The Study of Science and Religion

CHURCH OF SWEDEN
Research Series

Church of Sweden Research Series (CSRS) is interdisciplinary and peer-reviewed. The series publishes research that engages in topics and themes in the intersection between church, academy, and society.

Editor of the CSRS: Jonas Ideström

1. Göran Gunner, editor,
   Vulnerability, Churches and HIV (2009)
2. Kajsa Ahlstrand and Göran Gunner, editors,
   Non-Muslims in Muslim Majority Societies (2009)
3. Jonas Ideström, editor,
   For the Sake of the World (2010)
4. Göran Gunner and Kjell-Åke Nordquist,
   An Unlikely Dilemma (2011)
5. Anne-Louise Eriksson, Göran Gunner, and Niclas Blåder, editors,
   Exploring a Heritage (2012)
6. Kjell-Åke Nordquist, editor,
   Gods and Arms (2012)
7. Harald Hegstad,
   The Real Church (2013)
8. Carl-Henric Grenholm and Göran Gunner, editors,
   Justification in a Post-Christian Society (2014)
9. Carl-Henric Grenholm and Göran Gunner, editors,
   Lutheran Identity and Political Theology (2014)
10. Sune Fahlgren and Jonas Ideström, editors,
    Ecclesiology in the Trenches (2015)
11. Niclas Blåder,
    Lutheran Tradition as Heritage and Tool (2015)
12. Ulla Schmidt and Harald Askeland, editors,
    Church Reform and Leadership of Change (2016)
13. Kjell-Åke Nordquist,
    Reconciliation as Politics (2017)
14. Niclas Blåder and Kristina Helgesson Kjellin, editors,
    Mending the World? (2017)
15. Tone Stangeland Kaufman,
    A New Old Spirituality? (2017)
16. Carl Reinhold Bråkenhielm,
    The Study of Science and Religion (2018)

The Study of Science and Religion
Sociological, Theological, and Philosophical Perspectives

CARL REINHOLD BRÅKENHIELM

☙PICKWICK *Publications* · Eugene, Oregon

THE STUDY OF SCIENCE AND RELIGION
Sociological, Theological, and Philosophical Perspectives

Church of Sweden Research Series 16

Copyright © 2018 Carl Reinhold Bråkenhielm. All rights reserved. Except for brief quotations in critical publications or reviews, no part of this book may be reproduced in any manner without prior written permission from the publisher. Write: Permissions, Wipf and Stock Publishers, 199 W. 8th Ave., Suite 3, Eugene, OR 97401.

Pickwick Publications
An Imprint of Wipf and Stock Publishers
199 W. 8th Ave., Suite 3
Eugene, OR 97401

www.wipfandstock.com

PAPERBACK ISBN: 978-1-5326-1968-7
HARDCOVER ISBN: 978-1-4982-4617-0
EBOOK ISBN: 978-1-4982-4616-3

Cataloguing-in-Publication data:

Names: Bråkenhielm, Carl Reinhold, author.

Title: The study of science and religion : sociological, theological, and philosophical perspectives / by Carl Reinhold Bråkenhielm.

Description: Eugene, OR: Pickwick Publications, 2018 | Church of Sweden Research Series 16 | Includes bibliographical references and index.

Identifiers: ISBN 978-1-5326-1968-7 (paperback) | ISBN 978-1-4982-4617-0 (hardcover) | ISBN 978-1-4982-4616-3 (ebook)

Subjects: LCSH: Religion and science.

Classification: LCC BL240.2 B7 2018 (print) | LCC BL240.2 (ebook)

Scripture quotations are from the New Revised Standard Version Bible, copyright © 1989, Division of Christian Education of the National Council of the Churches of Christ in the United States of America. Used by permission. All rights reserved.

Manufactured in the U.S.A. 06/06/18

To Anders Jeffner—friend and mentor

Contents

Tables and Figures | ix
Preface and Acknowledgments | xi
Abbreviations | xvii

PART ONE: *Perceptions of Science and Religion in Sweden and Beyond*

Introduction to Part 1 | 1

1 Biology and Worldviews | 5
2 Swedish Public and Clergy on Biology and Religion | 21
3 Public and Clergy on Science and Religion | 39
4 Swedish Clergy on Science and Religion; a Cluster Analysis | 52
5 A Comparison between Sweden and the U.S. | 60

PART TWO: *From Söderblom to Jackelén: Swedish Twentieth Century Theology on Science and Religion*

Introduction to Part 2 | 77

6 Early Twentieth Century—Theological and Philosophical Transformations | 85
7 Theological Responses to Axel Hägerström | 94
8 A Model of Independence and Its Critics | 107
9 Lundensian Models of Contact | 117

10　New Philosophical Perspectives on Science and Religion | 132
11　From Generalities to Specifics | 154

PART THREE: *Main Models in the Dialogue between Science and Religion*

Introduction to Part 3 | 169

12　Models of Incompatibility | 175
13　Models of Independence | 205
14　Models of Contact | 222

Appendices | 265
Bibliography | 297
Author Index | 313
Subject Index | 319

Tables and Figures

Table 1   Summary of cluster-analysis of telephone-interviews 2006 and 2017. | 12

Table 2   Religion and biology in telephone interviews. | 23

Table 3   Science and religion among the general public and among believers in God in Sweden. | 41

Table 4   Interest to participate in courses about science and religion. | 48

Table 5   Subjective estimation of past change of view on science and religion—denominational affiliation. | 49

Table 6   Subjective estimation of past change of view on science and religion—gender. | 49

Table 7   Subjective estimation of future change of view on science and religion—denominational affiliation. | 50

Table 8   Overview of the CA-results. | 56

Table 9   Responses to the statement "Whenever science and religion conflict, religion is always in the right" in Sweden and the U.S. | 67

Table 10	Responses to the statement "Whenever science and religion conflict, religion is always right" and the question "Do you believe in God?" in Sweden and the U.S.	68
Table 11	Religion and biology in a national sample of the Swedish population and scientists.	70
Table 12	Scientists and public in the U.S. and Sweden on belief in God.	72–73
Table 13	Dimensions of science and religion.	146
Table 14	Six models of conflict between science and religion.	176
Figure 1	The place of clusters in a two-dimensional scheme	16
Figure 2	General statements on science and religion.	44
Figure 3	Ministers in Church of Sweden and pastors in the Uniting Church in Sweden on science and religion.	45
Figure 4	What does Swedish Clergy want to know about natural science?	46
Figure 5	The interplay between image of God, gender, denomination, and interest in science and religion.	58

Preface and Acknowledgments

Year 2016 marked the fiftieth anniversary of Ian Barbour's groundbreaking work *Issues in Science and Religion*. His work sparked the interest not only in the general issue about the relationship between science and religion, but also in specific issues in the theological relevance of cosmology, evolution, and genetics. Philip Clayton argues that Barbour's book could be described as the birth of a new phase in the field of science and religion. Other phases followed such as attention to "methodology (Barbour, Phil Hefner, Bob Russell, early Clayton and Murphy), constructive research programs (such as the 15-year Vatican/CTNS series on divine action), and violent rhetorical counter-attacks (the New Atheism, Intelligent Design)." According to Clayton, "we are entering a fifth phase of the evolution of science and theology, one in which persons of faith have both permission and mandate to do constructive work out of their particular cultural contexts and faith locations."[1]

This book is an example of such an effort to do constructive work out of a particular cultural context, namely contemporary Swedish religion and worldviews against the background of twentieth century theology with special consideration to its shifting understanding of the relation between science and religion. The main aim is to contribute to the analysis of the relationship between science and religion. Three main problems are in focus. The *first* concerns the opinions of the general public, ministers, pastors, students and scholars and scientists on this relationship and how their solutions can be interpreted and theologically assessed. The focus is

---

1. Clayton, "God in Process, World in Process," lines 1–15.

upon theses different groups in Sweden, but also upon the US and to some extent also other countries. The *second* problem concerns Swedish twentieth century theology and how the relationship between science and religion has been managed. And the *third* problem departs from the international philosophical discussion and how comtemporary proposals about the relations between science and religion can be described and assessed.

The work is part of a project supported by the John Templeton foundation. It was initiated in March 2015 and concluded in November 2017 and had the title "Science and Religion. A Platform for Dialogue and Education in Sweden." It was directed towards religious leaders in Sweden and branded Cusanus after the famous fifteenth century forerunner Nicolaus Cusanus. Main activities were courses, conferences, seminars, study material production, and other initiatives aimed also towards a broader audience in Sweden. The project also included studies of opinions on science and religion among Swedish religious leaders and the production of a baseline assessment report on the state and background of the dialogue on science and religion in Sweden. The present book builds upon this report, but extends beyond it to a broader field of theological and philosophical reflection on the relationship between science and religion.

Readers of this book from outside Sweden will recognize many parallels between his or her social, theological and philosophical environment and the Swedish situation. This recognition is a function of at least two different factors. The *first* is secularization, i.e. the emergence and domination of "the immanent frame" and declining percentages of churchgoing, prayer, and belief in God. Sweden is among the most secularized countries in the world. But secularization is an ongoing process in the whole Western world. Readers from other parts of the Western world are not foreign to these cultural developments. *Secondly*, and more importantly, Swedish theology (which will be in focus in Part 2) is part of a larger community of scholarship. Swedish theologians such as Nathan Söderblom, Anders Nygren, Anders Jeffner, Antje Jackelén, Mikael Stenmark, and others have been extensively published outside Sweden by works written in English, German, or French. Over 50 percent of the influential thinkers in the field of science and theology mentioned by Swedish clergy in a web-survey 2015, were theologians and philosophers from outside Sweden. So even if this is a book with special consideration to Swedish religious leaders and theologians, the questions are—as it says on the rear mirror of your car—"closer than they appear." This will be especially apparent in Part 3 of this book where different models of science and religion will be considered.

A comprehensive study of science and religion would involve studies from a manifold of disciplinary perspectives. The perspective from history is one, from political science is another, from ethics a third, and from religious studies a fourth. There are glimpses of these in this book, but the emphasis is on sociological, theological, and philosophical perspectives. One justification for attending to all these different aspects in one and the same work is they enlarge the perspective on the issue in a way that reflects the multifaceted phenomenon of science and religion. But another reason is shamelessly biographical. During my academic work and into my retirement, I have been working in all these fields, crisscrossing the boundaries between the sociology, theology, and philosophy of science and religion. The publication of this book is an opportunity to bring these fields together and explore their interrelations.

As suggested by the subtitle, the book has three different parts. The *first* concerns more sociological issues. Surveys and interviews mainly from 2006 up until 2017 will be presented and discussed. Some of these explore attitudes within the Swedish public and others are more specifically directed at religious believers and religious leaders. Of special importance are two national surveys about the process of biologization the developments of religious beliefs and other worldviews in Sweden. Special attention is also given to some other studies in 2015, which were addressed to the clergy of the Church of Sweden and the Uniting Church in Sweden (an ecumenical union between the Mission Covenant, the Baptist and the Methodist churches in Sweden). A special chapter deals with a comparison between Sweden and the U.S. when it comes to attitudes on science and religion. The *second* part of the book is more concerned with theological issues. It gives a broader general background of Swedish theological developments during the twentieth century with a focus on the general ideas about the relationship between science and religion as well as upon analyses of specific issues. In contrast, the *third* part is thematically oriented. The first theme concerns the model of incompatibility regularly challenging Swedish twentieth century theology and vitalized in the contemporary international discussion. The second is the model of independence and separation, dominating Swedish theology up until the 1950s, but not without proponents in the present. The third is the model of contact, which has been explored by American and British scholars.

Part 1 can be read independently of Part 2 and 3. It is also possible to proceed directly to Part 2 and 3 without first reading part 1. Part 2 and 3 are more closely related, but a reader with limited time and/or special interest in the philosophical aspects on science and religion can bypass Part 2 and concentrate on Part 3. In one important respect the three parts form a progressive argument. The sociological part uncovers specific worldviews, which are of significance in Sweden and beyond. Part 2 gives example of Swedish theologians suggesting alternative ways to think about religion and science. Part 3 develops this theme further in dialogue with the larger international community of philosophers and theologians.

Much of the book is based on primary studies and I have previously published some of the material. However, most of it is written for the purposes of the present book. Needless to say, important contributions come from other authors and research groups.

In the process of writing the book, several colleagues in Uppsala and Stockholm have influenced its shape and content. Assistan professor Johan Gärde at Ersta Sköndal University College has done a critical reading of Part 1 and helped to improve the text in many different ways. Furthermore, grateful acknowledgement goes to my colleagues at the Faculty of Theology, Uppsala University, professor Mattias Martinson and the doctoral seminar in systematic theology. Assistant professor Katarina Westerlund must be highlighted for her important proposal concerning the basic structure of the work. Professor Mikael Stenmark has been an esteemed partner in the Cusanus project and an encouraging advisor in the development of the present book. Twice I had the opportunity of discussing chapters in part 3 with the doctoral seminar of philosophy of religion. I'm especially grateful to assistant professor Ulf Zackariasson, who has given me valuable comments on Chapter 14. I have also received valuable comments to the last part of this chapter from Nordic colleagues such as professors Sigurd Bergmann, Marion Grau, Cristina Grenholm, Theodor Jørgensen, Kjetil Hafstad, and Tage Kurtén. Special gratitude should go to Aku Visala, who made me aware of Steven Horst's writings on cognitive pluralism.

During the course of the last ten years, I have had significant cooperation with Ingrid Friberg and her colleagues at IFS (Ingrid Friberg Samhällsinformation) in Stockholm. She has realized the practical details of most of the studies referred to in Part 1 and delivered advice and support. So has Jan Nylund (at Refina Information AB), whose statistical work was of special importance throughout Part 1.

The support of my wife, Lotta, is one of the explanations for the very realization of this book. With patience and care she has studied my manuscript and delivered significant critique and encouragement in a well-proportioned mix.

My colleagues in research department at the central office of the Church of Sweden have provided important comments to my work and paved the way for publication at Pickwick Publications. The staff of the publisher has patiently provided me with all necessary support during the process. My special gratitude goes to director Cecilia Nahnfeldt, the present editor of the Church of Sweden Research Series, Jonas Ideström, and the former editor Göran Gunner, who has given me invaluable assistance when it comes to giving the manuscript a formally acceptable shape. Needless to say, any lingering deficiencies are entirely my own responsibility.

Anders Jeffner has suggested some important improvements to chapter 11 of Part 2. His theory of fundamental patterns is of great significance especially in Part 3, but the application of the theory in the study of science and religion is my own responsibility. His personal and intellectual inspiration has been of continuing significance over the years. The gratitude expressed by dedicating this book to him is long overdue.

Carl Reinhold Bråkenhielm
Djursholm September 1st, 2017

Abbreviations

CA	Cluster Analysis
CMI	Creative Mutual Interaction (between science and religion)
CMP	Christian Mystical Doxastic Practice
CP	Cognitive Pluralism
CSR	Cognitive Science of Religion
EIFP	Existence Implying Fundamental Pattern
FP	Fundamental Pattern
HADD	Hypersensitive Agency Detection Device
IWV	Inventory of World Views
JTR	Jeffner´s Theory of Religion
MN	Methodological naturalism
MetaN	Metaphysical naturalism
MP	Mystical Perceptual Doxastic Practice
mWCS	Milder form of Worldview-Customized Science
NIFP	Non Existence Implying Fundamental Pattern
NOMA	Nonoverlapping Magisteria
PC	Principle of casual/explanatory closure
RAAS	Religion Among Academic Study
RASIC	Religion Among Scientists in International Context

RCT	Rational Choice Theory
RP	Religious Perceptual Doxastic Practice
RS	Religious Statements
SV	Standard View of Cognitive Science
SP	Sense Perceptual Doxastic practice
ST	Secularization Theory
WVS	World Values Survey

Part 1

Perceptions of Science and Religion in Sweden and Beyond

INTRODUCTION TO PART 1

In *Re-thinking Science. Knowledge and the Public in an Age of Uncertainty* (2001) Helga Nowotny, Peter Scott, and Michael Gibbons assumes an increasing interaction between science and society. A new kind of contextualized and context-sensitive science is emerging.

> One important change is the erosion of the collectivist belief-systems that characterize the science system and generate the norms which bind it together. The result is less "segregation" from, and more "integration" with, society . . . Scientists now share their once exclusive systems for communicating information with these "outsiders." One way of putting it is to say that the rising tide of individualism in society now has reached collectivist scientific communities.[1]

This development is connected with another equally important social process, namely, the rising levels of education and the broadened acquaintance with scientific and scholarly achievements. Scientists share their knowledge because it is in demand within the general public. Furthermore, the arrival of "knowledge society" requires new communicative skills, and particularly the skill to link up with public values and lifeviews. Nowotny et al. remind us that the Charter of the Royal Society did not want "to meddle with politicks, rhetoric, divinity." By the dynamics of the knowledge society, it

---

1. Nowotny, Scott, and Gibbons, *Re-thinking Science*, 102–3.

is increasingly difficult to avoid such meddling. In short, the last century has been undergoing a silent revolution from what Nowotny et al. call a "Mode-1" to a "Mode-2" society, more porous and no longer representing distinguishable domains—and from a Mode-1 to a Mode-2 science, where the social context no longer respectfully listens but "speaks back."

Contemporary dialogues between science and religion may be understood against this background—and so also social scientific research on how the general public (and specific social groups) perceive this relation. Looking back to the nineteenth century, theological opposition against the theory of evolution can be viewed as the first intimation of the larger change from a Mode-1 to Mode-2 knowledge production. US studies on the public perception of science and religion are still determined by this conflict,[2] which is less significant in the European, not to mention the Swedish context.

In a secularized society such as Sweden interest has shifted to the general study of worldviews, i.e., non-religous as well as religious. Many such studies have a special background in the more general theory of worldviews developed by professor Anders Jeffner at Uppsala university. Jeffner gained international attention for his work in philosophy of religion as well as his pioneering studies on worldviews in contemporary Swedish society. Jeffner's approach is holistic rather than atomistic. Worldviews (Sw. "livsåskådningar") consist of convictions about human beings, nature, and the world in combination with a basic values system and a basic mood of trust or fear, optimism or pessimism. These worldviews may be derived from science, but they may also in turn influence the presentation and organization of scientific results. Jeffner illustrated these intricate relationships in an analysis of *Biology and Religion as Interpreting Patterns in Human Life* (1999).

Chapter 1 begins with a presentation of Jeffner's research program and a number of recent studies within this paradigm. Many of them have been particularly focused on the significance of the biological and ecological sciences in contemporary formation of worldviews. Theories and results in these sciences have developed as interpreting patterns of human life in a broader sense. The presence of "biologism" and "ecologism" is indicated by different studies from the late 1980s and onwards. One study was made in 2006 in the form of telephone interviews with a representative sample of the Swedish adult population between 18 and 74 years old. This study was replicated in 2017. In another study from 2007 academically trained

2. See further chap. 2.

PART 1: PERCEPTIONS OF SCIENCE AND RELIGION IN SWEDEN AND BEYOND   3

biologists were interviewed about their science and their worldviews. Half of them clearly agreed to the statement "that all life is only the result of an unplanned development, built on the harvests of chance without meaning, purpose or plan." The other half was more hesitant and found it difficult to integrate randomness and chance in their worldview. Furthermore, 10 of the 35 respondents claim that a biology-induced loss of religious faith had occurred in their life.

Chapter 2 concerns opinions on biology and religion and presents a comparison between two representative Swedish samples. The first is drawn from the general Swedish public (in 2009) and the second consists of participants in a web-survey made among Swedish clergy (in 2015). Overall, Swedish clergy seems less prone to biologization in comparison to the Swedish public. Few Swedish ministers and pastors affirm a biological view of life in general. The most striking difference between Swedish clergy and Swedes in general concerns the perception of the relationship between science and faith in God. Only 2 percent of the clergy think it has become more difficult to believe because of the progress of science—in contrast to almost half of the general Swedish population.

Chapter 3 presents further results from the web-survey in 2015 among Swedish ministers and pastors. This survey contained some more specific questions concerning opinions about the relationship between science and religion. The respondents were asked to take a stand on different pre-formulated statements concerning science and religion. An overwhelming majority of Swedish clergy endorses the independence *as well as* contact between science and religion. There is generally an interest to participate in courses on science and religion. A minority of the overall sample (less than a third) foresees a change in their view of the relation between science and religion.

Chapter 4 presents the results of a cluster analysis of the responses in the web-survey among Swedish clergy, i.e., a statistical study of different segments within the total sample of respondents. The purpose was to describe how each of these segments differs from each other when it comes to the different philosophical and religious issues as well as social background factor. The respondents could be divided into six different clusters, where each one of the clusters displayed a particular worldview and personality-profile. There are interesting differences between the clusters. For example, cluster 5 is women-dominated, environmentally concerned, and interested in deeper knowledge about science and religion. Seventy two percent agree with the statement that nature is animated by a power that permeates all life. This is in clear contrast to cluster 2 with 0 percent in agreement that nature is animated and at the same time strongly affirming that humanity has sovereignty over the rest of creation. This group consists of a larger than average number of male pastors from small town areas. It must be noted that all clusters are equally opposed to biologism.

Chapter 5 gives an overview of different studies about science and religions in the U.S. These studies indicate a huge gap between American scientists and the American public concerning religious belief. Skepticism and atheism among American scientists contrast with the strong standing of religion among Americans in general. Comparing figures drawn from more or less dissimilar surveys and interviews are not maximally reliable. Nevertheless, comparing US and Sweden on public conception of science and religion reveals striking differences.

In the second half of the chapter, two questions are discussed. First, the presence of biologization among Swedish scholars in literature, political science, and physics is compared to the Swedish public. A comparison shows that overall scholars are less prone to biologization than the public. Secondly, an effort to a comparison between Sweden and the U.S. is made. In contrast to the U.S., scientists in Sweden are generally as secularized as the Swedish public. Swedish scientists and scholars still seem as religious as they are in the U.S. Why is there practically no difference between Swedish scientists and the general Swedish public on religion? If the proportions in Sweden would resemble those in the U.S., one would expect religious belief as practically nonexistent among Swedish scientists and scholars. Explanations why it is not are proposed and discussed.

1

Biology and Worldviews

There are certain things that most North Americans recognize as Swedish. They know about Ingmar Bergman, ABBA, Astrid Lindgren, and, possibly, Stieg Larsson. IKEA and Volvo are universal brands and without doubt well known also in the U.S. There is also one Swedish cultural institution that is clearly world famous: the Nobel Prize. It is awarded in five different categories of which three are in the sciences: physics, chemistry, and physiology/medicine (from 1968 also in economics).[1] The Nobel Prize provides an inroad to an important value in Swedish culture, namely, the value of science and technology. Sweden is not unique for the appreciation of the practical significance of science and technology. The citizens of most countries in the world completely agree that "science and technology are making our lives healthier, easier, and more comfortable."[2] This notwithstanding, Swedish citizens *are unique* for their trust in the *scientific method*. Over 80 percent believe that the scientific method is the best way to gain knowledge about reality.[3]

This trust in the scientific method also surfaces in some of the responses to items in the World Value Study (WVS, particularly in the responses to the so-called Wave 6, an international survey in 2010–2014). One of the items was the proposition that "we depend too much on science and not enough

---

1. *Images of Sweden Abroad*. The Nobel Peace Prize is awarded by a special Nobel committee appointed by the Norwegian parliament.

2. *World Values Survey*, Wave 6. Response to the item "science and technology are making our lives healthier, easier and more comfortable."

3. Ingemyr, *VoF-undersökningen*, 7. See also Bråkenhielm, *Världsbild och mening*, 239.

on faith." Over 60 percent of the Swedish respondents disagree—in stark contrast to only 25 percent in the United States. The difference to neighboring countries such as Germany and Estonia is equally striking.[4]

The appreciation of science and the scientific method comes to expression in different ways. One expression is in the individual worldview of Swedish citizens. Evidence for this is found in the worldview-studies at the Department of Theology, Uppsala University.

STUDIES OF WORLDVIEWS: CONCEPTS, AIMS AND METHODS

The study of worldviews as a specific academic field of study dates to the 1968–1976, when Anders Jeffner headed a research project—"Worldviews in Modern Society"—with support from the Swedish Tercentenary Foundation. This project produced several reports on the study of worldviews. One of the most lasting contribution of the project was a conceptual proposal made by Anders Jeffner himself, namely, a stipulative definition of worldview. Jeffner suggest that the concept has three components:

> The first is a basic value system: those values most important to someone which are retained for a longer period of time. The second factor can be called the picture of the world—or, why not, metaphysics. This deals with how we try to organize everything we know along a pattern that will make the world understandable, and how we assign human beings to a particular place in reality. And we believe, moreover, that there is a third factor, which must be noted and which is related to the other two. This factor can be described as a feeling toward life, which can stay with us for a long time and give a certain emotional color to the rest of our experiences. We call it basic mood. This can be, for instance, trust or fear, optimism or pessimism.[5]

Essentially, a worldview consists of three parts: (1) central values and norms, summarized more solemnly in ethics and morality, (2) a comprehensive picture of reality, and (3) a basic mood, optimism or pessimism, hope, or despair.

Jeffner's concept of worldview has proved fruitful in many empirical studies of worldviews. One significant study was initiated in 1986. It was an exploratory study with the aim to study worldviews disseminated

---

4. *World Values Survey*, Wave 6, v194 (2011).
5. Jeffner, "A New View of the World," 138.

among the general public of Sweden.⁶ An important result was that there two minority views. One group—about 8 percent—consisted of committed Christians. They had a specific comprehensive view of reality, but they were not significantly different from the majority in their value system or basic mood. Another minority (c. 25 percent) was found at the other end of the spectrum. This group distanced themselves from belief in God while professing a materialistic view of reality. In between these extremes was the majority view. It was dominated by a worldview with an anchorage in the biological-ecological context. But there was ambivalence between two ideological complexes. The first is a *biological deterministic* pattern. Genetic inheritance determines what becomes of a human being. There is little difference between humans and animals. Human beings are only bodies and matter. There are small possibilities for change. The other ideological pattern is *biological-personal*. Human choice and initiative are important. Humans are not only matter, organs, blood, and bones. Possibly, there is something beyond death.⁷

I will return to similar empirical studies shortly, but before that attend to some theoretical issues. One such theoretical issue concerns the very existence of worldviews as comprehensive pictures or ideas of the world as a whole. The existence of such comprehensive ideas is a central problem in cognitive science. Is there room for such comprehensive ideas?

The American philosopher Steven Horst distinguished between the Standard View of cognitive science (SV) and his own proposal, Cognitive Pluralism (CP). SV assumes concepts, beliefs and arguments as the central elements of the human cognitive architecture. In contrast, Horst argues that mental models are the fundamental units of understanding. "We *believe* propositions, but we *understand* things such as Newtonian mechanics, the game of chess, and the etiquette for dining in a restaurant, and we do so through having mental models of those domains."⁸ Mental models are (1) idealized in the sense that they are *domain-specific*, and (2) *good-enough* ways of thinking and reasoning about some particular set of purposes "without these being exact, unidealized, or context neutral ways of representing their targets." Horst summarizes:

> The basic thesis of cognitive pluralism is that the mind employs special-purpose models of parts, aspects, and features of the

---

6. Jeffner, "A New View of the World Emerging among Ordinary People," 137–45. See also, Jeffner, *Livsåskådningsforskning*, 42–45.

7. Jeffner, "A New View of the World Emerging among Ordinary People," 141; Jeffner, *Livsåskådningsforskning*, 43.

8. Horst, *Cogntive Pluralism*, loc. 182.

world, rather than (1) a single consistent, and integrated model of everything or (2) a long inventory of more specific and individual beliefs. If a "worldview" is construed as a comprehensive and consistent model of the world, then we possess nothing that answers to the description of a worldview.[9]

Horst hastens to add that adherents of CP need not be hostile to the project of unification as a regulative idea. But it is unlikely that such efforts of unification into a comprehensive and consistent model of the world will (ever) able to do so without "oversight, distortion or explaining away on the basis of its key concept".[10]

At this point, I will not digress into a further analysis into these basic and controversial issues of contemporary cognitive science. I will, however, return to the problem when elaborating the idea of models about science and religion (in the introduction to Part 3). At this point, I will bypass the problem of worldviews as a comprehensive picture of the world, and substitute Jeffner's definition with another definition. This revised definition links up with an earlier idea proposed by Jeffner, namely, that the cognitive component is delimited not by a set of comprehensive and general beliefs, but rather by those which are influenced by the central value system or the basic mood in a way that the individual person is prepared to accept.[11] I will depart from a definition, which does not rely on the subjective estimation of individual respondents, but rather *on statistical correlations between responses to certain propositions and value-judgements*. Such correlations will be taken as indications of a certain worldview.

One example of a worldview in this sense is biologism. In later research, Jeffner has suggested that contemporary discussions indicate that there is a "road from biological science to biology as an interpretative pattern of the whole human life."[12] This process can be called "biologization" and the resulting worldview "biologism." Theories and results in biological science move into the foreground in the understanding of life and human nature. Furthermore, they are correlated with certain value-judgments. Nature becomes normative for human behavior. Later studies have revealed more details about this worldview.[13] What are the indications of biologization? Here are some suggestions:

9. Horst, *Cognitive Pluralism*, loc. 1692.
10. Ferré, *Language, Logic, and God*, 163.
11. Jeffner, *Livsåskådningsforskning*, 18.
12. Jeffner, *Biology and Religion*, 10.
13. See, for example, Uddenberg, *Det stora sammanhanget*. Uddenberg payed close attention to related phenomenon of "ecologization" and "ecologism." See also Philipson,

1. Close relationship between animals and humans.
2. Nature is normative.
3. "Human beings are nothing but a heap of neurons" (Francis Crick).
4. Human personality is genetically determined.
5. Only that is morally right which leads to reproductive success.
6. Religion is an evolutionary by-product.

Are there empirical evidences for biologization in contemporary society and among scholars and scientists? What happens to biological theories when they are being transformed into interpreting patterns of human existence? These questions have been in the focus of eight different studies between 2006 and 2017.[14]

BIOLOGISM, RELIGIOUS BELIEFS, AND OTHER WORLDVIEWS

The first study (*Study 1*, 2006) was divided into three parts. In the *first* part the respondents were asked about their attitudes to a series of statements and instructed to indicate their position to each of them on a five-point scale from total agreement to total disagreement. In sum, twentyfive items were used, some of which were derived from the so-called New Environmental Paradigm Scale (NEP-scale, constructed by Riley Dunlap and K. D. van Liere).[15] Combined with other items used in Jeffner's study in the late 1980s, this resulted in an *Inventory of World Views* (IWV). Eight statements of these statements in the IWV were specifically formulated to measure the presence of biologism. Other statements were formulated to serve as indicators of other worldviews—for example, religious belief and ecologism. The *second* part of the survey consisted of ten propositions about different personality types. The respondents were asked to indicate to what degree the personality type described resembled themselves on a six point-scale from no resemblance to large resemblance. The *third* part consisted of items about age, education, occupation and residence. 503 respondents between 18 and 74 years old—representative of the Swedish adult population—was interviewed by telephone in May 2006.

In May 2017 *Study 1* was replicated together with four questions on the relation between science and religion (*Study 8*). The national sample

---

Med naturen som referenspunkt, and Bråkenhielm, "Environmentalism."

14. An overview of the studies is provided in appendix 1.
15. See Bråkenhielm, "Environmentalism," 49–66.

consisted of 500 respondents of the same age as those interviewed in 2006. Both studies were successful in the sense that the response-rate (compared to the net sample) was very good. 82 percent of the respondents participated in *Study 1* and 69 percent in *Study 8*. The interviewers reported that the respondents were engaged by the questions and that it was a very rewarding assignment to conduct the interviews.

Comparing the results, the overall impression is one of stability.[16] Two digits-changes are rare. But there are two notable exceptions. One concerns the decline of traditional religion and the other rise of environmental pessimism.

The decline of religion.—Responses to items about God, holiness, and the ensoulment of nature all indicate decline, even steep decline over the eleven-years period between the two studies. Most clearly, this comes to expression in response in item 29, which shows an increase of 25 percent (!) for the atheistic alternative (and from 19 to 11 percent decline for belief in a God to whom you can have a personal relationship). When it comes to item 25 about the existence of a god, a supernatural power or force, 48 percent disagreed in 2006, while 60 percent disagree in 2017. This might be compared to a similar figure from a study made in 1987, where 28 percent disagreed. If this trend continues, only around 10 percent of the Swedish citizens will have belief in God in the middle of this century.

It is sometimes argued that while Christian belief is declining in Sweden, religion is not.[17] Rather religion is transforming into a vaguer form of spiritualism.[18] But judging from the results of *Study 1* in comparison to *Study 8*, this is not case. There is no increasing sympathy for belief in an impersonal power (item 29). And there is also a sharp decline (from 45 to 36 percent) when it comes to belief in a spiritual power that permeates all life. Similarly, meeting holiness in nature is waning (item 18). This decline may seem surprising considering claims that "God is back" and the "new visibility of religion." To be sure, religion has come in focus especially since the terror attacks on the 11th of September 2001 and numerous similar events before and after. Religion has often been a factor involved, but this visibility has not been in its favor and may have contributed to its decline. Moreover, the whole notion of the "new visibility of religion" cannot be generalized to all countries and time periods. "Mia Lövheim and Alf Linderman find that the proportion of editorials on religion in Swedish newspapers has actually

---

16. For a detailed display of the results, see appendix 6.
17. Hamberg, *Gud*, 36.
18. See, for example, Thurfjell, *Det gudlösa folket*, chap. 1.

declined since 2005. Such findings are an effective antidote to general claims about the return of religion."[19]

Rising environmental pessimism.—Nearly reaching a two-digit change is also apprehensions of humanity facing a self-inflicted environmental disaster. It is now up to 67 percent (from 58 percent in 2006). Moreover, the whole increase is among those who are in *complete* agreement. Furthermore, those in disagreement about the statement that technology will be able to solve the problems with our limited natural resources (item 1) is up from 23 to 27 percent. Expansion of nuclear power (item 4) is down sharply with 15 percent. On the other hand, the number of those disagreeing with the statement that there are no limits to growth in a country like Sweden is down from 58 to 51 percent. In sum, these figures speak in favor of a rise in a domain-specific pessimism, i.e., pessimism related to environmental problems and the ability of human technology to solve these problems. This kind of pessimism might, of course, develop into a more general mood of gloom and despair. I will return to this shortly.

The responses to the different items were interpreted with the help of *cluster-analysis*.[20] Six such clusters emerged in *Study 1* from 2006 and they can be found also in *Study 8* from 2017 (80–90 percent likelihood that a given response-pattern in 2006 reappears in 2017). Table 1 gives an overview of the profiles of the different clusters and their sociodemography.

19. See Repstad's review of Hjelm, *Is God Back?*, 433.
20. For a description of the method, see appendix 4. The method is further presented in chap. 4.

	Cluster 1: Traditional religion	Cluster 2: Spiritualism	Cluster 3: Biologism	Cluster 4: Anti-environmentalism	Cluster 5: Skepticism	Cluster 6: Environmental pessimism
Share 2006	10	7	19	15	13	9
Share 2017	7	4	14	15	17	12
Dimensions (items in parenthesis):						
Human sovereignty (10, 11, 14, 17)	++	–	+			
Belief in God/meaning, holiness (18, 23, 25, 29)	++	+	–		– –	
Tradition & family (82, 84, 87, 88, 89)	+	+	+		– –	
Environmental concern (5, 12, 13, 15)	+	++	+			++
Hedonism & egoism (81, 83, 85, 86)	–	– –	+			
Technology & science (1, 2, 3, 8)	–	– –	+		+	
Socio-demography 2017	+ Women + Univ. edu. + Sick leave (2017)	+ Women in rural areas (2006) + 18–34 years (2017)	+ Men + Low educ. + Small city (2017)	None	+ 18–34 years + Men + Big city + Univ. edu.	None

Table 1: Summary of cluster-analysis of responses to telephone-interviews concerning worldviews and biologization in May 2006 (*Study 1*, n=503) and May 2017 (*Study 8*, n=500). ++ indicates a positive agreement which is much stronger than the avarege of the whole sample, + indicates a weaker agreement, – – indicates a much stronger disagreement, and – indicates a weaker disagreement. Percent.

The overall tendency is (1) the decline of religion and spiritualism, and (2) the rise of skepticism and environmental pessimism. (1) is in line with theories of secularization. However, my results are not in support of the common opinion that it is not religion as such that is declining, but only its outward form. No movement from traditional religion to spiritualism can be discerned. Rather, skepticism and environmental pessimism is on the rise. The relationship between the rise of these two clusters requires further study.

The largest cluster—*skepticism*—merits a closer look. Skepticism is negatively determined by items affirmed by members of the religious cluster. In fact, it is its antipode not only when it comes to the religious statements denied, but also when it comes to its sociodemographic profile. Young well-educated men from the big cities are overrepresented within this sample. They are less concerned than the average of Swedes when it comes to tradition and environment, and their hope that technology will solve the problems with our limited natural resources is fading (item 1, from 51 percent in 2006 to 41 percent in 2017).

The emergence of *anti-environmentalism* (cluster 4) goes contrary to the findings in the original worldview studies of the 1980s. Possibly, the wave of environmentalism has created a kind of environmental back-lash. Throughout the survey, they are low on indicators of environmental concern. A basic mood of resignation and lack of engagement is visible in this cluster. This trend could be linked to political changes from internationalism to nationalism finding expression in the rapid rise of the Sweden Democrats, i.e., the immigration hostile party, which now attracts nearly 20 percent of the Swedish electorate. They score higher than average on indicators of hedonism and egoism. Interestingly, this cluster has is no special sociodemographic profile.

Contrary to my expectations, it does not appear that Swedes in general are moving towards *biologism*. Cluster 3 is shrinking from 19 percent in 2006 to 14 percent in 2017. This may be a part of a larger trend from the 1980s. Looking at just one item, namely, item 20 about humans being only matter and body, agreement seem to be declining from about 40 percent in 1987 to a little over 25 percent in 2017. This finding falsifies the hypothesis about the increasing significance of biologism, even if it is still among the major worldview alternatives in Sweden. One curious anomaly is found in this cluster. One would expect that emphasis on the biological perspective would go hand in hand with a *devaluing* of human sovereignty in nature. But this is not case. In fact, the results show the opposite. Possibly, the members of cluster 5 affirm human sovereignty not in the religious, but rather in the social-darwinian sense. This could also explain the relatively high score on item 6 (human have unique qualities which no other creature has).

The cluster I have called *environmental pessimism* (cluster 6) is also on the rise (from 9 to 12 percent). They are not unlike cluster 5 (skepticism), but more environmentally concerned and at the same time pessimistic about the future. Another difference is that the pessimists are much less inclined to affirm that human beings consist only of matter and body. In fact, only 13 percent do, while almost 80 percent of the skeptics. This is difficult to understand, but it is possible that the pessimism of cluster 6 stems from the contrast between the strong affirmation of the unique qualities of human beings and their equally strong conviction that environmental changes for the sake of humanity creates serious problems (item 15).

Traditional religion and *spiritualism/ ecologism* is now at the low end of the six clusters discerned in the surveys. Commenting on the results in the study made in 1986, Anders Jeffner suggested that a vitalization of Christianity might come through a combination of Christian belief with environmental concern and at the same time providing a metaphysical anchorage for human rights and human dignity.[21] Efforts in this direction are evident and the papal pronouncements as well as ecumenical initiatives and episcopal letters can serve as examples. But there is no evidence that this has halted secularization.

Needless to say, the demise of traditional religion as well as spiritualism and the rise of skepticism and environmental pessimism cry out for an explanation. It was not the purpose of the studies in question to provide such an explanation. But some suggestions can nevertheless be considered.

In a very general sense, it is possible to link the changes between 2006 and 2017 to the rise of so-called postmodernism and the increasing distrust in the "grand narratives." This trend of distrust would be consistent with the decline of religion (cluster 1) and spiritualism (cluster 2) as well as the rise of skepticism (cluster 5) and environmental pessimism (cluster 6).

Another possible explanation is the rise of neo-atheism. The publication of Richard Dawkins *The God Delusion* in 2005 (Swedish translation in 2006) and its wide dissemination and influence springs to mind as one possible factor behind the rapid rise of skepticism. During the last decade, the Swedish Humanist Association has been a strong voice against religious belief. One of their main arguments has been that there has been and will always be a strong conflict between science and religion. This might be an explanation for the rapid rise of skepticism. Without discounting this factor entirely, I find it less likely to be among the main ones. The reason is that there is no increase in the conflict model of science and religion advocated by Dawkins and other neo-atheists. No significant increase is registered in response to the relevant item (item 28).[22]

---

21. Jeffner, *Livsåskådningsforskning*, 44–45.
22. See further below in chap. 3.

A more credible factor behind the decline of cluster 1 and 2 and the rise of cluster 5, could be the spectacular acts of terrorism performed by Jihadist religious groups—especially since 2001. This violence has received a lot of attention in the media, where a large part of the public gets news about religion and religiously inspired conflicts. This could have fostered a more general alienation from religion. While this seems intuitively probable, there is no way to verify this on the basis of the studies discussed.

There is one further item, which deserves to be highlighted. It is optimism and pessimism as measured by item 5 about humanity facing a self-inflicted environmental disaster. Such pessimism is to some extent correlated with skepticism and to an even stronger degree when pessimism is measured in terms of future economic growth (item 16). Furthermore, there is a clear correlation between skepticism and affirmation of the randomness of biological evolution (item 11). This indicates that skepticism can be linked to what Anders Jeffner termed "basic mood." Members in the cluster of skepticism have a tendency to affirm a pessimistic basic mood in contrast to members of the religious and spiritualistic clusters (1 and 2).[23] In 2006, 51 percent of those belonging to the skeptical cluster affirmed that humanity is approaching an environmental disaster. In 2017, this figure has increased to almost 75 percent. The pessimistic mood has also increased in the other clusters, but not as strongly as among the skeptics.

To summarize the results of the two studies, it is fruitful to single out two dimensions providing a field of possible combinations. The first dimension runs from environmental concern to environmental indifference, and the second from belief in God to atheism. Crossing these dimension, a scheme displayed in Figure 1 emerges with the clusters representing various positions.

---

23. Another study—*Study 5*—was launched in the spring of 2012 in the context of a larger project on environmental values among Swedish high school students (born 1994, n=1196). Once again a selection of items from the IWV was used. A factor analysis of the results shows (1) that 1/3 of the students have an integrated worldview and (2) that this group is divided in three different subgroups, of which one may be called religious (15 percent of the total sample), the second ecological (10 percent), and the third biological (10 percent). A follow-up study in 2014 indicated the presence of a fourth paranormal group affirming statements about precognition, spirits, and ghosts.

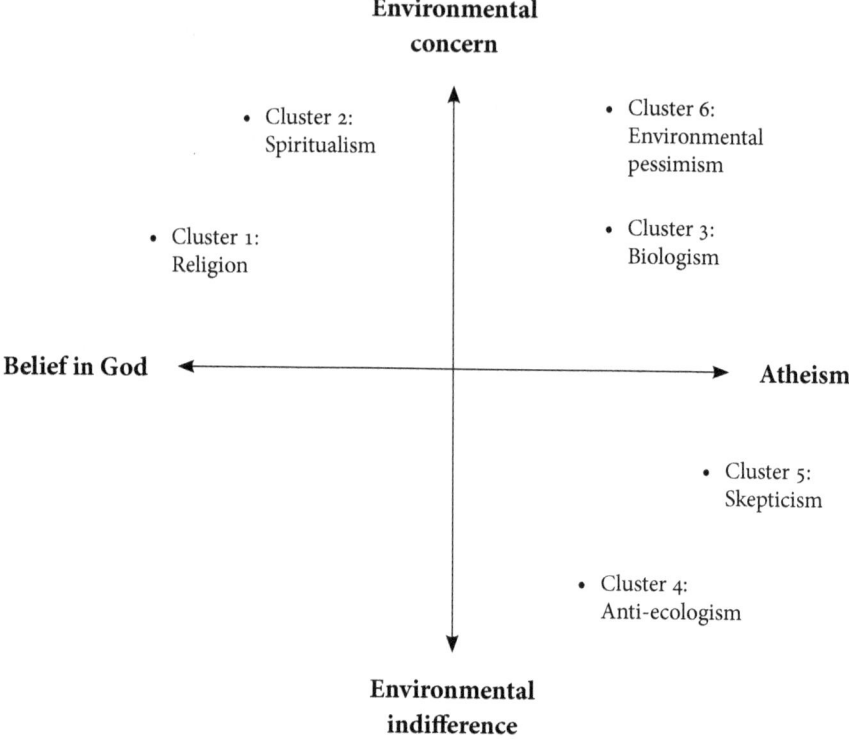

Figure 1: The place of clusters in a two-dimensional scheme of belief in God and environmental concern.

The most noticable fact about this overview is that there is an empty field in southwest. There is no cluster even close to a combination between belief in God and environmental indifference. This highlights an interesting difference between Sweden and USA. According to a recent PEW-study made in 2014, 50 percent of the American population say the Earth is warming primarily due to human activity but only 28 per cent among white evangelicals agree. 40 % of the mainline Protestants agree.[24]

BIOLOGISM AMONG SWEDISH BIOLOGISTS

One of the main purposes of another study—*Study 2* (2007)—was measuring biologism among Swedish academics trained in biology. To explore this, 36 biologists representing student, teachers and scholars were being asked to indicate their positions in respect to some of the statements in the

24. Funk and Alper, "Religion and Views on Climate and Energy Issues."

IWV presented to the general public. (Needless to say, this group is not representative for the whole group of biologists, but with this reservation a comparison might nevertheless be interesting.) A selection of the questions of the IWV, which were put to the general public in *Study 1* was also put to the biologists in *Study 2*. Furthermore, they were asked some open questions about worldviews and biology.

If we compare the answers to the standardized items from the general public in *Study 1* with the answers from the biologists in *Study 2*, a number of interesting differences between the general public and the trained biologists appear (discounting age- and gender-differences).

- Biologists are more prone to dissent from a response and either choose the don't know-alternative—or not give an answer at all.
- Biologists are more definitely non-anthropocentric—34 of 36 disagree completely with the claim that plants and animals exist only for the sake of humans (54 percent in the general public). All 36 of the biologist completely disagree with the claim that humanity exists to rule over the rest of nature.
- More surprisingly, biologists are less prone to biologization in the sense that they are clearly more critical than the public to (1) the statement that human beings are largely determined by genetic inheritance, and (2) the statement that life has no other meaning than reproduction.
- On the other hand, biologists are in sympathy with a biological worldview in the sense that human beings are only body and matter. Here the public is more critical than the biologists.
- Biologists are more clearly opposed to traditional faith in God, but also to more pantheistic ideas about a power or spirit that pervades all nature. A clear majority of them claims that Christian belief in creation is inconsistent with the theory of evolution. Over 25 percent affirmed that they encounter the holy in nature, but it should be noted that almost as many declined to respond to the item at all.

Study 2 also provides us with a large material in the form of responses to some open questions (without any formulated response alternatives). Here are some examples of those open questions:

- What is most important in life?
- What are your reflections on life as whole?
- Have your scientific and biological knowledge affected your attitudes to life—and if so—in what way?

- Can human beings completely be explained biologically?
- Can biological knowledge give us guidance when it comes to basic moral issues (for example that we ought to strive for justice and take care of the sick)—and in that case which?
- Are biological theories about human beings consistent with the idea that human beings have a soul?
- In a widely-used textbook a leading biologist has claimed that the theory of evolution states "that all life is only the result of an unplanned development, built on the harvests of chance without meaning, purpose or plan." Describe your attitude to this claim and if it corresponds to your own opinion.

The responses to the last question is of special interest. It is rather easy to classify the answers and the sample is clearly divided in two groups. The *first* group consists of those who clearly and without reservation accept the claim "that all life is only the result of an unplanned development, built on the harvests of chance without meaning, purpose or plan." One of the biologists writes:

> Couldn't have said it better myself, agree to a 100 percent.

Another biologist writes:

> I believe rock hard on evolution and that all organisms adjust to the life-space which they have been given.

Most of those who are convinced, express their conviction in emotionally strong terms. This is rather sharp contrast to most of those belonging to the *second* group, who in one way or the other are hesitant. In addition, their hesitation is linked to their view of human beings in general. Humans have consciousness, and we can plan our lives. Furthermore, biological science raises questions about chance. Many of the respondents have clear difficulties to integrate the notion of chance and randomness with their general view of the world.

GENERAL OBSERVATIONS

There is one general observation about the results of *Study 2*. On the one hand, there is an expected skepticism against religion. This becomes evident in the responses to various questions, but it is especially clear in the answers to the question "Have your scientific and biological knowledge affected your attitudes to life—and if so—in what way?" Ten of the

35 respondents claim that such a biology-induced change has occurred. And it can most adequately be described as *loss of faith*. One of those—a scholar—writes the following:

> I'm brought up in a free-church environment and I have for a long time called myself a Christian. My scientific view of life has changed all this and induced me to turn away from faith, religion and religions in general.

Another respondent writes:

> Totally! I was a "believing Christian" before I opened my eyes to natural science. I went to church, was going to confirmation and prayed at nights before I went to sleep. Now 10 years later I have given up most of my old views and realized that we are not as special as we think. I have begun to take things a little less serious and I'm open for most things in life.

What is the character of the Christian belief, which is left behind? There is no clear answer to this question. There are, however, some indications. Traditional anthropocentrism of Christian belief is perceived to be superseded by the biocentrism of evolutionary theory. But there are also indications that the former Christian belief of many of the respondent is understood as existing on the same level as scientific belief. What is left behind is Christian belief in the form of an "empirical theism"—to borrow a term from Anders Jeffner.[25] It is a worldview, according to which religious beliefs should be supported by the same kind of arguments that support scientific claims. Such a form of Christian belief is discarded, when a superior—and possibly simpler and natural—explanation is found.

Alongside the clear skepticism against Christian belief, there is also another tendency, which must be mentioned to be fair to the material. At the outset of the questionnaire, the respondents were asked how often they reflect on the big existential questions of life: "what is most important in life? How should we think about life as a whole? How does it feel to live?" One third of the respondents do not reflect on such issues very often, but two thirds think about those things often or daily. Some of those who are uninterested are uninterested because they think it is meaningless to reflect on questions, which have no answer. A biology student writes:

> I cannot think that I reflect upon something. I realized already as a young person that there are questions, which nobody could answer. What happens after death, the end of the universe etc.

---

25. Jeffner, *Filosofisk religionsdebatt*, chap. 2.

> Why should I devote a lot of time to issues to which there are no answers?

But most of the respondents have another attitude. "How often do you reflect on existential issues of life?" One respondent—a biology teacher—gives the following answer:

> Every day, many times a day. There are so many things, which we are used to and patterned which I perceive better after such reflections.

This is an interesting response. Many postmodernists argue that people in our late modern society are no longer interested in the overall perspectives and that they have resigned when it comes to try to understand the world as a whole. This does not seem to be the case in the responses from the respondents. On the contrary, many of them use biology to gain an interpretative pattern of life—even if they find it difficult to accommodate chance and randomness in their personal life.

SUMMARY

In this chapter the focus has been upon different worldviews such as religion, spiritualism, biologism, and skepticism among the general Swedish public and social subgroups within Sweden. The main result from a study made in 2017 (*Study 8*)—compared to a similar study in 2006 (*Study 1*)—shows that religion and spiritualism are decreasing, but so is biologism (from is 19 to 14 percent). Skepticism is increasing and in 2017 up from 14 to 17 percent of the Swedish population.—When the same survey in 2006 was directed to academically trained biologists, the result shows that they are less prone to biologization than the Swedish public. In another study (*Study 2*) 36 biologists (scholars, teachers, and students) responded to a set of open questions about their worldviews and biology. Half of them clearly agreed to the statement "that all life is only the result of an unplanned development, built on the harvests of chance without meaning, purpose or plan." The other half was more hesitant, but found it difficult to integrate randomness and chance in their worldview. Furthermore, 10 of the 35 respondents claim that a biology-induced loss of religious faith had occurred in their life.

2

Swedish Public and Clergy on Biology and Religion

Secularization is often described in terms of differentiation. The sociologist José Casanova has underlined this several times, for example in his *Public Religion in the Modern World* (1994):

> [T]he central thesis of the theory of secularization is the conceptualization of the process of societal modernization as a process of functional differentiation and emancipation of the secular spheres—primarily the state, the economy, and science—from the religious sphere and the concomitant differentiation and specialization of religion within its own newly found religious sphere.[1]

In the context of science and religion, the thesis of secularization as differentiation could imply the presence of a divide between Christian communities and the larger public. Such a divide is also implied in the concept of the Christian community as a "cognitive minority"[2] in separation, tension or even conflict with science. What is the evidence for such a divide? This is an empirical question. A comparison between *Study 3* (made in 2009), directed to the general public, and *Study 6* (made in 2015, directed to religious leaders) could contribute to an answer.

---

1. Casanova, *Public Religion*, 19. Note that Casanova has certain misgivings about this understanding of secularization.
2. Laeyendecker, "The Church as a Cognitive Minority," 71–81.

METHODOLOGICAL QUESTIONS

One of the purposes of *Study 3* and *Study 6* was to estimate the presence of biologism in comparison to other worldviews.[3] In the foregoing chapter, I presented some studies indicating the influence (albeit declining) of biologism. How is this worldview related to the worldview of Swedish minsters and pastors? And what are the attitudes of Swedish clergy to other existential issues?

Study 3 was launched in 2009 and based on 502 telephone interviews with adult Swedes between the age of 18 and 74. It contained thirty-six items and thirty-two of them were derived from IWV.[4] The respondents were asked to state agreement or disagreement on a scale from 1 (totally agree) to 5 (disagree completely). *Study 6* was launched in the spring of 2015 and directed to ministers in the Lutheran Church of Sweden and pastors in the Uniting Church in Sweden. It contained five different sections, of which the first included ten of the items used in *Study 3*. Section 2, 3 and 4 concerned science and religion, and section 5 consisted of items about age, gender, etc. (will be discussed in Chapter 3.[5]) Methodological issues are presented in the introduction to appendix 3.

RESULTS

Table 2 presents a comparison between the responses of the two samples chosen from the general public of adult Swedes (*Study 3*, 2009) and from Swedish clergy (*Study 6*, 2015). The percentages consist of the sum of those who completely (5) or partly (4) agreed to the statement in *Study 3* in 2009 (n=502) and the sum of those who chose 8–10 on the ten grades-scale used in the *Study 6* in 2015 (n=1492).

---

3. See above in chap. 1.

4. Item 2, 5, 6–11, and 20–23 coincided with items in *Study 1* and *Study 6*. See Appendix 2.

5. The web survey is found in Appendix 2. The items are found in section 1 of the survey. The Uniting Church of Sweden was formed in 2011 between the Mission Covenant Church of Sweden, the Baptist Union of Sweden, and the United Methodist Church of Sweden.

	Swedes in general (2009), n=502	Ministers and pastors (2015), n=1492
(1) Human beings have certain unique qualities, which no other creatures have.	81	91
(2) Humanity is going towards a self-inflicted environmental disaster.	61	48
(3) We must respect the order of nature irrespective of how this affects human welfare.	76	54
(4) Humans would be much better off if they lived a simple life without so much technology.	39	17
(5) Biology gives us the complete answer to the meaning of life.	24	1
(6) Humanity exists to govern over the rest of creation.	15	39
(7) The more science advances, the more difficult is it to believe in a God who created the world.	48 (2008)	2
(8) Humans consist only of body and matter.	25	2
(9) Nature is filled by a spiritual force permeating all life.	42	43
(10) Life has no other meaning than to further the human species.	22	1
(11) What becomes of a human being is mostly dependent on her biological heritage.	16	1

Table 2: Religion and biology in telephone interviews to a national sample of the Swedish population between 18 and 74 years in 2009 (*Study 3*) and to ministers in the Lutheran Church of Sweden and pastors in the Uniting Church in Sweden in 2015 (*Study 6*). Percentage.

It should be added that the clergy responding to the web survey in 2015 also were given the opportunity to make their own comments to the statements in a special comment field. Most of the respondents did this, more or less extensively. Some of these comments were in the form of requests for clarifications of one or more statements in the survey. This is hardly surprising. The study was directed to a well-educated group with reflected views concerning the issues raised in the survey. Similar reactions occur in surveys directed to medical doctors and lawyers.

It should also be added that a more limited number of respondents were given the opportunity to develop their views in a more extended telephone interview. The results of this study will not be discussed in the present context.

The results of the surveys will be presented in the following way: *first*, I will focus on the general question of *science and religion*, i.e. item 7. *Secondly*, attention will be directed to the responses to items 5, 8, 10, and 11, which all concern aspects of *human nature*. *Thirdly*, I will highlight the view of *nature* in the two samples and particularly attend to items 3, and 9. *Fourthly*, there are the items about the relationship between *humans and (other) animals*, i.e. items 1 and 6, and *fifthly*, questions about *environmental* issues, i.e. items 2 and 4.

Science and religion

In contemporary sociology of religion, the nature and explanation of religious change is a much-disputed issue. The so-called secularization paradigm is often contrasted to rational choice theory,[6] according to which the demand for religious "goods" is fairly stable and rooted in our need for salvation and eternal life. Religions thrive by competition between different religious firms seeking to attract religious consumers. This new trend in sociology of religion notwithstanding, the classical secularization paradigm is still influential. Religious decline may not be inevitable as claimed by classical theorists, but it is certainly visible on the European scene. According to Steve Bruce, there are at least two factors behind this development: individualism on the one hand, and science and rationality on the other. These factors remove many of the purposes of religion and render many of its beliefs implausible.[7] But " . . . it is science that has the most deadly implications for religion."[8]

Whether this version of the secularization paradigm is adequate can be debated. Nevertheless, it seems to be a popular paradigm within the larger Swedish public. *Study 3* in 2009 contained an item concerning science and religion in general. Almost 50 percent believes that the advance of science makes it more difficult to believe that God created the world. Results from a similar study made by the *Swedish Association for Public & Science* confirm

---

6. See, for example, Stark and Finke, *Acts of Faith*; Bruce, *Choice and Religion*.
7. Quoted in Davie, *Religion in Modern Europe*, 24.
8. Stark and Finke, *Acts of Faith*, 61.

this figure. This study was also directed to the general adult public in Sweden. In response to the question whether a scientific view of the world could be reconciled with religious belief 47 percent answered "no." 43 percent answered in the positive.[9] The most common answers among those who thought that science and religion are compatible were the following:

- Two completely separate things—thus there is never a problem (most women respond so).
- Science and religion are two alternative approaches.
- Science is not enough. Where it cannot go any further, religion kicks in. Some respondents point specifically to science exploring God's creation and evolution, which God initiated. (Most men are in this response category).
- There are scientists who are religious, so it must go.[10]

The most common answers among those who thought that science and religion are incompatible were:

- They are two completely different things that do not go together.
- I do not believe in God
- Religion has negative consequences.
- The nature of science is to resist religion, which rests on faith and feelings.
- Faith is unscientific.
- Darwin and the theory of evolution make it impossible.[11]

A similar item was contained in *Study 6* directed to the Swedish clergy. Nearly 1500 respondents were asked about their opinion to the following statement: the more science advances, the more difficult is it to believe in a God who created the world. Only 2 percent agreed; 90 percent disagreed. *This indicates a huge gap between religious leaders and the Swedish public when it comes to the perception on the general relationship between religion and science.*

---

9. Hermansson, *Vetenskap att tro på?*, 17.

10. Ibid., 19.

11. Ibid., 18. On the last response about the faith—or non-faith—of scientists, see Ecklund, *Science vs. Religion*. Ecklund surveyed nearly 1,700 scientists and interviewed 275 of them. She found that most of what is commonly believed about the faith of elite scientists is wrong. Nearly 50 percent of them are religious. Only a small minority are actively hostile to religion. See further below in chap. 6.

The previously mentioned survey made by *Swedish Association for Public & Science* gives a clue to an interpretation of this gap. A large majority (71 percent) of those who believe in God also deny that the more science advances, the more difficult is it to believe in a God who created the world. *In contrast, 72 percent of the non-believers affirmed the statement.*[12] The perception of the general relationship between science and religion seems to reflect already held beliefs or disbeliefs in God.

An ambiguity in the statement of item 7 in *Study 6* should be noted. A pastor in the Uniting Church in Sweden can serve as an example:

> Somewhat unclear and leading questions. For example, question 7 . . . I don't know if this question is directed to me or to what I believe others are experiencing. I personally would like to mark 1 (disagree completely), but in regard to society a 7 (agree somewhat).

There are, however, other comments, which indicate that the item 7 was taken as an item about the perception of the impact of science on his or her faith (and not about what "others are experiencing"). Such an interpretation is supported by another study from 1999, where a representative sample of the adult Swedish population was asked about *their* attitude to science. 48 percent agreed that the more science advances, the more difficult is it to believe in a God who created the world.[13] This is close to the result in the study made by the association *Public & Science* in 2008–2009. It indicates that item 7 in *Study 3* measures the perception of the impact of scientific progress on the personal religious faith of the respondents. *The overwhelming majority of Swedish clergy do not perceive the progress of science to make it more difficult for them to believe in God.*

It is interesting to compare these results with some results in the Finnish Science Barometer from 2007.[14] The question had the following text: "What is your attitude towards the following statement on science and research: The scientific world view is not incompatible with religion." About one third of the respondents (n=1068) answered in the positive, while about almost 40 percent disagreed, i.e., were convinced that the scientific worldview is incompatible with religion. There seems to be no great difference in this question between Finland and Sweden.

---

12. Hermansson, *Vetenskap att tro på?* 17.
13. Bråkenhielm, *Världsbild och mening*, 238.
14. *Finnish Science Barometer 2007.*

Commenting on the report, the Finnish researchers notice very little change compared to the previous Science Barometers in 2004 and 2001.[15] But a more noticeable change has taken place regarding another statement, i.e., the statement that "belief in science has become a modern form of religion which is guiding people's sense of values in the wrong direction." About 25 percent agreed and 20 percent disagreed. These figures should be compared to those from 2007, i.e., 33 percent and 30 percent respectively. First, these figures seem to indicate a kind of polarization between those who are in sympathy with scientific values and those who are not. Secondly, the sympathies for the scientific worldview are growing. "This is the second largest change in the results to be seen in the extensive statement section of the questionnaire."[16] One possible explanation could be that sympathy for the scientific world view positively perceived as an alternative to religion is growing in the younger generation. One indication in favor of this hypothesis is that younger persons affirm the theory of evolution to a greater degree than older persons. (Furthermore, denial of the theory of evolution shows a significant correlation with trust in the church.[17])

The nature and purpose of human life

Four statements concern human nature, i.e., item 5, 8, 10, and 11. I will present the results and comment on the responses to each of these statements in turn beginning with item 8.

Item 8 concerns a central issue in philosophical and theological anthropology, namely, the interpretation of human nature. Are there truths about human beings other than those pertaining to their physical attributes? 25 percent of the general public affirms that humans consist only of body and matter. Only 2 percent of Swedish religious leaders concur. Still there is a noticeable gap between the Swedish public and the Swedish religious leadership, even if not as obvious as is the case in the former issue about the general relationship between science and religion.

---

15. Ibid., 28.
16. Ibid. No big changes are found in the barometers of 2010 and 2013.
17. *Summary of the Finnish Science Barometer 2013*, 16–17. The claim that "humans have evolved over millions of years from other, earlier species of animal" meets widespread, but not unanimous approval. Seven out of ten agree, but one in eight (12 percent) do not. The breakdown does not differ from the *Finnish Science Barometer 2010*. Differences in the responses according to population group are notable; the younger the respondent, the more convinced they are of the validity of the theory of evolution. The link to education is just as clear: the higher the education level, the stronger the belief that we share our origins with animals. The respondents who have the strongest trust in the church are widely represented amongst those that deny the theory of evolution.

Two important things should be noted about the response from the general public. *First*, there is an obvious gender difference when it comes to responses to the statement that human beings consist only of body and matter. 49 percent of the men affirm this statement, but only 19 percent of the women! Theories about gender differences may provide different explanations to this "gender gap." *Secondly*, when it comes to male response there is a significant change compared to male responses to the same question in 1999. The figure for masculine responses was then 33 percent—a significant increase in 16 percent in ten years. There is a gap between the genders when it comes to views on the nature of human beings—and an increasing gap at that.

The affirmation or denial that human beings are nothing but body and matter is related to the question of human consciousness and its relationship to the brain. In the history of philosophy, it has been framed as the problem of body and soul. The most influential solution was radical dualism: the soul is separable from the body, and the person is identified with the former. It was also taken as an essential part of Christian theology. This has been seriously questioned in modern theology and one source of this is Old Testament scholarship claiming that "human beings must be understood in their fully integrated, embodied existence."[18] This claim is reflected in some of the comments to the web-survey to Swedish clergy. A minster in the Church of Sweden comments:

> Question 8 is impossible to answer from biblical terminology. The human being is body. The human being is soul. The human being is also a spiritual creature. The questioner seems colored by a more dualistic view of reality.[19]

This view can be compared to the view of Nancey Murphy, who was one of the editors of the volume *Whatever Happened to the Soul?* Murphy distinguishes between four views of the human person:

1. *Radical dualism*: "the soul (or mind) is separable from the body, and the person is identified with the former."

2. *Holistic dualism*: the person is ultimately to be identified with the whole, whose normal functioning is as a unity.

---

18. Green, "Bodies—That Is, Human Lives," 158.

19. On this, see also the *Catechism of the Roman Catholic Church*, paras. 362–68, making use of the Aristotelian terminology of the soul as the form of the body. On St. Thomas's complicated argument on body and soul, see McInerny and O'Callaghan, "Saint Thomas Aquinas, " section 8.

3. *Nonreductive physicalism*: "the person is a physical organism whose complex functioning, both in society and in relation to God, gives rise to "higher" human capacities such as morality and spirituality."

4. *Eliminative/reductive materialism*: the person is nothing more than a physical organism, which ultimately will be explained by natural science.[20]

The minster's response could be interpreted as an affirmation of 2 and a denial of 1. In comparison, Nancey Murphy and other contributors to *Whatever Happened to the Soul?* make a plea for non-reductive physicalism, while arguing that reductive materialism is unacceptable to the Christian and that radical dualism is incompatible with science.[21]

Furthermore, the results of the survey among the general public suggest that a strong minority is in sympathy with reductive materialism. 25 percent affirm that human are nothing but body and matter and 24 percent affirms that biology gives us the complete answer to the meaning of life (item 5). This is further indication of the earlier mentioned trend of biologization. It is not surprising that only 2 percent of the Swedish clergy concurs. A pastor from the Uniting Church in Sweden comments:

> Are there things, which are not biological? Even philosophy and philosophical thoughts are part of our biological nature, so also belief in God. They are also the result of chemical reactions in the brain. But this disproves only a traditional view of the world and of God; it does not disprove the existence of God.

This comment links up with the former issue about the human person, and invites the further question how it is (1) that the belief in God is the result of chemical reactions in the brain is not incompatible with (2) the existence of God, nor for that matter with (3) the conviction that the meaning of life is found through belief in God.[22]

Those who affirm that biology *completely* answers the question about the meaning of life can be interpreted as proponents of a particular model of the relation between science and religion. This model has been described by Mikael Stenmark as a model of substitution according to which the domain of science should be expanded to cover the area of religion and provide us with a new worldview with—suggested by Edward O. Wilson—the evolutionary epic as a the central myth. In another context, Wilson writes:

20. See Murphy, "Human Nature," 24–25.
21. Murphy, "Nonreductive Physicalism."
22. On the question, how belief in God might be an answer to the question of the meaning of life, see Nagel, *Secular Philosophy and the Religious Temperament*, 3–18.

> We are obliged by the deepest drives of the human spirit to make ourselves more than animated dust, and we must have a story to tell about where we came from, and why we are here. Could Holy Writ be just one of the first literate attempts to explain the universe and make ourselves significant within it? Perhaps science is a continuation on new and better-tested ground to attain the same end. If so, then in that sense science is religion liberated and writ large.[23]

Whether the model of substitution is an adequate account of the relations between science and religion depends on whether science can perform all—or at least most—of the functions previously performed by religion. This does not seem very likely. Could science in the face of death and suffering give consolation in the same way as religion? Furthermore, could it give a credible answer to the question why there is anything at all, rather than nothing?

16 percent of the general Swedish adult population affirms that what becomes of a human being is mostly dependent on her biological heritage (item 11). The figure for Swedish clergy is 1 percent. Still it should be noted that many ministers and pastors underline the significance of genetic inheritance.

How far does the significance of genetic factors go? What is the weight of environmental factors? The studies of twins are undoubtedly relevant since identical twins have the same genetic equipment. Such studies suggest the genetic factor is more significant than commonly thought.[24] But—as Anders Jeffner underlines—"this is far from demonstrating that our personalities are totally genetically determined." Jeffner continues:

> To come to such result we must not only explain how the brain is formed through the interplay of the genes and the material environment but also that the mental states of the brain are not causally active—that our lives are not formed partly by free choice.[25]

Explanation of the formation of the brain lies within the domain of science, but the idea about the total determination of all mental states does not. Jeffner argues, "it can never acquire the same degree of certainty as my insight that I have decisions and purposes which are something more than

---

23. See Wilson, *Consilience*, 6; Wilson, *On Human Nature*, 201; Stenmark, "Naturvetenskap och religion," 79–95.

24. See Jeffner, *Biology and Religion*, 16, 28; with a reference to Scarr, "Theories for the 1990s," 1–19.

25. Jeffner, *Biology and Religion*, 16.

obedient companions to chemical processes in the brain."²⁶ Such insights are often dismissed as "folk-psychology" and superseded by a more scientific theory of human consciousness.

22 percent of the respondents representing the general Swedish population agree to the statement, "life has no other meaning than to advance the human species" (item 10). This claim is another indication of biologization, which we have discussed earlier.²⁷ The affirmation invites further questions. Is it to be interpreted as a flat denial of any other meaning to life than reproduction? Or is it rather to be understood as an emphasis on biological meaning without denying that there are other things beyond biology? I would suggest the latter, but a justification for this interpretation would require further studies based on open questions in longer interviews.

The claim "that life has no other meaning that to advance the human species" has almost no adherents among the Swedish clergy. This is hardly surprising even if the advancement of the human species is significant in Jewry and Christianity and often supported with reference Genesis 1:28: "Be fruitful and multiply, and fill the earth and subdue it." However, this reference is not only challenged in environmental theology; it is traditionally also supplemented by the idea of human beings as created in the image of God, not in the sense of dominating over the rest of creation, but finding the meaning of life in relationship to God. This is one possible explanation to the response of Swedish clergy to this item.

In conclusion, Swedish clergy forms a rather homogenous group when it comes to questions of philosophical and theological anthropology. The results indicate that ministers and pastors in Sweden are at variance with a strong minority (about 15 percent) of the general Swedish population in sympathy with biologism. The clergy stands in between a group of adherents to biologism and another group supporting non-biologism but not particularly interested in religion.

What will be the future development? One hypothesis is that new discoveries about the brain and the technological advances of artificial intelligence will strengthen biologism and reductive materialism. According to the results in chapter 1, this has not happened during the last ten years, even if there is evidence for in decline in more holistic view of human nature. Possibly, Christian and other religious communities might strengthen the resistance to a reductive view of human nature. The decline of biologism notwithstanding, no such development can be discerned in the data.

26. Jeffner, *Biology and Religion*, 16.
27. See above in chap. 1.

Nature

Two items in the telephone survey to the general Swedish public in 2009 and web-based survey to Swedish clergy in 2015 concern nature. As mentioned earlier, nature has been an important theme in many Swedish worldview studies.[28] The overall picture that emerges in these studies can be described in the following way:

> One could speak of the emergence of an ecological life-philosophy. Central to many persons is the idea of the unity of life. Human beings are but a thread in the web of nature. Humans and animals are equal—at least in principle. Health—the health of the body—is the most important thing in life. Why? Because this is the only life I have and I want to make the most of it. This worldview is permeated by a basic attitude of optimism . . . Somewhere there is an unclear sense of something more, that reality is greater and richer than the part of it that we can discern without senses and capture through our concepts.[29]

This worldview has been studied in more detail in various studies[30] and one aspect concerns the relationship between humans and animals. I will treat this issue separately in the next section. Two aspects of this ecological worldview are highlighted in two items, one of which concerns the spiritual aspect of nature (item 9) and the other the order of nature (item 3).

The responses to item 9 are in line with other studies. 42 percent of the sample of the general Swedish public affirms that nature is filled by a spiritual force permeating all life. In a study made in 1999, the same figure was 62 percent (n=1282).[31] This suggests that this spiritual aspect of the ecological worldview is weakening in Sweden. It seems to have an equally strong standing among the Swedish clergy. 43 percent affirms the statement. However, there are interesting gender and regional differences. Clergy from Stockholm and Uppsala have a stronger tendency to affirm the statement than clergy from other cities and less populated areas. Furthermore, women affirm the statement more strongly than men.

Item 9 is related to an important question in Christian theology, namely, the concept of God and God's relationship to the world. Those ministers and pastors, who affirm the statement that nature is filled by a spiritual force permeating all life, may suggest that God is immanent in the

---

28. See further Jeffner, "A New View of the World," 137–45.
29. Bråkenhielm, "Christian Tradition," 28–29.
30. See, for example, Bråkenhielm, "Environmentalism," 49–66.
31. See Bråkenhielm, *Världsbild och mening*, 232. In another study from spring 1997, 69 percent affirmed the statement.

world (and nothing beyond the spiritual force permeating all life). One of the respondents comments:

> I believe that the Spirit of God permeates all living, breathes in everything. If that is what you mean or not, I cannot decide.

A pastor in the Uniting Church in Sweden is of another opinion:

> I believe that the Spirit of God acts in nature. But not that nature in itself is permeated by a power. I don't believe that nature has a "soul."

These two religious leaders may both dismiss pantheism, i.e., the view that God and nature are one and the same (substance). In what way to they differ? What is the difference between the view that the Spirit of God acts in nature and that the Spirit of God "breathes in everything?" One possible interpretation is that the personal qualities come in the foreground when one speaks of God as agency. Agency requires a distinction between the person who acts and that towards which an action is directed. A minster in the Church of Sweden suggests that God and the holy Spirit of God "is not only in the created, God is more than creation." A pastor explicitly denies that "every human/animal/nature should have God inherent in themself" and supports this with a number of quotes from the Bible. Another respondent distances himself from the thought that nature is permeated with a soul, but affirms that everything is "leavened" by God. What is the difference here? Furthermore, the responses reveal a difference between those who affirm God as an independent self or God's aseity[32] and those who want to downplay this attribute and emphasize God's omnipresence. This is difficult to understand—in what way is aseity and omnipresence mutually exclusive?

One possibility is that these differences have to do with tension between a transcendent view of God and a more immanent. This is suggested by the sociologist of religion Thorleif Pettersson. Among other things, Pettersson referred to a sermon by the former archbishop of Sweden, K.G. Hammar, when he declined speaking about God "above" preferring the idea of God "at the level of eyes."[33]

Item 3 involved a statement that has been used in many surveys in order to measure the normative weight of things or processes considered to be "natural." The statement was used already in an interview study made in 1988, when almost 80 percent agreed (completely or partially) to the statement that we must respect the order of nature irrespective of how

---

32. From Latin *"a se esse,"* being from oneself. See Hick, *Philosophy of Religion*, 7.
33. Pettersson, "Svensken och religionen."

this affects human welfare. This figure has changed very little over time; in the telephone interviews to the representative sample of the Swedish adult population in 2009, 76 percent agreed more or less strongly. Swedish clergy is not far behind with 48 percent. Needless to say, this clear affirmation of the integrity of nature does not imply that humans have no right to interfere with nature. Of course, humans have such a right. But it stands against what could be expressed in the form of a precautionary principle. UN World Charter for Nature states, "when potential adverse effects are not fully understood, the activities should not proceed."[34] Defining and applying such a principle involves a host of problems, but an elaboration of this goes beyond the scope of the present study.

In conclusion, it is to be noted that the gap between the Swedish public, which was evident concerning science and religion, is still present but less evident when it comes to human nature, but almost negligable when we compare attitudes to nature.

Humans and animals

Animals are part of nature, but our relationship to them involves special considerations. Empirically many studies into this area have been performed. In one of the pioneering worldview studies in 1988, respondents were asked "Do you think that we ought to pay greater respect to human beings than to other living creatures?" A clear majority responded with a resounding "no." Anders Jeffner suggests that the most likely explanation of this opinion was that the world was perceived as an ecosystem where all components are mutually dependent.[35] It is this ecosystem, which has a normative standing. Of course, one may ask if the stability of the ecosystem is not overemphasized; nature is not static but dynamic and the role of humans in this dynamic is sometimes, but not always destructive.

The general trend from the responses to the former items about nature continues in the responses to item 1 and the statement that human beings have certain unique qualities which no other creatures have. A clear majority within the sample of the Swedish public (81 percent)—not to mention the sample of the clergy (91 percent)—affirm that human beings do indeed have certain unique qualities that no other creatures have. The interesting question is, of course, which these unique qualities are. Considering the history of philosophy, one natural answer would be the soul, but as we saw earlier, soul-language is not without problems for Swedish clergy. According to other studies, soul appears to be a forgotten concept. When asked about the reason

34. For an interpretation and a critical analysis, see Sunstein, *Laws of Fear*.
35. Jeffner, "A New View of the World," 139–40.

for the higher value of humans over animals, many persons responded that humans have a larger brain. Most respondents lacked a language to describe the mental world. Jeffner writes, "you cannot come further from Platonism and traditional theological anthropology."[36]

Item 6 asked for a response to the statement that humanity exists to govern over the rest of creation. It is hardly surprising that this was an idea foreign to the Swedish general public. Only 15 percent agreed completely or partially (still more than the 9 percent who affirmed the statement in 2006). Equally expected was that Swedish clergy were more positive—39 percent. But there are interesting differences. *First*, a gender gap is visible. Only 30 percent of the women agree, while 47 percent of the men concur. *Secondly*, there is also an age gap. Younger minsters and pastors (age 26–41) are less likely to agree than ministers and pastors between the age of 62–79 (33 percent vs. 49 percent). Moreover, and *thirdly*, there is a confessional gap. Pastors in the Uniting Church in Sweden are more ready to affirm that humanity exists to govern over the rest of creation than minsters from the Lutheran Church of Sweden.

It should be emphasized that many of the respondents expressed reservations about the expression "rule" in item 6. They insist that the "rule" should be interpreted to mean "responsible for," "caring," or "stewardship." If you have this in mind, the distance to the rest of the population is less clear. A minister in the Church of Sweden puts it as follows:

> No, not control, we must cultivate and care for it (1 Gen 2:15). In the Bible there is indeed "rule" but that does not mean that we shall devastate and exploit as dictators.

The different possibilities of interpretation may have prompted some of the respondents to be drawn to the middle option (i.e. 5 on the 10-point scale). A minister in the Church of Sweden expresses this clearly:

> Depends on what you mean by control. Man is part of creation and belongs to it, but because we have got the brainpower we have a much greater responsibility than other living creatures have. If "rule over" means stewardship my answer is 10. If "rule over" means manage as one feels like my answer 1. Now I place myself instead on 5, because I think the question is fuzzy.

In summary, the conclusion is that the criticism of the classical Christian worldview as a hierarchy with God at the top, human beings in the middle with animals and nature at the bottom, has had an impact among today's priests and pastors, and esepecially among younger female ministers in the

---

36. Jeffner, "A New View of the World," 141.

Lutheran Church of Sweden. Philosophical criticism of human's unique position in the sense of having unique characteristics is not visible. Contemporary religious leaders in Sweden maintain the unique nature of human beings consistently and unreservedly.

Environmental issues

Environmental issues have received a lot of attention in Swedish Christianity. The Uniting Church in Sweden has issued a special policy on Peace and Sustainability[37] and in 2014 the Bishops' conference of the Church of Sweden issued a much-discussed Bishops' letter about the climate.[38] "God has empowered humanity with the spiritual, psychological and material resources even to face the environmental crises," writes Anders Wejryd, former archbishop in the Church of Sweden.

How does Swedish clergy view environmental issues? *Study 3* and *Study 6* contained two different statements on this topic, the first of which concerned the possibility a self-inflicted environmental disaster (item 2) and the second about a simpler life without technology (item 4).

Is humanity approaching a self-inflicted environmental disaster? 48 percent of the Swedish clergy affirms this dark vision of the future. The differences between Lutheran ministers and the pastors in the Uniting Church, and between different age groups or regions were small. Optional comments were few, but some say that a disastrous development is not inevitable.

If compared with the results of *Study 3* directed to the Swedish public in general, one can conclude that pessimism is even stronger. Approximately 60 percent agreed completely or partly in the claim that we are moving towards a self-inflicted environmental disaster. In an interview study published in 1995, Nils Uddenberg wrote that most people nevertheless pronounce a sort of "optimism in the gallows."[39] There is a clear gender difference; women are more likely to affirm that humanity is moving towards a self-inflicted environmental disaster.

Are there any connections apocalyptic beliefs and the Christian doctrine of the world end? There is no trace of this in the comments field. Just one . . . A minister in the Church of Sweden writes:

> Everything created is finite. If the end of everything comes through an environmental disaster remains to be seen.

---

37. *Policy för fred och omställning.*
38. *A Bishops's Letter about the Climate.*
39. Uddenberg, *Det stora sammanhanget*, 83.

Another environmental issue raised by item 4 that humans would be better if she lived a simple life without so much technology. Only 17 percent of the sample of the Swedish clergy agreed on this, i.e., much less than the proportion of the sample of the general Swedish population did in 2009 (39 percent). A minster in the Church of Sweden is representative of what was said about the matter in commentary field.

> I agree if I think about how technology is used now. For example, it produces so many different machines that are thrown in the trash in a few years. Technical developments are crunching natural values. And human beings do not feel good to sit too much in front of the computer. But the technology itself is not necessarily a negative thing; it brings also very positive things.

In a similar way other members of the Swedish clergy pointed out that technology is both good and bad. A pastor of the Uniting Church stressed that "we could not cope with the continued survival by returning to some sort of romanticized Antiquity."

An important issue in the sociology of religion has been on the Christian faith promotes or prevents an environmental commitment. Earlier studies on this question point in different directions.[40] The survey on Swedish clergy in 2015 gives no grounds for any definite conclusions, but the general impression is that environmental issues engage today's religious leaders.

SUMMARY

In important respects, Swedish clergy concurs with a view of nature and humanity that is embraced by Swedes in general. In particular, this is the case for issues such as view of the unique qualities of human beings, respect for the natural order and the fears of a coming environmental disaster. There is also a significant part of the Swedish public who affirms that nature is animated, which nearly half of the Swedish clergy also affirms. A significant part of the Swedish population is skeptical of modern technology, and it also applies to some, but not most, priests and pastors.

40. See, for example, Hope and Jones, "The Impact of Religious Faith." "Muslim and Christian participants' opinions about climate change and CCS technologies were shaped by the importance of environmental stewardship and intergenerational justice. Both groups had relatively low perceptions of urgency for environmental issues, particularly climate change, due to beliefs in an afterlife and divine intervention . . . Secular participants expressed anxiety in relation to environmental issues, especially climate change. Lack of belief in an afterlife or divine intervention led secular participants to focus on human responsibility and the need for action, bolstering the perceived necessity of a range of technologies including CCS" (from the abstract of the article).

The responses to other items reveal striking differences. A significant minority of Swedes in general is convinced that human beings consist only of body and matter, or that biology provides the complete answer to the meaning of life. But very few Swedish ministers and pastors affirm such biologism. The most striking difference between Swedish clergy and Swedes in general concerns the perception of the relationship between science and religion. Only 2 percent of the clergy think it has become more difficult to believe in God, because of the progress of science—in contrast to almost half of the general Swedish population.

3

Public and Clergy on Science and Religion

The Lutheran state-church in Sweden has been gradually weakened during the nineteenth century, but it was not without power after the Second World War. Different events during the 1950s placed this sacred institution under strain. This process and its repercussions has been described by the British historian Hugh McLeod. In general, McLeod distinguishes between three levels of explanations for religious change in the "long sixties:" the long-term processes of modernization (level 1), immediate social changes (level 2), and specific events, movements and personalities (level 3).[1] One important part of the religious change of the Swedish 1950s had to do with level 3, i.e., specific events, processes and personalities. McLeod rightly emphasizes the impact of the intellectual attack on Christianity that was launched by Ingemar Hedenius, professor in philosophy at Uppsala University between 1947 and 1980. His book *Tro och vetande* (Faith and Reason) was published in 1949,[2] but some material was already published in the liberal newspaper *Dagens Nyheter* and in the social democratic journal *Tiden*. When the book was published, the heat was already on. Hedenius' articles and books received much criticism (not least from bishops and theologians), but the subsequent verdict was that Hedenius won the debate and that he swept the floor with his ecclesial and theological opponents. McLeod concludes:

---

1. McLeod, *The Religious Crisis of the 1960s*, 257.
2. Hedenius, *Tro och vetande*.

Church-going was already very low in Sweden, but the debate encouraged atheists and agnostics to 'come out' in a way that in other countries happened more often in the 1960s and 1970s.[3]

There are many anecdotes about the impact of Hedenius' [separate] book. One comes from my Swedish-teacher in high school. I wrote an essay on Hedenius in 1962 voicing certain criticism. He remarked that he found it interesting that a person had found it justified to question one of his—as he put it—"domestic gods." Another witness is Thure Stenström, now retired professor in literature, in the late 1940s student in Uppsala. He testifies to the significance of Hedenius; in the beginning of the 1950s, Hedenius thoughts were discussed in every student dorm in Uppsala.[4]

The impact was not only deep, but also lasting. In the beginning of the new millennium, new books on the debate emerged, and in the fall of 2002 the public discussion was revived in the national newspaper *Svenska Dagbladet*. A dozen of articles by scholars and philosophers were published and the public attention was wide-ranging. Fifty years since the original debate, the heat was on again.

STUDIES ON THE PUBLIC PERCEPTION OF SCIENCE AND RELIGION

These debates on faith and reason provide one important context for the related question of science and religion in Sweden. An interesting study on the public's perception of science and religion was made in 1999, i.e., some years before the revival of the debate on faith and reason in 1949–1951. A survey was sent out to a representative sample of the Swedish population (n=2000) between the age of 20 and 55 with a response-rate of 54 percent.[5] In one of the items, the respondents were asked about three different statements about science and religion. The study was replicated in 2017 (*Study 8*). The general results as well as the partial distribution among those who believe in a God and those who do not are displayed in table 3.

---

3. McLeod, *The Religious Crisis of the 1960s*, 55.

4. Lundborg, *När ateismen erövrade Sverige*, 12.

5. Görman, "Svenskars uppfattningar om relationen mellan naturvetenskap och religion," 34–38.

Science and religion	All respondents		Believe in a God with whom you can have a personal relationship (19 percent of the total sample)		No belief in any god, supernatural power or force (15 percent of the total sample)	
Year of study	1999	2017	1999	2017	1999	2017
. . . have no points of contact	24	22	14	17	35	36
. . . complement each other, express different aspects of one and the same reality	57	56	80	74	23	32
. . . are in conflict	19	22	7	9	42	32

Table 3: Science and religion among the general public and among believers in God in Sweden. Görman 1999 (n=2000, 20–55 years) and *Study 8* (n=500, 18–74 years).[6] Percentages.

Different comments are called for in connection with table 3. *First*, there are those who affirm that there is no point of contact. This view is well known in theological and philosophical thinking,[7] but seemingly also has some resonance in the general public, especially among atheists. *Secondly*, the idea of complementarity between science and religion is the most common understanding of the relationship both among the general public and affirmed by an overwhelming majority of religious believers. The interpretation of this alternative is not entirely clear and has been discussed by philosophers and theologians.[8] *Thirdly*, 7 percent of the religious believers (9 percent in 2017) affirm that science and religion are in conflict. How is this to be interpreted? Possibly, this minority of religious believers affirms that religion—presumably Christianity—is true, and that different Christian beliefs contradict scientific theories or results (for example, the theory of evolution), which

---

6. Görman, "Svenskars uppfattningar om relationen mellan naturvetenskap och religion," 34–38.

7. See further below, chapter 8, on the view of Anders Nygren.

8. The theory of complementarity between science and religion is discussed by Holte in *Människa, livstolkning, Gudstro* (see further below, in chapter 12). See also MacKay, "'Complementarity' in Scientific and Theological Thinking," 232 and 237. See also Mortensen, *Teologi og naturvidenskap*, 76–79.

should be rejected.[9] It might also be assumed that most of the 19 percent of the general Swedish population (22 percent in 2017) responding that there is a conflict, presuppose the truth of science and those scientific results indicate that the central beliefs of Christianity are false.

The differences between 1999 and 2017 are surprisingly small. But some figures stand out. Atheists responding in 2017 are less prone to affirm a conflict between science and religion (32 percent) than those responding in 1999 (42 percent). This may have a simple explanation. Görman's study in 1999 did not include the oldest age group between 56 and 74 years included in *Study 8*. Moreover, this age group is somewhat less inclined to affirm a conflict between science and religion, even if the difference is not statistically significant.

Another national survey was made in 2009 by the Swedish society Vetenskap och Allmänhet (VA, Public & Science).[10] The formulation of the questions was somewhat different, but a partial comparison with Görman's survey from 1999 and *Study 8* is nevertheless possible. The respondents were asked to give their opinion about the possibility of reconciling religion with science. 47 percent answered in the negative and 43 percent in the positive.[11] These figures indicate a significant increase for the conflict position. It should be noted that the level of education made significant difference to the answers. Those with a high school education tended to affirm the harmony between science and religion more often (39 percent) than those who had only elementary school (31 percent), and those with a university degree more often than those with high school education.

SWEDISH CLERGY ON SCIENCE AND RELIGION

The previously mentioned web-based survey directed to Swedish ministers and pastors in March 2015 (*Study 6*[12]) contained different items about the science and religion, the knowledge and interest of the clergy concerning science and its relation to religion. After the questions about biology and biologism, some other questions were formulated about (1) the general conceptions of the relation between science and religion, (2) the

---

9. It should be noted that religious believers in Sweden have no problem combining their beliefs with high esteem of science. The figures from Sweden in the WVS-studies from 2006 and 2011 show that believers tend to value science not as much as non-believers. But almost half of those who believe in a personal God do not think that the progress of science makes it harder to believe in God. (23 percent neither agree nor disagree and only 28 percent agree more or less completely.)

10. See Vetenskap & Allmänhet, https://v-a.se.

11. Hermansson, *Vetenskap att tro på?*, 17.

12. See table 3.

knowledge about natural science, (3) about the scientific areas of special interest to the respondents, (4) the interest to participate in courses and conferences on science and religion, (5) whether any change of view on science and religion had occurred during the respondent's professional life and, (6) whether such a change would occur in case of increased knowledge about scientific research. The survey concluded with some questions about gender, age, habitation, and denominational affiliation.[13]

General perceptions of the relations between science and religion

The *first question* listed six different statements concerning the general relationship between science and religion. The respondents were asked to declare their position. These different statements were:

1. When religious faith and science contradicts each other, religion is always right.
2. Natural science and religion can be united because they are two very different perspectives on reality with different questions and purposes.[14]
3. Natural science and religion are two different activities, but they can nevertheless come into contact with each other and affect each other's content.[15]
4. It is important to reinterpret the Christian/Jewish/Muslim faith so that it at all points is consistent with the scientific worldview.
5. Christian/Jewish/Muslim faith rests on a scientific foundation.
6. Religion should to a greater degree influence the direction and methods of natural science.[16]

The main result of the responses to the six items is displayed in Figure 2.

---

13. See further Appendix 3.
14. This statement reflects the model of independence or separation.
15. This statement has been termed the model of contact.
16. The formulations of these statements were derived from a taxonomy proposed by Mikael Stenmark. See further below, chap. 10.

Figure 2: General statements on science and religion. N=1492. Mean values.

Statement	Mean
When religious faith and science contradicts each other, religion is always right.	3.1
Science and religion can be united because they are two very different perspectives.	8.3
Science and religion are two different activities, but they can come into contact with each other.	8.2
It is important to reinterpret religious faith so that it is consistent with the scientific worldview.	4.1
Faith rests on a scientific foundation.	3.6
Religion should to a greater degree influence the direction and methods of natural science.	3.4

Mean value: 1-10. 1= Disagree entirely, 10 = Agree entirelly.

The main result is clear. Expressed in percentages, an overwhelming majority of Swedish clergy endorses the independence and contact between science and religion (c. 80 percent). Only a minority concurs with the idea that religion is right when it contradicts science (7 percent), that it rests on a scientific foundation (13 percent) or that it should influence the direction and methods of natural science (14 percent).

One thing is puzzling. It is that alternative 2 and 3 does not differentiate the responses. The respondents who affirm alternative 2 about science and religion being a unity because they are entirely different perspectives also affirms alternative 2 that science and religion albeit different activities can come into contact with each other. This may be due to deficiencies in the wording of the statement trying to express the model of independence and the model of contact, respectively. Another interpretation is that the respondents are thinking about certain religious beliefs when answering item 2 and other beliefs when answering item 3. For example, there are certain religious and scientific beliefs are entirely independent of each other. Alternative 2 may direct attention to the belief that is true that Jesus is God incarnate has nothing to do with physics or chemistry. On the other hand, there are other religious beliefs for which it is true that their truth-value is (at least partly) determined by certain scientific results and theories. For example, the belief that God created the world is affected by evolutionary theory, because evolutionary theory explains—at least in part—how God created the world.

It should be noted that in most instances there are no big *gender differences* in the responses to these general questions about the relation between science and religion. However, there is one exception; there is a significant difference between men and women when it comes to the statement that it is important to reinterpret religious belief so that it at all points it is in accord with the scientific worldview. *Within Swedish clergy, significantly more men dissent from the importance of a reinterpretation than women do.* One hypothesis of this difference is that environmental concerns are more significant for women than for men and that these concerns—particularly for younger women—has provided an inroad not only for interest in science and religion issues, but also for reconsiderations of classical interpretation of Christian belief.

Similarly, there are some smaller differences *between different age groups*, for example when it comes to the response to the statement that natural science and religion can be united because they are two very different perspectives on reality with different questions and purposes. Respondents in older age groups are more likely to affirm this model of independence than respondents in younger age groups. Younger persons are also more open to the statement that Christian/Jewish/Muslim faith rests on a scientific foundation.

Furthermore, the *denominational differences* between ministers in the Church of Sweden and pastors in the Uniting Church in Sweden are small—but not altogether absent. This is clarified in Figure 3.

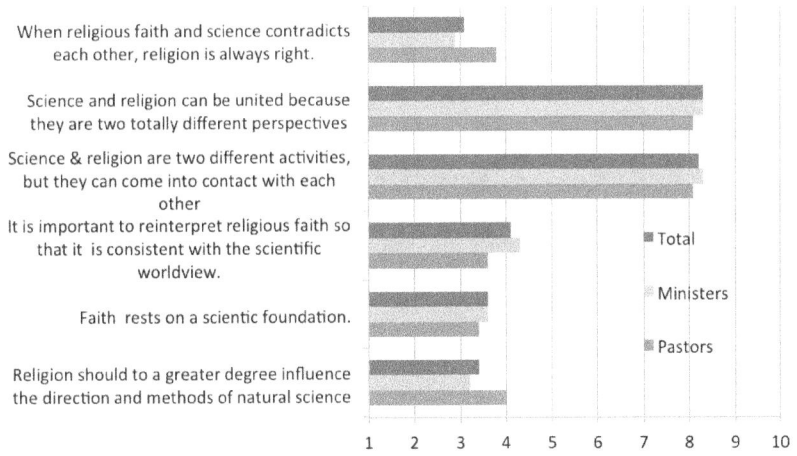

Figure 3. Ministers in Church of Sweden and pastors in the Uniting Church in Sweden on science and religion. According to denominational affiliation. Mean values.

There are insignificant differences when it comes to the emphasis on science and religion being different perspectives nevertheless in contact with each other. However, on three other items some differences are discernible. Pastors (54 percent) are less inclined than ministers (67 percent) to disagree with the statement when religious faith and science contradicts each other, religion is always right. Furthermore, more pastors (65 percent) disagree with minsters (54 percent) on the statement that it is important to reinterpret the Christian/Jewish/Muslim faith so that it at all points is consistent with the scientific worldview. Pastors (56 percent) do not as strongly as ministers (67 percent) disagree with the statement that religion should to a greater degree influence the direction and methods of natural science. On this item, there is also a significant difference between women (51 percent disagree) and men (61 percent), but no difference between age groups. In sum, ministers in the Church of Sweden seem a bit more liberal than their colleagues in the Uniting Church in Sweden.

Subjective estimation of scientific knowledge

The respondents were also asked to make a subjective estimation of their knowledge about natural science and the significance of such knowledge in their role as ministers or pastors. Figure 4 gives an overview of the mean values of the responses to the three different items in this section.

Figure 4. What does Swedish Clergy want to know about natural science? Mean values.

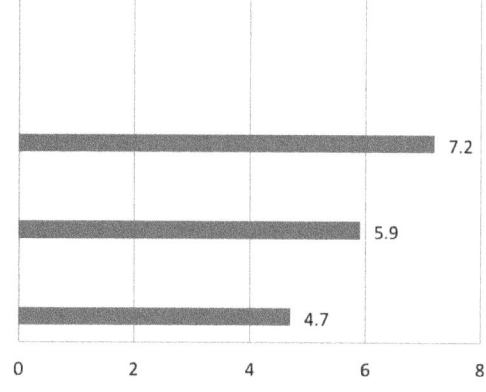

The main result is the emphasis on the importance of deeper knowledge in science. This has priority in comparison to knowledge about one's own religious tradition. Presumably, most clergy think they have sufficient knowledge in their own tradition in comparison to their knowledge in science and religion. There are, however, some interesting differences. Male clergy puts more emphasis on the significance of one's own tradition than female clergy does. Ministers in the Church of Sweden show a similar tendency in comparison to pastors in the Uniting Church in Sweden. Younger pastors emphasize the significance of scientific knowledge to a higher degree than elder pastors.

Science-areas of interest

Another question of the survey was an open question about the scientific areas of special interest to the respondents: "have you in your activity as minister or pastor met any special questions about science and its relationship to religion?" Over two-thirds (1142) the total sample of 1492 respondents formulated a response (albeit mostly in a sentence or two).

A first analysis shows that questions about evolution and creation are most frequently mentioned followed by questions about cosmology and the Big Bang-theory. It should be noted that ethical questions about life and death, issues of reproduction (abortion), and climate change are high on the list of important questions. Further down are issues on prayer and miracles, the problem of evil, death, eschatology, and eternal life. For some questions of stewardship and the place of human beings in creation are important. Surprisingly, few mention questions about human consciousness and almost no one issues about cognitive science and theories on the origin/development of religion. Questions about intelligent design and creation are of no great interest among Swedish clergy.

Interest to participate in courses and conferences

Furthermore, the survey contained an item about the interest to participate in courses and conferences where scientists present their science and engaged in a dialogue with representatives of different religions. Four alternatives were given: yes willingly, yes, possibly, doubtful, and no. 51 percent of the total sample responded that they willingly would do so and 37 percent possibly. Only 9 percent were doubtful and 4 percent answered in the negative. Table 4 gives a summary of the results:

Responses in percentage	Total sample	Ministers in the Church of Sweden	Pastors in the Uniting Church
Yes, willingly	51	55+	37-
Yes, possibly	37	35	43+
Doubtful	9	8-	14+
No	4	3	6

Table 4: Interest to participate in courses about science and religion where scientists present their science and engage in a dialogue with representatives of different religions. Percentage. "+" indicates that the percentage is significantly higher than the percentage of the total sample; "–" indicates that the percentage is significantly lower.

There are no gender differences here, but a clear difference between ministers in the Church of Sweden and pastors in the Uniting Church in Sweden: ministers are more interested than pastors to participate in courses on science and religion

Past change of view concerning science and religion

The question referred to a past change during the professional activity of the respondents. It called for a subjective judgment and, needless to say, the reliability of such estimations can be questioned. This notwithstanding, it is of a certain interest to get an insight into such evaluations. For example, a high frequency of perceived change could indicate increasing significance of the issue.

Four different alternatives were given in the response to this item: yes, to a high degree, yes, to some degree, no, hardly and no, not at all. Table 5 gives a summary of the percentages.

Responses in percentage	Total sample	Ministers in the Church of Sweden	Pastors in the Uniting Church
Yes, to a high degree	4	4	5
Yes, to some degree	37	35	45+
No, hardly	43	44	40
No, not at all	15	17	9-

Table 5: Subjective estimation of past change of view on science and religion—denominational affiliation. "+" indicates that the percentage is significantly higher than the percentage of the total sample; "–" indicates that the percentage is significantly lower. Percentage.

The main result is that the respondents perceive no high degree of personal change in view on science and religion. One interpretation is that the whole issue of science and religion has not gained visibility among other concerns in the professional activity of Swedish clergy. This interpretation is supported by scattered remarks in the commentary field of this and other items, that the issue of science and religion was important 15 or 20 years back in time, but since then it has been resolved and is now deleted from the theological agenda. This is, of course, a misperception, and the source of this misperception is difficult to discern.

Pastors in the Uniting Church in Sweden seem to perceive a change in their views on science and religion "to some degree" and choose this alternative significantly more often than ministers in the Church of Sweden. The percentage of respondents from the Uniting Church choosing the alternative "No, not at all" is also fewer than the ministers in the Church of Sweden (9 percent). The reason for this difference is difficult to discern.

It must be added that there is a small, but nevertheless significant gender difference in the responses to the question about change of view concerning science and religion. This is clarified in Table 6.

Responses in percentage	Total sample	Men	Women
Yes, to a high degree	4	6+	2-
Yes, to some degree	37	43+	29-
No, hardly	43	38-	50+
No, not at all	15	13-	19+

Table 6: Subjective estimation of past change of view on science and religion—gender. "+" indicates that the percentage is significantly higher than the percentage of the total sample; "-" indicates that the percentage is significantly lower. Percentage.

These results raise a larger issue about gender differences revealed in *Study 6* as a whole. As will become clearer in the analysis of the telephone interviews, women in general seem less interested in the issue of science and religion than men. This might be reflected in the response to question 6. Interestingly, however, younger women minsters and pastors displaying environmental concern and putting less emphasis on human uniqueness, *emphasize the importance of courses and conferences on religion and science* (question 5).

Possible future change of view

Finally, the respondents were asked to make an estimation of future change on science and religion in case of increased knowledge about scientific research.

Responses in percentage	Total sample	Ministers in the Church of Sweden	Pastors in the Uniting Church
Yes	28	27	33
No	30	32	25-
Doubtful	41	41	43

Table 7: Subjective estimation of future change of view on science and religion—denominational affiliation. "+" indicates that the percentage is significantly higher than the percentage of the total sample; "-" indicates that the percentage is significantly lower. Percentage.

A minority of the sample (less than a third) foresees change in their view of the relation between science and religion. This may indicate that the view on the relationship is dependent on *apriori* theological considerations, i.e. a general view that science and religion belong to two different realms. Such opinions are not easily changed. This is in line with earlier discussed results indicating that this model of independence (natural science and religion can be united because they are two totally different perspectives on reality with different questions and purposes) has a strong standing among Swedish clergy. Presupposing this model, it is to be expected that increased scientific knowledge is not likely to change the perception of the relationship.

In the responses, *gender differences* are barely noticeable. This is not surprising on the hypothesis that the responses to both this and the former question reflects preconceived theological assumption about science and religion being two different realms of reality without any contact with each other.

Similarly, there are no big differences in the responses of different age groups. However, one figure stands out. Respondents in the youngest age are more likely to expect their views on the relation between science and religion to change if they were to receive new knowledge about scientific research. This is in line with the earlier result that the younger generation of Swedish clergy is less inclined than their elderly colleagues to accept the model of independence and separation.

SUMMARY

One important context for the public perception of science and religion, was the debate caused by Ingemar Hedenius and his book *Faith and Reason* in 1949, regularly revived during subsequent decades. Empirical studies show that the most common view is one of complementarity, but also that background attitudes to religious belief determine your view on the relationship between science and religion. Religious believers tend to minimize the conflict between science and religion, while atheists are inclined to emphasize it. This tendency is less clear in recent studies. The growth of atheism during the last decades must have other sources than growing perception of a fundamental conflict between science and religion.

When it comes to opinions on science and religion among Swedish clergy the main result is clear. An overwhelming majority of Swedish clergy endorses the independence *and* contact between science and religion (c. 80 percent). It should be noted that minsters and pastors do not seem to differentiate between two different models of science and religion, i.e., the model of contact and the model of independence. There might be multiple explanations for this, for example deficiencies in the wording of the statement trying to express the model of separation and the model of contact respectively. Generally, Swedish clergy appears as a rather homogeneous group. About 80 percent endorse both independence *and* contact between science and religion. In comparison to the general public, members of Swedish clergy do not consider science to be a problem for religious belief.

There are some interesting differences among Swedish clergy when it comes to denominational affiliation, gender and age. For example, there are significantly more male than female clergy who dissent from the importance of a reinterpretation of religious belief. When it comes to interesting subjects in the area, questions on evolution and cosmology are frequently mentioned as well as ethical questions about life and death, issues of reproduction (abortion) and climate change. Surprisingly, few mention questions about human consciousness and almost no one highlights issues about cognitive science and the origin/development of religion. Questions about intelligent design and creation are rarely mentioned. Furthermore, there is generally an interest to participate in courses on science and religion, but pastors in the Uniting Church in Sweden are generally a little less interested than ministers in the Church of Sweden. A minority of the sample (less than a third) foresees change in their view of the relation between science and religion.

4

Swedish Clergy on Science and Religion; a Cluster Analysis

In this chapter, I will move to a summary of the results in *Study 6* of Swedish clergy in 2015. This will be done by means of a cluster analysis (CA), which has been introduced in Chapter 1.[1] CA is especially useful when there is a wealth of material and no particular background hypothesis. It should be noted that CA results in a description of structures in the material, but is silent about the explanation of these structures.[2]

PURPOSE AND METHOD

The purpose of CA is twofold, i.e., first, to study if there are different groups within the total sample of clerical respondents with different opinions on philosophical and religious issues, and secondly, to describe how each of these groups differ from each other when it comes to these issues as well as social background factors.

All responses on social background data as well as to items where a response on the 10-degree scale was offered were used as in-data (i.e. responses to Question 1–11 in section 1, Question 1–6 in section 2, Question 1–3 in section 3, and the questions in section 6 and section 7).[3] Initially, the data were analyzed by factor analysis. The main result of this analysis is that

---

1. See further Appendix 5.
2. On CA, see Dell Inc., *Statistics: Methods and Applications*.
3. All questions are to be found in Appendix 3.

the responses to the questions could be divided according to six different factors, where the responses to each question were mutually correlated.[4]

An analysis of the data resulted in that 1139 (76 percent of the total sample of 1492) of the respondents could be divided into six different clusters, where each one of the clusters displayed a particular worldview- and personality-profile. The remaining 353 respondents (24 percent of the total sample) were placed in an "other-group" (cluster 0) without any distinct profile. The final clustering can be described as *intermediately stable*. The probability that a respondent at reiterated cluster analyses ends up in the same cluster is 82 percent. The mean stability value ranges between 70 and 90 percent. The main impression of the statistical expert is that the whole group of respondents is relatively homogeneous and that the differences between the clusters are relatively small.

PRESENTATION OF THE GENERAL PROPERTIES OF THE SIX CLUSTERS

In general, the sample reflects are rather homogenous group, which come to expression—among other things—in its almost unanimous rejection of biologism. However, within this frame, there are several interesting differences. I will present the different clusters and label each cluster according to some significant characteristics. The labels reflect my personal impression of the main characteristics of the cluster in question.

Ministers for integration and contact with science

The *first* cluster (6 percent of the total sample,) is the smallest of the six clusters, containing 94 respondents. This means that some caution is advised when drawing detailed conclusions from such a limited sample. Remembering this, it should first to be noted that there is an over-representation of ministers in the Church of Sweden in this cluster (84 percent in comparison to 75 percent in the total sample). The minimal overweight of women in the cluster (51 percent, 42 percent in the total sample) is not statistically significant and neither are any of the other social background factors.

The cluster displays an interesting combination of positions on issues related to science and religion. First, it should be noted that the traditional theological doctrine of the sovereignty of humankind in creation is downplayed in combination with a strong emphasis on the importance to make a reinterpretation of religious belief so that it at all points is consistent with the scientific worldview. A significant number (38 percent, 10 percent in the

---

4. See further Appendix 5.

total sample) of the respondents in this cluster also agree to the statement that Christian/Jewish/Muslim belief rests on a scientific foundation. To a stronger degree than the total sample, they also tend to agree to a dialogue/integration view of the relation between science and religion. It is also of importance to notice that they show a higher environmental concern, i.e., 10–20 percent higher than the average of the total sample. Finally, the members of the cluster consider themselves to have sufficient knowledge of science and its relation to religion, but at the same time, they consider deeper knowledge of science and its relation to religion to be great significance for their professional work as religious leaders. They would be interested to participate in future courses on science, even if they don't foresee their position on such issues to change. The image that emerges is a minister who seeks integration between science and Christian beliefs, albeit a stranger to the minority of Swedes that are prone to biologism.

Pastors with strong roots in "old time religion" and foreign to modernity and science.

The *second* cluster has a very different profile. It is twice as large as the first cluster (12 percent, 185 respondents). Cluster 2 also puts a stronger emphasis on traditional religion with a clear affirmation of the sovereignty of humankind, priority on deepening the knowledge on one's own religious tradition and very little interest in reinterpretation of religious belief so that it at all points is consistent with the scientific worldview. A strong minority (14 percent, 4 percent in the total sample) agrees with the statement that when religious belief and science contradict each other, religion is always right. Furthermore, environmental concerns and values are downplayed and the members of the group are especially foreign to the idea that nature is animated with a power that permeates all life (0 percent, 43 percent in the total sample!). They show less sympathy than Swedish clergy in general to the models of dialogue and integration and are mostly foreign to the idea that scientific knowledge is of significance in their work as religious leaders. A large group in the cluster is in consequence not very interested in future courses on science and religion (30 percent, 51 percent in the total sample). There are more pastors than the average in the total sample (35 percent, 24 percent in the total sample), but many ministers in the Church of Sweden share the general outlook on issues in science and religion. Men in small town areas of Sweden dominate the cluster.

Traditional male pastors with environmental concern.

The *third* cluster shares many features with cluster 2. It encompasses 8 percent of the total sample, i.e., 114 respondents. As the members of cluster 2, they are rooted in traditional Christian belief and suspicious of theological reinterpretations to accommodate to the scientific worldview. There is a stronger presence of pastors in the Uniting Church in Sweden. More importantly, they are more environmentally concerned than the members of cluster 2 and a large minority of cluster 3 emphasize that religious faith rests on a scientific foundation. They would have some interest in participating in future courses on science and religion, but most of them are doubtful if they would change the opinion on the issues. A representative of the group is a Swedish religious leader—possibly from the Uniting Church in Sweden—firmly identifying with classical Christian convictions, but concerned about environmental issues and of some interest in the dialogue between science and religion.

"Modern" clergy with technological orientation and critical of anthropocentrism.

The *fourth* cluster—13 percent of the total sample (194 respondents) —lacks a clear profile when it comes to social background data. In contrast to cluster 2 and 3, they are critical of traditional Christian anthropocentrism. Interestingly, they are also less prone to environmental concerns and particularly to the idea that nature is animated (just 1 percent in agreement!). More than any other group, they are foreign to the idea that humankind would be better off if we could live a simple life without much technology. This could be taken as an expression of sympathy to technological development (albeit combined with a modest interest in science and its relationship to religion). What emerges is a minister or pastor of more practical orientation, clearly in tune with modernity and acquainted with the critique of anthropocentrism in modern theology.

Women clergy speaking for environmentalism, ensoulment of nature, and critical of anthropocentrism.

The *fifth* cluster is the largest of the six clusters emerging from the analysis (19 percent, 283 respondents). It is the most interesting cluster, primarily because not a single member of this cluster agrees with the classical Christian doctrine that humankind exists to rule over the rest of creation. Many of them still affirm the idea that human beings have characteristics that no other creature has, but in contrast to all the other clusters, there is a strong minority of 23 percent of the cluster, who do not agree to this statement (all the other clusters agree to about 90 percent or more). Another characteristic of the group is that its members emphasize environmental values and are in strong agreement with the proposition that nature is animated by a power that permeates all life. Furthermore, it is interesting to note that this cluster contains significantly more women from urban areas. In contrast to the male-dominated cluster 2, they admit insufficient knowledge of science and its relationship to religion and acknowledge that deeper knowledge would be of great significance to their work as religious leaders.

Clergy in favor of anthropocentrism and ensoulment of nature.

The *sixth* cluster (18 percent, 269 respondents) is like cluster 4 in that it lacks a clear profile when it comes to social background data. However, the members of the group are at variance with the members of cluster 4 in that they strongly affirm the classical Christian anthropocentrism and at the same time express agreement to the ensoulment of nature—even if most of them disagree with other environmental ideas such as living life without so much technology or believing in an environmental apocalypse. The main results of the cluster-analysis are presented in the Table 8.

	Cluster 1 n=94 (6 %)	Cluster 2 n=185 (12 %)	Cluster 3 n=114 (8 %)	Cluster 4 n=194 (13 %)	Cluster 5 n=283 (19 %)	Cluster 6 n=269 (18 %)
(1) Anthropocentrism	–	++	++	–	– –	+
(2) Biologism						
(3) Environmentalism	+		+	– –	+	–
(4) Coherence with science	+	–				+
(5) Faith rests on and should influence science	+		+			
(6) Interest in deeper knowledge about science	+	–			+	
Background data	+ course interest + minister	– course interest + pastor + small town + male	– course interest + pastor		+ course interest + minister – small town + female	

Table 8: Overview of the CA-results. "+" indicates that response to one or more of the different questions within the factor was at least 10 percent higher than the mean percentage of the total sample. "–" indicates that response to one or more of the different questions within the factor was at least 10 percent lower than the mean percentage of the total sample. "++" and "– –" indicate 20 percent higher/lower than the mean percentage of the total sample.

COMPARATIVE ANALYSIS

First, cluster 2 is in clear contrast to cluster 5. There are gender, denominational, and demographic differences, but also different degrees of interest to participate in courses and conferences on science and religion (31 percent in cluster 2 vs. 64 percent in cluster 5). Furthermore, the male-dominated cluster 2 puts significantly less emphasis on the importance of such knowledge for religious leaders. It is not wholly transparent how these differences can be explained, but a part of an explanation may be that the members of cluster 5 consider themselves to be less knowledgeable in science and it relationship to religion (only 8 percent) than the members of cluster 2 (22 percent). It should also be noted that only 2 percent of the members of cluster 2 agree with the importance to reinterpret religious belief so that it is consistent with the scientific worldview. The figure on that score for cluster 5 is 14 percent. Both these things might explain a big part of the difference between the two clusters concerning the interest in courses and conferences as well as the importance of such knowledge for religious leaders.

Secondly, the contrast between cluster 2 and cluster 5 also raises the question of gender and how gender determines the perception of issues on science and religion. In 1997, Ann Pederson and Mary Solberg noted that "even today very few women are active in theology and science, and the number of feminist women (or men) is even fewer."[5] Cluster 5 with a strong presence of women clergy with interest to know more about science and religion might indicate that a significant change has occurred during the last 20 years.

Thirdly, and moving back to table 8, one other factor of importance might have to do with environmental concern. The members of cluster 2 show a much stronger commitment to environmental values. This commitment might "spill over" to an emphasis on issues of science and religion. It is of special interest to notice that there is total absence of agreement in cluster 2 with the statement that nature is animated by a power that permeates all life. In cluster 5, 72 percent agree with this statement! Here we may have of one of the more significant factors behind the interest in issues for science and religion. *Belief in God as the Spirit of nature goes hand in hand with concern for the relation between science and religion.* This might also explain the significance that members in cluster 1 attaches to issues in science and religion. 66 percent in this cluster affirms the ensoulment of nature and 73 percent affirm the importance for religious leaders of deeper knowledge in science and religion. This is the highest figure in all of the six clusters.

In conclusion, the cluster analysis suggests that significance of the image of God for interest in dialogue between science and religion. It is

---

5. Pederson and Solberg, "CyberFlesh."

possible that many of the respondents that find this dialogue important—for themselves as well as for religious leaders in general—because this dialogue deepens explorations into the spiritual dimensions of human life as well as the universe as a whole. Furthermore, two other factors are involved. It is gender and denominational affiliation. The overall impression is interplay between four different factors as illustrated in Figure 5.

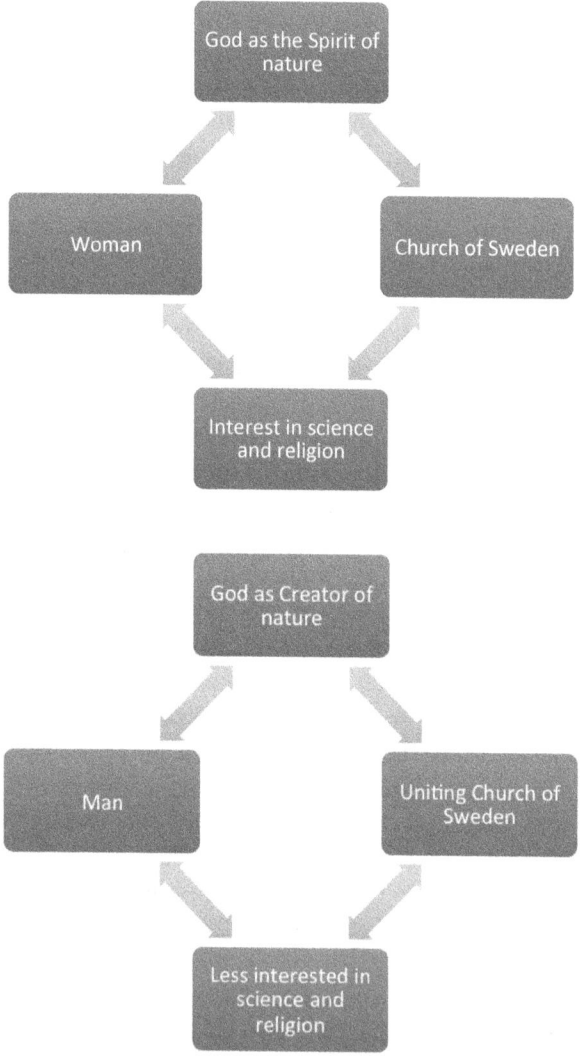

Figure 5: The interplay between image of God, gender, denomination, and interest in science and religion.

SUMMARY

The responses in the web-survey among Swedish clergy were subjected to a cluster analysis, i.e., a statistical study of different groups within the total sample of respondents with different opinions on philosophical and religious issues. The purpose was to describe how each of these segments differs from each other when it comes to these issues as well as social background factors. The result shows that that 1139 (76 percent of the total sample of 1492) of the respondents could be divided into 6 different clusters, where each one of the clusters displayed a particular worldview- and personality-profile. Cluster 1 is a rather mixed cluster with a small dominance of men in the Church of Sweden downplaying the sovereignty of human beings in creation, but underlining the significance of science and religion as well as showing environmental concern. Cluster 2 is more traditional with an overrepresentation of male pastors from small town parishes in the Uniting Church in Sweden. They put strong emphasis on human beings as the masters of creation and are not particularly interested in issues in science and religion. Cluster 3 also shows an overrepresentation of pastors (men and women), but in contrast to the previous cluster questioning the sovereignty of human beings in creation, more environmentally concerned and interested in science and religion. Cluster 4 is more undefined and somewhat skeptical of human sovereignty and less environmentally concerned. Cluster 5 seems to be the opposite of cluster 2: women-dominated, environmentally concerned and wanting deeper knowledge about science and religion. Cluster 6 is in turn more anthropocentric, but with an interest in the dialogue between science and religion. It must be noted that all clusters are equally opposed to biologism.

In conclusion, women clergy in the Church of Sweden are in more sympathy with belief in God as the spirit of nature and interested in the dialogue between science and religion, while male Pastors in the Uniting Church have a more traditional image of God and are not as much engaged by science and religion. But it should be noted that these correlations are rather weak.

5

A Comparison between Sweden and the U.S.

Sociological studies on opinions concerning the relation between science and religion is a field of growing interest. A crude distinction can be made between studies of *public* perceptions, studies of *religious* perceptions and studies of *scientific* perceptions. In the foregoing chapters, some studies of the Swedish general public and religious leaders in Sweden were presented. In the present chapter, I will (1) focus on a comparison between and Swedish academic students, teachers, and scholars (relying on *Study 2*, 2007, and *Study 4*, 2009), on the one hand, and American academic students, teachers, and scholars, on the other.[1] These studies are a mixture of quantitative surveys and qualitative interviews. I will also (2) compare the general public, religious believers, and scientists in the U.S., and in Sweden. Some of these studies are only partially commensurable and the comparative analysis should be regarded as hypothetical and as a point of departure for a more thorough and methodologically well-prepared comparison.

One thing should be noticed at the outset. My own studies as well as the US studies are mostly of an explorative nature. They do not depart from a specific social scientific theory about the relation between science and religion. This need not be a deficiency, but it would nonetheless be desirable to complement accomplished studies with studies having a more explicit and elaborated theoretical point of departure. One such point of departure could be contemporary theories of cognitive science. In Chapter 1, I briefly referred to Steven Horst and his theory of cognitive pluralism (CP). One of

---

1. See Appendix 1.

the major claims of CP is that mental models are the fundamental units of understanding. Applying this in a study of the perception of science and religion, could mean that awareness of conflict or coherence is not what should be expected but rather independence and separation. But further theoretical reflection on the implication of cognitive science for social scientific studies of science and religion are surprisingly scarce.

STUDIES OF THE PUBLIC PERCEPTION OF SCIENCE AND RELIGION IN THE U.S.

Gallup USA has made several studies into opinions about the theory of evolution. These studies began in 1982 and the results are rather stable. 42 percent of the American population say "God created humans in present form." [2] In contrast, Pew Research Center in Washington DC finds only 31 percent saying that humans and other living thing have "existed in their present form since the beginning of time." It is less likely that there is a difference in the sampling of the respondents. Differences in the wording of the statement is a more likely explanation. The Gallup formulation is somewhat more open to different interpretations than the one used by Pew Research, which does not mention God or makes explicit the time span human beings have existed in their present form.

Pew Research has other studies on the relation between science and religion in the U.S. during the last ten years. These publications include both historical overviews, scholarly discussions, and surveys. One important survey was published in 2009 and a more recent one 2014.

One result in the study 2009 concerned the perception of a conflict between science and religion. A comparison was made between (1) those with no religious affiliation (attended a religious service at least once a week) and (2) those with no religious affiliation. Interestingly, the group of religiously observant only 48 percent acknowledged a conflict between science and religion, while 68 percent of those with no religious affiliation did.[3] Pew Research also had a question about of a conflict between science and *one's own* religious beliefs. Only 15 percent of the unaffiliated perceived a conflict, while 46 percent of the religiously observant did so. This suggests that the religiously observant acknowledging a conflict between science and

2. Newport, "In U.S., 42 percent Believe Creationist View of Human Origins," lines 15–22. The polls concerned views on the origin and development of human beings and was based on telephone interviews conducted May 8–11, 2014, with a random sample of 1028 adults, aged 18 and older, living in all 50 U.S. states and the District of Columbia.

3. *Public Opinion on Religion and Science in the United States,* 2009 (http://www.pew-forum.org/2009/11/05/public-opinion-on-religion-and-science-in-the-united-states/).

religion are quite aware that this conflict involves their own religious beliefs. This leads to the larger question how it is possible to manage such a conflict. On the other hand, such a larger question presupposes that it is awareness of a conflict between science and religion that needs to be explained. *It could, however, be the case that it is the opposite that needs explaining, i.e., an affirmation of a harmony or coherence between science and religion.*

The researchers of Pew Research Center also asked all those who affirmed a conflict between their own religious beliefs and science, which specific scientific theories conflicted with their religion. 36 percent mentioned conflict over the creation of the universe, evolution, and Darwin, 24 percent broad differences over the belief in God, facts vs. beliefs, miracles, and view of man as "in charge," and 11 percent views about the beginning of life. It is evident that the conflict between religious belief and the theory of evolution is still influential in American culture.

It should also be noted that many persons disbelieving the theory of evolution might argue that there is no conflict between their religion and science, because *true* science provides an alternative to the established theory of evolution. This may also explain why as many as 31 percent of the total sample say that humans and other living thing have "existed in their present form since the beginning of time." Presumably, many of them still believe that this is not in conflict with *true* science.

STUDIES ON SCIENTIFIC AND RELIGIOUS PERCEPTION OF SCIENCE AND RELIGION

Significant studies into religious beliefs of American *scientists* have been made during the last century. One classical study was conducted by James Leuba in 1914. Leuba concentrated on two religious beliefs, i.e., belief in God and in immortality. He asked 1000 randomly selected scientists and found that 42 percent believed in a supreme being and 58 percent disbelieved or had no definite beliefs about the existence of God. In a new study from 1933 the figure of disbelieving and doubting scientists had risen to 67 percent. In an effort to replicate the study in 1997, Edward Larsson and Larry Witham found that little has changed from 1933; about 60 percent of their sample of scientists expressed doubt, agnosticism, or disbelief in a personal God.[4]

---

4. Larson and Witham, "Scientists are Still Keeping the Faith," 435–36. In another survey from 1969—assigned by the Carnegie Commission—showed that 35 percent of scientists did not believe in God. Trow et al., *Carnegie Commission National Survey of Higher Education*.

Leuba also made another study into the beliefs of "greater scientists of superior knowledge, understanding and experience" with 400 respondents in his sample. He found that their doubt or disbelief was even more pronounced with over 70 percent expressing doubt, agnosticism, or disbelief in a personal God. Larson and Witham replicated this study in 1998 and draw their sample from members in the National Academy of Science (NAS). They found that the figure now had risen to 93 percent; only 7 percent expressed their belief in a personal God![5] Larson and Witham cited the Oxford University scientist Peter Atkins commenting the 1996 survey: "You clearly can be a scientist and have religious beliefs. But I don't think you can be a real scientist in the deepest sense of the word because they are such alien categories of knowledge." For a reliable interpretation of these figures, it must be kept in mind that the response rate of Larson's and Witham's study in 1998 was a little over 50 percent. Secondly, the wording of the questions about belief in God and immortality is of relevance. Larson and Witham used the same wording as Leuba in 1914. For example, the respondents were asked to state their response to the following three statements:

1. I believe in a God in intellectual and effective communication with humankind, i.e., a God to whom one might pray in expectation of receiving an answer. *By "answer," I mean more than the subjective psychological effects of prayer.*
2. I do not believe in God *as defined above.*
3. I have no definite beliefs regarding this question.[6]

Leuba in 1914 (and his replicators Larson and Witham in 1997 and 1998) had a very specific concept of God in mind. Other options could have been asking for the respondents' belief in a divine creator or simply the existence of a transcendent being.

ECKLUND'S STUDIES

Elaine Howard Ecklund has made the most recent and extensive contributions to the study of what scientists "really" think about religion and the relationship between science and religion. During four years of intensive research between 2005 and 2008, she conducted a survey of nearly 1700 natural and social scientists as well as on-to-one conversations with 275

---

5. Larson and Witham, "Leading Scientists still Reject God," 313.

6. For the wording in comparison to the wording of other surveys as well as a thorough analysis of the results, see Pappas's dissertation, "Is There a Belief in God and Immortality?" Appendix A.

of them (Religion Among Academic Scientists, RAAS). Her main conclusion was that "much of what we believe about the faith of elite scientists is wrong." And she continued:

> The "insurmountable hostility" between science and religion is a caricature, a thought-cliché, perhaps useful as a satire on groupthink, but hardly representative of reality.[7]

To be sure, some scientists rejected religion, but not necessarily *because* they are scientists. Others lived with a closeted faith, became boundary pioneers or came to practice spiritual atheism with a new kind of individual spirituality—without God.

These *general* results notwithstanding, the general conclusion of a classical study made by James Leuba and the replications made by Edward Larsen and Larry Witham still stands: scientists are generally much less religious than the U.S. public at large. Ecklund found that 34 percent of the scientists affirmed "I do not believe in God," 30 percent were agnostics ("I do not know if there is a God and there is no way to find out"), 27 percent have belief in God with explicit reservations and only 9 percent said, "I have no doubts about God's existence." When the general population was presented with similar alternatives, only 2 percent said, "I do not believe in God," while 63 percent affirmed "I have no doubts about God's existence."[8]

There are—finds Ecklund—three main reason why at least one third of American scientists do not believe in God.[9] The *first* reason is because science trumps religion. Science—and only science—gives us reliable knowledge and not religion. The *second* reason is that religion let them down. Their doubts about religion were neglected or even quenched. Almost 40 percent were raised in Protestant homes. Less than half of these still identified themselves as Protestant. Many social scientists left religion after reading the founding fathers of social science (for example, Emile Durkheim) and becoming convinced that religion is nothing but a social construction. Others had bad experiences of religion, were troubled by the problem of evil or the negative consequences of religion in society. A *third* reason was that the religion of their parents was "only significant as a label rather than a matter of regular practice."[10]

Ecklund's study represents a clear progress in relation to earlier surveys and provides us with a richer and more nuanced picture of American

---

7. Ecklund, *Science vs. Religion*, 5.
8. Ecklund, *Science vs. Religion*, 16.
9. Ecklund, *Science vs. Religion*, 17–27.
10. Ecklund, *Science vs. Religion*, 25.

scientists and their reflections on science and religion. In the present context, another and more recent study deserves special attention. Religious Understandings of Science (RUS) study was presented in 2014 and based on a survey utilizing the research tool KnowledgePanel@. It consisted of 50000 adult members and for the present survey 16746 panelists were asked to participate. 10497 responded positively, i.e., a response rate of nearly 63 percent, which Ecklund noted as an extraordinarily high response rate for modern survey research.[11] An intentional oversampling of science panelists was realized, representing occupational sectors such as computer and mathematics, architecture and engineering, life, physical and social sciences etc. After some further exclusions, the total sample amounted to 9138 of which 574 were identified as scientists.[12]

The sample of RUS consisted of a variety of religious tradition including 3 percent from non-Christian religions and 16 percent atheists, agnostics or "nones." It should also be noted that in addition to the survey, 315 telephone interviews were conducted. Ecklund's and Scheitle's report is largely focused on the subsample of evangelical protestants.

Looking at a number of selected religiosity measures, Ecklund and Scheitle concludes that the responding scientists do not differ significantly from the general population "although they are slightly less likely to consider themselves very religious, attend religious services weekly, and pray several times a day." Ecklund and Scheitle continue:

> We see larger differences between the scientist and general populations when looking at the frequency of reading of religious texts and the likelihood of having no doubts about God's existence. Evangelicals, on the other hand, show much higher levels of religiosity on all measures when compared to the general population. Evangelicals who are also scientists are similar to evangelicals in general, but they are more likely to say that they are "very religious."[13]

Of special interest in the present context is the perception of the religion-science relationship. Ecklund and Scheitle distinguish between four different positions:

- Conflict ... I consider myself to be on the side of religion
- Conflict ... I consider myself to be on the side of science

---

11. Ecklund and Scheitle, "Religious Communities."

12. The special problem of defining a scientist is discussed at length, see Ecklund, *Science vs. Religion*, 4–5.

13. Ecklund, *Science vs. Religion*, 10.

- Independence... they refer to different aspects of reality
- Collaboration... each can be used to help support the other [14]

Some of the results are—considering earlier studies—not surprising. Atheists, agnostics, and "nones" are overrepresented among those affirming there is a conflict between science and religion. Contrary to expectations, almost 50 percent of the Evangelicals stand for collaboration between science and religion and that "each can be used to help support the other" (average of all respondents was 38 percent). But more than twice as many Evangelicals as the average of all respondents (14 percent) affirm that there is conflict between science and religion and consider themselves to be on the side of religion. Notwithstanding that this is not the position of the majority of Evangelicals, it is for many and it is largely tied to views on human origins. Affirming the view that "God created the universe, the Earth and all life within the past 10,000 years" was twice as likely among Evangelicals in general (43 percent). 9 percent of the scientists were of the same opinion and 32 percent of the Evangelical scientists![15]

These figures illustrate that the question of human origins still haunts the dialogue between science and religion in the U.S. Ecklund and Scheitle conclude with some comments on the future of this dialogue:

> ... in the grand scheme of the conversation on religion and science, it would serve us well to consider what everyday religious people think—not only the most vocal voices and visible debaters. It would also serve us well to remember the topics on which there is greater agreement or openness to collaboration, such that a framework of trust and mutual respect might develop between these two communities. If this framework can be strengthened from the perspective of ordinary Americans, then controversial debates over specific topics, such as human origins, have a chance to become productive spaces of dialogue and learning.[16]

COMPARING THE GENERAL PUBLIC IN SWEDEN AND THE U.S.

It is difficult to find a comparison between these data from the U.S. and the situation in Sweden. But there are some sources for a reliable comparison.

---

14. Ecklund, *Science vs. Religion*, 15.
15. Ecklund, *Science vs. Religion*, 23.
16. Ecklund, *Science vs. Religion*, 26.

One is the World Values Survey (WVS), an international and on-going study of values and beliefs around the globe.[17] WVS is an important and open source for knowledge not only for the public perception of science in different countries, but also for opinions on the relationship between science and religion. One item in the WVS, Wave 6, (2010–2015) is especially relevant. The respondents were asked if they strongly agree, agree, disagree, or strongly disagree with the following statement: "Whenever science and religion conflict, religion is always right." Table 9 shows the results of responses to this question given to a representative sample of the adult population in Sweden and United States in 2011.

	"Whenever science and religion conflict, religion is always right." Percentage. Numbers may not sum to 100 percent due to rounding.	
	Sweden	U.S.
Strongly agree	2	15
Agree	6	24
Disagree	35	37
Strongly disagree	52	22
Missing	1	–
No answer	1	2
Don't know	4	–
(N)	1206	2232

Table 9: Responses to the statement "Whenever science and religion conflict, religion is always in the right" in Sweden and U.S. (2011).

The results highlight an important difference between Sweden and United States, but an even more important difference is revealed when responses to the present item are correlated with the responses to another item, namely, belief in God. This is clarified in Table 10.

---

17. For further information about this research project, see World Values Survey, http://www.worldvaluessurvey.org/wvs.jsp.

"Whenever science and religion conflict, religion is always right."	"Believe in God?" Yes	
	Sweden	U.S.
Strongly agree	3	17
Agree	13	26
Disagree	46	41
Strongly disagree	28	14
Missing	1	-
No answer	2	2
Don't know	6	-
(N)	494	1967

Table 10: Responses to the statement "Whenever science and religion conflict, religion is always right" and the question "Do you believe in God?" in Sweden and U.S. (2011). WVS Wave 6 (2012). Numbers may not sum to 100 percent due to rounding. Percentage.

Only 16 percent of Swedish believers in God agree and agree strongly that whenever science and religion conflict, religion is always right—in the U.S. 43 percent do. The result reflects the difference standing of religion in the respective countries. Measured with different parameters (church attendance, prayer, importance of belief in God etc.) religion has a weaker standing in Sweden than in the U.S. In this perspective, the differences displayed in Table 10 are not surprising. Moreover, it is to be expected that in a country where religion is weaker and science is stronger, religious believers are more inclined to give science priority in the case that science and religion are in conflict.

PUBLIC AND SCIENTISTS IN SWEDEN

U.S.-studies into public and scientific attitudes to (science and) religion show that scientists—and particularly elite scientist—are more foreign to religious beliefs than the general public. The studies of Leuba, Larson and Witham suggest a wider gap between public and scientist than the studies of other scholars (for example, Ecklund), but there is still a huge difference. Over 60 percent of the general public affirmed, "I have no doubts about God's existence," while only 9 percent of the scientist, concurred.

Studies into beliefs of scientists have also been made in Sweden, but the approach has been different. U.S.-studies are generally done against the background of Christian beliefs. Christianity forms the point of departure.

Terminology and categories are derived from the context of Christian belief. In contrast, many Swedish studies are made in a society where Christian beliefs and customs have receded into the cultural background. To be sure, cathedrals and churches still stand in the center of cities and villages—but more as a symbol of ages past than as expressions of contemporary public attention and personal commitments. Studies into public attitudes must take account on these cultural realities and accommodate surveys and interviews accordingly. For example, the concept of worldview has been substituted for the concept of religion. Examples of studies with this point of departure have been given in the first part of this book. Anders Jeffner initiated such studies already in the late 1980s and other have followed.[18] One central question has been the presence of biologization. For this purpose, an *Inventory of Worldviews* (IWV) was constructed and used in different studies (as earlier described in Chapter 2). IWV was also used in 2009 with the purpose of comparing biologization in the Swedish public with biologization among Swedish scientists.[19]

The data for this comparison has been collected in two different ways. In May-June 2009 telephone interviews were made with a representative sample of the Swedish adult population (*Study 3*). The sample was drawn from the adult population between the ages of 18 to 74. 502 respondents were willing to participate with a response rate of 78 percent. Simultaneously, a web survey was set up directed to teachers, researchers and students in three university disciplines, i.e., literature, political science and physics (*Study 4*).[20] Contact was taken with provosts and student counselors at the university of Stockholm, Gothenburg, Chalmers University, and Uppsala University. Relevant departments at these universities were also contacted with the purpose of informing staff and students about the web survey and give them the link to the web site. At the same time an advertisement was published in professional journal for teachers with the reach of c. 100 000 readers. Respondents were weighted according to gender, age and subject in order to avoid overrepresentation of a particular group. All in all, 233 respondents were assembled.

I have described the methods of these studies in some detail to make clear that the sample of teachers, researchers and students undergoing academic education may not be entirely representative of Swedish scientists. The study still gives an indication and should be understood as the first step to a more reliable comparative study between the Swedish public and Swedish scientists and scholars. Bearing this in mind, the following comparison between Swedes in general and Swedish scientists can be made.

18. See Jeffner, "A New View of the World," 137–45.
19. See Appendix 1.
20. See Appendix 1.

	Swedes in general (2009), n=502	Scientists (non-biologists) (2009), n= 233
(1) Human beings have certain unique qualities, which no other creatures have.	81	73
(2) Humanity is going towards a self-inflicted environmental disaster.	61	55
(3) We must respect the order of nature irrespective of how this affects human welfare.	76	54
(4) Humans would be much better off if they lived a simple life without so much technology.	39	23
(5) Biology gives us the complete answer to the meaning of life.	24	13
(6) Humanity exists to govern over the rest of creation.	15	7
(7) Humans consist only of body and matter.	26	26
(8) Nature is filled by a spiritual force permeating all life.	42	26
(9) Life has no other meaning than to further the human species.	22	17
(10) What becomes of a human being is mostly dependent on her biological heritage.	16	11
(11) Biology teaches that only the strongest survive.	32	22
(12) There is a God, a supernatural power or force.	32	21

Table 11: Religion and biology in telephone interviews to a national sample of the Swedish population between 18 and 74 years in 2009 (*Study 3*), and scientists (in literature, political science, and physics) in 2009 (*Study 4*). Totally agree + Agree somewhat. Percentage.

The overall result is clear. Swedish scientists are less prone to biologization than the Swedish public. One explanation of this could be that the scientists were largely chosen from the human and social sciences. But this explanation seems less likely, since *Study 2* directed to trained biologists—albeit not representative—showed that they also are less prone to biologization. This suggests that scientific and scholarly training in general is likely to counter the tendency to make a move from biological science to biologism.

Taking a more detailed look at the figures, it could be noted that the first four questions are largely questions about value and questions 5–12 concern philosophical or religious issues. When it comes to issues of value, the difference between Swedish public and Swedish scientists is small. Scientists are not as categorical as the public, but except for Question 4, there seems to be no large disagreements between the values of the public and the scientists. Among the philosophical and religious questions, there is a clearer disagreement—especially when it comes to question 5, 6, and 8. Scientist are more skeptical to the statement that biology gives us the complete answer to the meaning of life, that humanity exists to govern over the rest of creation and that nature is filled by a spiritual force permeating all life.

The responses to question 7 are that humans consist only of matter and body is interesting. A significant minority of about 25 percent of both the scientists and the public agree. *But over half of both samples disagree.* When it comes to this question on human nature, there is a significant overlap between scientists, the public, and religious leaders.[21] Such a broad agreement is also surfaces in response to question 9 (life has no other meaning than to further the human species). A majority (58 percent) of the scientists disagree, half of the Swedish public, and almost all religious leaders! A similar agreement surfaces in the response to the statement that what becomes human being is mostly dependent on her biological heritage.

In contrast to these fairly substantive agreements, there is a more obvious contrast when a comparison is made between the responses to the statement that biology teaches that only the strongest survive. There is a fair amount of sympathy for this idea within the public (32 percent), but only 22 percent of the scientists agree.

Finally, the responses to the two more explicitly religious statements are worth considering. The Swedish public is largely in sympathy to the statement that nature is filled by a spiritual force permeating all life. But among the scientists is quite the contrary. A large majority is opposed (48 percent) and only a minority of 25 percent agrees. Interestingly, the difference is not as pronounced when it comes to belief in God (32 percent among the public, 21 percent of the scientists). In fact, 53 percent of the public disagree and only 47 percent of the scientists. 32 percent of the scientists can be described as agnostics.

---

21. See above, chap. 2.

SCIENTISTS AND PUBLIC ON SCIENCE AND RELIGION: A COMPARISON BETWEEN SWEDEN AND THE U.S.

These figures are of interest in comparison to the responses in the U.S. The Swedish studies have not been replicated in the U.S., but it is possible to make a preliminary comparison between science and religion with the help of other data.

In general, the studies in the U.S. are clear on one point—scientists do not affirm belief in God to the same degree as the public. In fact, there is a large difference also in Ecklund's RAAS study of natural and social scientists in 2005. 63 percent of the general U.S. public said that "I have no doubts about God's existence"—but only 9 percent of the scientists. Moreover, 34 percent of the scientists affirmed "I do not believe in God"—compared to only 2 percent of the general public.

Table 12 shows the results of a comparison between the public in U.S. and Sweden and between scientists and public. I have compared those who in one way or another express firm belief, i.e., those that affirm "I have no doubt in the existence in God" in Ecklund's study and those who agree/disagree fully in the studies in Sweden.

	U.S. scientists (n=1646)	U.S. population (n=2812)	Swedish scientists (n=233)	Swedish population (n=502)
Firm belief in God	9	63	15	20
Weak belief in God[22]	19	21	6	13
Firm denial of God	34	2	37	34
Weak disbelief in God[23]	8	10	10	19
Agnostics (don't know)	30	4	32 (13 percent neither affirm nor deny + 19 percent don't know)	14 (13 percent neither affirm nor deny + 1 percent don't know)

22. In the U.S. sample this is measured by the sum of the percentage of positive response to the statements "I have some doubts, but I believe in God" and "I believe in God sometimes." In the Swedish sample this is measured by the percentage affirming *to some extent* (2 on a five point-scale) that "there is a god, a superior power or force."

23. In the U.S. sample this is measured by the sum of positive response to statements "I believe in a higher power, but it's is not God." In the Swedish sample this is measured by the percentage denying *to some extent* (2 on a five-point scale) that "there is a God, a superior power or force."

Table 12: Scientists and public in the U.S.[24] and Sweden[25] on belief in God. Percent.

This is, admittedly, a somewhat unorthodox way of comparing figures. Until a more reliable study comes along, it may still be worthwhile to ponder on these figures. At least three observations can be made: (1) Belief in God in Sweden is generally *much less* frequent than in U.S. (33 percent vs. 94 percent), *but* (2) there is *no dramatic difference* between the *low* proportion of religious belief among Swedish scientists and U.S. scientists (21 percent vs. 28 percent). (3) The difference between scientists and the public on belief in God in Sweden is *dramatically weaker* than it is in the U.S. (only 12 percent more believers in the Swedish public than among Swedish scientists, but 66 percent more believers in the U.S. public than among U.S. scientists.)

A THEORETICAL EXPLANATION OF DIFFERENCES BETWEEN SWEDEN AND THE U.S.

There might be different explanations for these differences. One possible explanation to (1) is the so-called *Rational Choice Theory* (RCT) or the *New Paradigm for the Sociological Study of Religion*. The RCT emerged in the 1990s and can be regarded as an alternative to the classical Secularization Theory (ST). Briefly summarized, RCT goes like this. The demand for religious "goods" is fairly stable, rooted in a universal need for salvation and eternal life. Religion thrives by competition between different religious firms seeking to attract religious consumers offering assurances in these matters. Religious monopolies—for example in Scandinavia—lost their incentives to develop their "product" and actively design their message to attract believers. Moreover, in alliance with the state religious competitors to the Swedish state church were repressed and persecuted. Eventually, they voted with their feet and many of them moved to North America (as the Swedish author Vilhelm Moberg described in his immigrant-epos). In the U.S. religious freedom produced an innovative climate for religious competition between a variety of denominations and churches.[26] Therefore, religious beliefs and practices declined in most parts of Europe and particularly in Scandinavia, but thrived in the free market of religious "firms" in the U.S.

ST originated in the European context. It has its roots in nineteenth century sociology and has been forcefully defended by—for example— the British sociologist Steve Bruce. He argues that two main factors have

24. See Ecklund, *Science vs. Religion*, 16. Sampling described on 160, and 210n10.
25. *Study 3* and *4*.
26. See, for example, Davie, *Europe: The Exceptional Case*, 15–17, 42–44.

contributed to the decline of religion. "Individualism threatened the communal basis of religious belief and behaviour, while rationality removed many of the purposes of religious belief and behaviour and rendered many of its beliefs implausible."[27] This kind of explanation seems especially relevant for the scientific community and could be interpreted as a factor contributing to the low proportion of both Swedish and U.S. scientists expressing firm belief in God. Let me give some comments to this explanatory hypothesis.

At a first glance, it might seem that neither RCT nor ST would be able to explain (2), i.e., that there is no significant difference between the percentage of believing scientists in Sweden and U.S. One would assume that the combined effect of a repressed market, individualism, and strong beliefs in human rationality would result in fewer believers among Swedish scientists than in the U.S. Or to put in in the opposite way: the thriving competition among religious "firms" in the U.S. would be expected to counteract the individualism and rationality in the scientific context and stimulate religious beliefs in the scientific community. But here we may find a clue to an explanation why religious belief is so low among U.S. scientists. Scientists are simply too few to be of any interest for the competing religious "firms." There is a big profitable market of non-scientists—and they prefer a religion of a totally different kind than the one that would attract scientists! In Sweden, it is rather the opposite. Religion has been tightly controlled by the state and pruned, for example, through the education of Lutheran State Church-ministers at secular universities. This may to some extent have resulted in a religion more attractive—at least marginally so—for the scientific community. The unregulated religious market in the U.S. contributes to the wider gap between the percentage believing persons in general and believing scientists. At the same time the tightly regulated religious market in Sweden contributes to a more "refined" and intellectual religion, which is less attractive for the broader population. In sum, based on RCT and ST in combination, we might have an explanation to (3), i.e., that there is such a huge difference between the large percentage of religious believers in the U.S. and the percentage of believing scientists, while that there is a much smaller gap in Sweden. This difference is the effect of the unregulated religious market in the U.S. and regulated religious market in Sweden.

These reflections link up with the previous parts of this book about the views of Swedish clearly and theologians on science and religion. A general result of this analysis was that a majority in these groups favour a model of independence. Nathan Söderblom formulated one version of this idea by suggesting that science and religious lie on two different planes

---

27. Bruce, *From Cathedrals to Cults*, 230.

(although he acknowledged without explaining that "[t]hat the two planes do not lack connection the one with the other.")[28] Studies of the Swedish population at large seem to favor a similar kind of model—and even more so the portion of the Swedish population that describe themselves as believers. There is no evidence that this can be explained with reference to the views of Swedish clergy—nor to the views of Swedish theologians (at least not only). They may rather be an expression of secularization in the sense of a growing tendency to differentiate (and marginalize) religion from other spheres of society.[29] It might also be the case that Swedish clergy, Swedish theologians, and the Swedish population at large walk hand in hand—but for different reasons.

The standing of the model of independence in the U.S. had been analyzed in Elaine Ecklund's latest study in 2011–2012. 9422 scientists from eight global regions were interviewed about science and religion.[30] This study shows that the independence model enjoys sympathies in the U.S. too (29 percent of a representative sample of the U.S. population according to Ecklund[31]). A more detailed analysis shows that these figures are hard to interpret[32], but it seems clear that affirming a conflict *and siding with religion* is very much stronger in the U.S. than in Sweden. In the WVS-study 2012 quoted in the beginning of this study only 8 percent of the Swedish sample agreed with statement "Whenever science and religion conflict, religion is always right" in comparison to 41 percent of the U.S.-population. Swedes would go for the opposite that science is right—or for an independence

---

28. See below, chap. 8. Anders Nygren made a similar concession to a contact model between science and religion in his earlier writings. See Nygren, *Anders Nygren's Religious Apriori*, 123–25.

29. See above chapter 2.

30. Ecklund, "Religion among Scientists in International Context," 4.

31. Ecklund, "Religion among Scientists in International Context," 6.

32. This is verified by another study, namely, Hill, *National Study of Religion*. One of the items was intended to measure the adherence to Steven Jay Gould's NOMA-model. The respondents were asked to state their views on the statement: "Science is about facts and religion is about faith. The two do not overlap." The atheistic evolutionists stand out from the rest. "A majority tends to agree with this statement (62 percent), while only 32 percent and 26 percent of theistic evolutionists and creationists agree." Hill, *National Study of Religion*, 30. Further analysis showed that the formulation of the statement might have been conflated with an adherence to the conflict model. "[A]theistic evolutionists who claim to agree with NOMA are actually *less* likely to believe in this compatibility then atheistic evolutionists who don't agree with NOMA." Hill, *National Study of Religion*, 31. Interestingly, in a Swedish study referred to earlier both those who argued against and those who argued *for* the compatibility between science and religion said that they are "two completely different things." The independence model in a scholarly context is not always what it is in a public context.

model (of sorts). This might be interpreted as an indicator of my thesis that the religious attitudes in Sweden are much more pruned and disciplined by an ideological market regulated by schools and universities and softening the impact of secularization among scientists. In U.S. the unregulated religious market has resulted in successful religious "firms" in opposition to science—and especially to the science of human origins.

Finally, it might be interesting to put the percentage of believing scientists in a global perspective. Elaine Ecklund's RASIC-study from 2011–2012 shows that the general public in France, United Kingdom and USA is much more religious than the scientific community. But the general global pattern is more diverse. " . . . in India, Hong Kong, and Taiwan, when comparing scientists to the local population, a higher proportion of scientists participate regularly in religious services.[33]" Furthermore, in Turkey there is no difference. What is the explanation for this difference? Could it be that science as it is practiced in USA, UK, France, Italy (and Sweden) science is more frequently framed by a naturalistic worldview—and that this is the reason for the relatively low percentage of believing scientists in these countries? If so, there is reason to believe that it is not science *as such*, which has this eroding effect on religion or religious practices. It is rather science combined with a naturalistic worldview, which has this effect. This would require a modification of the classical secularization theory. The secularizing effect of science was a result of a particular historical process. When science was exported to other global regions, it has gradually been framed by other cultural contexts. This may require another kind of analysis for these contexts than the classical secularization theory has provided for Europe and the U.S.

---

33. Ecklund, et al., "Religion among Scientists," figure 1.

Part 2

From Söderblom to Jackelén

Swedish Twentieth Century Theology on Science and Religion

INTRODUCTION TO PART 2

This second part of this book is an analysis of efforts to do theological work on science and religion in a particular cultural context. My source material is Swedish twentieth-century academic theology. "Twentieth century" should be taken in "longer" sense, i.e., from the end of the nineteenth century to the present, i.e., the beginning of the twenty-first century. The aim is twofold. The *first* is to give a historical background to the sociological description of contemporary attitudes to science and religion among Swedish clergy given in Part 1. Needless to say, contemporary attitudes are shaped by several different factors, but the theological heritage of the twentieth century is one important factor among many other. The *second* aim is to provide an inroad to Part 3 about the philosophical analysis of science and religion. In their reflection on science and religion, Swedish theologians of the twentieth century are in different ways and different degrees informed by philosophical ideas. This will become apparent and serve as a foreplay to the philosophical discussion in Part 3.

MODELS OF THE RELATIONSHIP BETWEEN SCIENCE AND RELIGION

A more precise question guiding the content of this second part of the book can be formulated in the following way: *what are the conceptions of the relationship between science and religion among twentieth century Swedish theologians* from Nathan Söderblom to Antje Jackelén *and how can these models or theories be philosophically and theologically assessed?* This question is descriptive, analytical, and critical. It concerns the content, significance and tenability of different models or theories.

Ian Barbour has proposed one of the most influential model of relating science and religion. He distinguishes between four models: conflict, independence, dialogue, and integration.[1] *Conflict* between science and religion is the main model among scientific naturalists and Biblical literalists. *Independence or separation between science and religion* has been suggested by—for example—theologians and philosophers influenced by Wittgenstein's theory of differing language-games. *Dialogue* is the hallmark of philosophers and theologians discerning common issues in the intersection between the two fields, but also of those discerning methodological parallels. *Integration* is emphasized among thinkers developing a theology of nature, but also among those seeking a more radical reinterpretation of both fields in terms of an overarching philosophical perspective (such as process-philosophy).

Many other taxonomies have followed Barbour's proposal. Most of these exist in the tension between what can loosely be described as (1) *models of independence* or *separation* and (2) *models of contact and interplay*.[2] The model of separation interprets science and religion as two totally separated spheres with no contact or overlap. The model of contact acknowledges that science and religion interact with each other, come into interplay and affect the shape with each other. Mikael Stenmark underlines that a model of contact can be focused on (a) harmony and consilience, (b) consistency, (c) tension, or (d) conflict.[3] Emphasizing (c) tension or (d) conflict between science and religion does not amount to the idea of a deep-seated conflict, which in principle is impossible to resolve. Such a more radical idea is advocated by proponents of a different third model, namely, the *model*

---

1. Barbour, *Religion in an Age of Science*. For an extensive overview of different models of the relation between science and religion, see Russell, "Dialogue, Science and Theology."

2. See Bråkenhielm, *Att bekänna människans värde*, 32–35. For a different terminology, see Bråkenhielm, "Religion och vetenskap." Mikael Stenmark has independently refined and developed the concepts of the science-religion relationship in, for example, Stenmark, *How to Relate*; Stenmark, "Ways of Relating."

3. If (d) conflict is claimed it may be resolved through correction of scientific statements, through correction of religious statements or through correction of both. See Simonsson, *Face to Face with Darwinism*, chap. 2; Stenmark, "Ways of Relating," 266.

of incompatibility, which states that there is an irreconcilable opposition between science and religion.[4]

Mikael Stenmark has pointed out that science and religion is a multilevel or multidimensional relationship.[5] They are formed by certain epistemologies and beliefs, theories and stories, but they can also be described as social practices with certain specific goals. I shall return to these social practices and goals later in connection with Stenmark's contribution to the study of science and religion.[6]

In the chapters of this second part of the book, Stenmark's terminology will be used as an analytical tool to clarify the position of different Swedish theologians. However, it should be noted that specific theologians—as Swedish clergy—often waver between different models. It should also be noted that the question of the relationship between science and religion is often perceived from a general perspective and more seldom as the result of an analysis of the relation between specific scientific results and specific theological doctrines. Part 3 of this book provides a more elaborated analysis of the models of conflict, independence and contact.

SUMMARY OF PART 2

Part Two begins with *chapter 6*, which shortly describes the background in nineteenth century of Swedish theology. It was to a large degree developed in the form of confessional Lutheran theology. Towards the end of the century heavy attacks against Christian beliefs emerged, mostly in the form of critique against Biblical literalism, but also from proponents of the scientific worldview. *Nathan Söderblom* (1866–1931) (who later became Archbishop of the Church of Sweden) developed one important line of apologetics, but unlike other Swedish theologians Söderblom never engaged the Swedish philosopher *Axel Hägerström* (1868–1939). Hägerström is famous for his rejection of Swedish nineteenth century philosophical idealism, primarily in the form of *Christopher Jacob Boström's* (1797–1866) subjectivism. Hägerström argued that reality is the spatial-temporal context of experience—and nothing more. This placed him in direct opposition to

---

4. It should be noted that Stenmark in later writings has added a fourth model, the model of substitution according to which the domain of science should be expanded to cover the area of religion and provide us with a new worldview with—suggested by Edward O. Wilson—the evolutionary epic as the central myth. Wilson, *On Human Nature*, 201; Stenmark, "Naturvetenskap och religion."

5. See Stenmark, *How to Relate*; and below, chap. 10, table 16.

6. See below, chap. 10.

metaphysical religion, i.e., religious belief claiming access to the "essence of reality" beyond space and time.

Chapter 7 concerns some of the main theological responses to Hägerström. One prominent example can be found in the thought of *Emilia Fogelklou* (1878–1972). She followed Söderblom in his emphasis on religious experience with Henri Bergson as a common source of inspiration. *Torsten Bohlin* (1889–1950) took his point of departure in the thought of Søren Kierkegaard. Here the emphasis is on a personally acquired faith in contrast to scientific knowledge. A related line of thought is represented by *John Cullberg* (1895–1983), who like Bohlin ended up as bishop in the Lutheran Church of Sweden. Cullberg was inspired by continental existentialist philosophy in the form of the I-Thou-thinking of, for example, Martin Buber. Another effort to come to grips with the challenge of Hägerström is found in the philosophical theology of *Hjalmar Lindroth* (1893–1979). Lindroth admired Hägerström's critique of subjectivism, but argued that Hägerström himself fell prey to metaphysics, when he affirmed that the *only* reality is the spatial and temporal context of experience.

The focus of *chapter 8* is a well-known Lundensian theologian, namely, *Anders Nygren* (1890–1978). One of Nygren's main ideas was his theory about fields of experience (or contexts of meaning): the theoretical, the ethical, the aesthetical, and the religious. The ethical, aesthetic, and religious fields of experience are atheoretical and Nygren draws a line of demarcation between the theoretical and the atheoretical. In consequence, Nygren draws a firm line of demarcation between science and religion. *Ingemar Hedenius* (1908–1992)—professor of practical philosophy, Uppsala University—is well-known for his sharp and energetic objections to Christian theology in general and Lundensian theology specifically. He defended a radical interpretation of Nygren according to which religion and Christian belief should be understood as containing no truth-claims. *Ragnar Bring* (1895–1988)— Anders Nygren's colleagues in Lund—argued that Hedenius' interpretation of Nygren was unfair. Bring made an important supplement Nygren's theology by identifying Hägerström's *temporal and spatial context of experience* with the theoretical-scientific context of meaning. Bring argues that the intention of Nygren by providing religion with a specific context of meaning was *not* to deny that religious beliefs could refer something real, but to deny that they refer to something real in the sense of something which belongs to the spatio-temporal context.

Chapter 9 concerns a number of theologians who take quite a different tack on the whole issue of science and religion. In contrast to Nygren, they are skeptical against all efforts to disjoint religion from other fields of human existence such as science. One representative of this kind of thinking is *Harald Eklund* (1901–1960). He was strongly opposed Anders Nygren's theory of religion and argued that religious belief refers to something that is to accessible to theoretical analysis. His main argument is empirical. One other example is the dissertation of Tord Simonsson (1926–), bishop of Strängnäs 1982–1989. His work represents a significant step forward in the discussion on science and religion in Sweden. In his analysis, Simonsson gave concrete historical evidence for the existence of a common field of interest between Christianity and science. During the 1960s Swedish philosophy of religion shifts attention to the English-speaking analytic tradition. One example is Eklund's successor *Hampus Lyttkens* (1916–2011). Lyttkens had broad fields of interest and played an important role for the contact between British and Swedish philosophy of religion. Lyttkens went further than Eklund and studied the possible parallels between scientific and religious hypotheses with the help of theories and concepts from philosophy of science. *Urban Forell* (1930–)—later professor in the philosophy of religion at the University of Copenhagen—made several contributions in the field of theological ethics, but he also explored the question of acts of God and scientific explanations. *Lars Haikola* (1947–) made other contributions in the area of science and religion, for example in an interesting dialogue with the British theologian Maurice Wiles about divine action.

Chapter 10 changes the focus to theologians at the Faculty of Theology in Uppsala. They took their inspiration from English-speaking analytic philosophy and developed a theoretical framework for understanding the relationship between science and religion. One example is *Anders Jeffner* (1934–)—professor in theological and ideological studies 1976–2000—and his theory of fundamental patterns (FP). FPs involve the adoption of a certain attitude "to the totality of what I experience in life." Fundamental patterns can be falsified by science, but they can never be verified by science, because by the very nature of fundamental FPs, they go beyond science. *Ragnar Holte* (1927–), professor in ethics in Uppsala until 1992, is internationally known for studies of the history of theology, but is also of significance for developing the idea of complementarity between science and religion. *Jarl Hemberg* (1935–1987) got his basic training in philosophy of religion in Uppsala, but moved into theological ethics and became the successor of the well-known Gustaf Wingren (1910–2000) in Lund. Hemberg passed away in the middle of his career in 1987. In a couple of scholarly articles, he took an interest in the relationship between science and religion. *Mikael Stenmark* (1962–) became professor in philosophy of religion at the Faculty of Theology, Uppsala University, in 2008. Stenmark has made significant analyses of different aspects of the study of science and religion. Of particular significance are his efforts to engage the general problem how to relate science and religion as well as explorations into the nature and tenability of scientism.

Chapter 11 has the title *From Generalities to Specifics* and highlights more analyses of different specific problems in the interplay between scientific theories and Christian doctrines. Tensions and conflicts are clarified, and solutions suggested. This trend is visible in a number of dissertations from the middle of the 1990s up until the present. Seven different examples are presented and discussed. The *first* comes from the middle of the 1990s and was defended in 1994 by *Anders Nordgren* (1953–). His stated aim was to investigate the relevance of biological theory of evolution for the philosophical treatment of problems related to mind, rationality, morality and religion. The *second* example is *Antje Jackelén* (1955–). She has been a productive scholar since the middle of the 1990s with 8 monographies, over 70 articles in scholarly journals and the editor of over 20 books. Jackelén has a unique voice in the studies on the relationship between science and religion. She has emphasized the significance of theological hermeneutics, especially in dealing with Christian eschatological thought. *Thirdly*, Åke Jonsson (1943–) published his doctoral thesis, in 1999. The method of the thesis was to compare the Swedish author Gunnar Edman's texts with theologies of contemporary professional theologians who develop their theology in dialogue with scientists. A *fourth* example is *Åsa Nordén* (1935–) with her doctoral thesis 1999. This philosophical study is right at the center of studies on science and religion. Nordén illustrates the indirect relevance of science to theology with reference to three different scientific theories, namely, the theory of Big Bang, the theory of the fine-tuned universe and quantum-theory. *Fifthly*, Eva-Lotta Grantén (1961–) presented her doctoral thesis in 2003. The thesis concerned the relevance of biological results and theories for Christian ethical discussion of human altruism. In a subsequent publication from 2013 the focus is on the Lutheran doctrine of original sin in an evolutionary perspective. *Sixthly, Anne Runehov* (1955–) defended her thesis in 2004. The purpose of her thesis was to investigate the potential of contemporary neuroscientists to explain religious experiences and the relevance of these explanations for the veridicality of religious experiences. Lastly, there is *Lotta Knutsson Bråkenhielm* (1954–) and her dissertation in 2016. Knutsson Bråkenhielm distinguishes between different critical as well as constructive philosophical interpretations of religious belief. She arrives at the conclusion that *certain*—but not all—theories in cognitive science may have relevance for the rationality of *certain*—but not all—forms of religious beliefs.

EARLIER RESEARCH

The main question of this second part of the book—*what are the conceptions of the relationship between science and religion among twentieth century Swedish theologians and how can these models or theories be philosophically and theologically assessed?* —has not been the object of earlier studies. To be sure, there are several studies which have been concerned with the writings of Swedish twenties century theologians and some of which have discussed issues related to conceptions of the relationship between faith and reason with occasional reference to the relation between science and religion. The most important are Martin Fries work 1948 and Jarl Hemberg's dissertation 1966.[7] The value of these two works is, however, limited by their concentration on the issue of metaphysics and whether a particular theologian is "guilty" of metaphysics or not. Nonetheless, they are resources for the present study in that they contain analyses of works, which are important for my interest in the conceptions of science, and theology developed by Swedish twentieth century theologians.

Arne Rasmusson's contribution is valuable for giving a broader background to the social, cultural and institutional context of Swedish twentieth century theology.[8] This article is a valuable complement to the present analysis, partly because it presents Swedish theologians not relevant in the present context of the relationship between science and religion (for example, Einar Billing, Gustaf Wingren, and J.A. Eklund). Last, but not least, an essay of Anders Jeffner deserves a special mentioning.[9] Despite its character of short overview, it contains many valuable insights into Swedish twentieth century theology and it has been a significant guide for the present study.

---

7. Fries, *Metafysiken i modern svensk teologi*; and Hemberg, *Religion och metafysik*.

8. Rasmusson, "A Century of Swedish Theology." Many thanks to Håkan Möller who has drawn my attention to this article.

9. Jeffner, "Teologin inför vetenskapens utmaningar."

6

Early Twentieth Century—Theological and Philosophical Transformations

Throughout the nineteenth century, Swedish theology was largely in opposition to the historical-critical view of the Bible, which was dominating the development of continental theology. Liberal voices (for example, from the historian Erik Gustaf Geijer, author Viktor Rydberg and philosopher Pontus Vikner) were largely heard from outside theology. In the beginning of the twentieth century the Faculty of Theology in Uppsala was dominated by theological conservatism. This was challenged by *Nathan Söderblom*. He became professor at the faculty in 1901 and Archbishop of the Church of Sweden in 1913. Two of his most important works in the present context are *The Nature of Revelation* (in Swedish 1903) and *Religionsproblemet inom katolicism och protestantism* (1910).[1]

Söderblom writes that the second volume of *Religionsproblemet* should be regarded as an attempt to sketch the religious break-through in spiritual works of the then-present era. This sketch had a remarkable impact on many intellectuals at the time it was published. Folke Holmström—a prominent Söderblom-scholar in the 1930s—writes that the work stands as a monument over the ability of Nathan Söderblom to "open up the exclusive and suffocating seclusion that for a long time characterized Swedish theology, to turn the windows wide open for the fresh breezes from the European centers of culture."[2] One of his students—Emilia Fogelklou—witnessed the

---

1. For Söderblom's life and work, see Sundkler, *Söderblom*; and Jonson, *Söderblom*.
2. Holmström, *Uppenbarelsereligion och mystik*, 239.

same and the Swedish author Sven Stolpe writes that the book almost had the character of a revelation.[3]

THE ENDURING LEGACY OF SÖDERBLOM

Nathan Söderblom is primarily known as the Archbishop of Sweden and the initiator behind the Ecumenical meeting in Stockholm in 1925. But he was also a respected religious scholar particularly in the study of world religions. Furthermore, he engaged natural science not primarily by reflection on particular scientific results or theories, but rather through suggesting a particular model by which the relationship between science and religion could be interpreted. This model reemerges in different shapes and forms throughout the twentieth century Swedish theology. It can be described as a model of independence—with a twist.

Söderblom provides a natural starting point for a study on twentieth century Swedish theology in general and a study of conceptions on science and religion more specifically. One entry to his theological thinking is his *Nature of Revelation*.[4] Central in this work and in Söderblom's thinking as a whole is the idea of God's revelation. Revelation—in the general sense—occurs where God comes within the orbit of human inner experience. Söderblom makes a clear distinction between revelation and science.

> In both we are concerned in some sense with reality. In science it is, however in a very abstract way. That is to say, our knowledge gathers up out of reality, that which can be arranged within the causal continuity and such as constitutes the condition necessary for the adaption of nature to our purposes. Of course this scientific view of the totality with its hypotheses stands in an intimate connection with reality. The two planes do not lack connection the one with the other. But the task of science is not simply theoretical, not simply to know, but to rule. The plane of revelation is different. Here we penetrate deeper than we are able to do with the intellect alone . . . we seek to penetrate into and catch the message of reality itself. The truth is that God himself presses in upon us.[5]

What is Söderblom's reason for making a distinction between two planes of reality—and a corresponding distinction between two kinds of knowledge—in the first place? *His answer presupposes the classic distinction*

---

3. Ibid., 238nn13–14. See also Andrae, *Nathan Söderblom*, 220.
4. English translation 1933—originally published in Swedish in 1903.
5. Söderblom, *The Nature of Revelation*, 135–36.

between absolute or ultimate reality and empirical or relative reality. Söderblom derived this distinction from ontological idealism of Henri Bergson, the basic tenets of which are common with the Plato of Sweden, Christopher Jacob Boström and his pupil Vitalis Norström.[6]

So what is the ultimate reality? In another work[7] Söderblom argues that the only way to find an answer to this question is to seek an analogy in our everyday experience. Basically, there are only two alternatives. Either ultimate reality is understood in analogy with *physical* nature or it is understood in analogy with *living* nature, life.[8] Is the universe only a piece of matter—only larger? Or is it like a living universal organism?

Söderblom argues that a physical interpretation of the fundamental nature of reality is inadequate. He has—as far as I can see—three arguments. In the first he follows a kind *reduction in absurdum*-argument. Let us suppose that reality ultimately is like a physical piece of matter. It then follows that there are no differences—other than quantitative ones—between the large and small, between the biggest galaxies and the smallest living cell. You really learn nothing about the *qualitative* nature of reality. But this is absurd.[9] The second argument can be summarized as a critique against the reductionism of the physical interpretation. Is it really possible to reduce life and consciousness to mere physical reality? It is not possible to reach an exhaustive understanding of life and organic processes by reducing them physics and chemistry.[10]

Furthermore, Söderblom has a third argument against a special form of the physical interpretation. A physical interpretation of reality gives ultimate priority to a mechanistic and deterministic explanation of reality. But such an explanation cannot help us to explain how new phenomena can emerge in the process of biological evolution. Söderblom refers to Henri Bergson and his idea of creative evolution.[11] Bergson's book *L'Evolution creatrice* appeared in 1907 and Söderblom held seminars on Bergson in the autumn of 1909. Bergson's dualism between the quantitative, mechanical, conceptual, intellectual on the one hand and the qualitative, living, intuitive runs like an Ariadne-thread through the second volume of *Religionsproblemet*.[12]

6. See further below.
7. Söderblom, *Religionsproblemet*. See, especially, chap. 17.
8. Söderblom, *Religionsproblemet*, 404.
9. Söderblom, *Religionsproblemet*, 404.
10. Söderblom, *Religionsproblemet*, 405–6.
11. Bergson, *L'evolution creatrice*.
12. Söderblom, *Religionsproblemet*, 413–14: "If existence is living, it implies a struggle between opposites, not a monism."

All these arguments for the shortcomings of a physical interpretation of ultimate reality prompt Söderblom to opt for the second alternative. Life is far from being an enigmatic and perhaps noble parasite in one or many rooms of the universe; it is a more enlightening revelation of what reality ultimately is than anything in the physical universe. *Life holds the key to understanding fundamental nature of reality.*

There is an important corollary to this idea. If life is revelatory of the fundamental nature of reality, higher forms of life are *even more* revealing. Ultimate reality is most clearly discerned in human existence—and among human beings it is most clearly encountered among artists, geniuses, founders of religion, and ultimately and most deeply in Jesus Christ. Jesus Christ is the qualitatively highest peak of life—at least on this planet.[13] Moral consciousness here coincides with the experience of eternity. In an ethically colored experience of eternity, God is revealing God's self.[14]

In sum, Söderblom presents conception of the relation between science and religion according to which there is a separation between the two. This separation is explained with reference to an ontological distinction between two planes of reality. Science—in the form of its results and theories—refers to empirical reality, whereas religious phenomena—in the form of revelatory experiences—are experiences of absolute or ultimate reality, which is God.

But it is important to note that Söderblom makes an important qualification of his two-planes-theory. He acknowledges "[t]he two planes do not lack connection the one with the other." This connection is visible in his theory of miracles. On this issue, he consistently defended two different points.[15] (1) Miracles are not to be described as contra-causal events, i.e., events that go against the laws of nature, and (2) they are, rather, lawful events, which the eye of faith are interpreted as a specific revelation of God's will. "In the measure that faith is conscious of the presence of God and sees a divine purpose in that which is happening, the event is unveiled as a miracle and viewed as a revelation."[16]

---

13. It interesting to note that Söderblom is open to the possibility of life in other parts of the universe. Söderblom, *Religionsproblemet*, 404.

14. Söderblom, *The Living God*, 96, 98. See also Jeffner, "Teologin inför vetenskapens utmaningar," 142.

15. For an overview, see Ehnmark, "Nathan Söderbloms ställning," 83–98. Söderblom's earliest writing on this topic occurred already in *Jesu bärgspredikan och vår tid*. In his last work, Söderblom acknowledges that the new physics has superseded the old, but he nevertheless affirms that there are no breaks in the causal chain. See Söderblom, *The Living God*, 435. Compare this to his statement in Söderblom, *The Nature of Revelation*, 149.

16. Söderblom, *The Nature of Revelation*, 197–98 (with a reference to the Swedish

This theory of God's action is based upon that which Alvin Plantinga describes as the *Laplacean picture of the causal closure of the physical deterministic universe*.[17] Direct interventions of God in the natural world are impossible. God acts within the natural laws. Söderblom writes "God uses the natural processes of events to freely enforce his intentions."[18] The point of entry lies outside the realm of science in the "inner room" of the soul[19] and especially in the creative soul of the genius.[20]

"THE PLATO OF THE NORTH." THE PHILOSOPHY OF CHRISTOPHER JACOB BOSTRÖM

Söderblom opened new perspectives when he arrived in Uppsala in 1901. A similar process occurred in philosophy. Philosophical idealism dominated the philosophical scene in Sweden during the second half of the nineteenth century. "Plato of the North," Christopher Jacob Boström stood at the center. His principal work was translated into English and published in 1962[21]. In a review of the translation, Ninian Smart suspects that Boström's absolute idealism is only of historical interest—even if "the idealistic Adam has been stirring strangely of late."[22] A short summary of his doctrine will suffice in the present context.

Boström's doctrine rests on the basic tenets of ontological idealism. The ultimate reality is not the material world of the senses. Material reality is ultimately a world of spiritual and eternal ideas beyond space and time. But in contrast to Plato, these ideas are not abstract entities, but persons. The highest person is God who perceives reality as it is and not as humans who perceive reality as a sensual reality in time and space. This sensual reality can be described as a phenomenal, distorted and relative reality as opposed

---

theologian N. J. Göransson.

17. See Plantinga, *Where the Conflict Really Lies*, 84–90.

18. Söderblom, *Bärgspredikan*, quoted in Ehnmark, "Nathan Söderbloms ställning till problemet tro och vetande," 89.

19. Paul Sabatier spoke about "un Dieu intérieur" or even "un Dieu tout intérieur." In a famous sermon from 1907, Söderblom gives an interpretation of Sabatier's "Dieu intérieur" under the theme *The Inner Guest*: "The guest within cannot be met in the outer rooms of the soul. Do not neglect the guest in you innermost room! You make ready your house for all kinds of other guest. You are so eager to show them attention and courtesy. But meanwhile the Supreme Guest who awaits you within is forgotten." Cited in Sundkler, *Nathan Söderblom*, 79–80.

20. Söderblom, *The Nature of Revelation*, part III.

21. Boström, *Philosophy of Religion*.

22. Boström, Review of *Philosophy of Religion*, 381.

to the noumenal, undistorted, and absolute reality of God. God is the light of the world, which enlightens itself and everything else.

The aim of human persons is to work their ways towards realizing their union with God. This striving requires more than a single life for its completion and Boström was open to the Hindu doctrine of the transmigration of souls. Smart summarizes this peculiar doctrine in the following way:

> In all this he was influenced by Plato—but the Forms become persons; by Leibniz—but God is not creator of the best of all possible worlds, and the monadology is not deterministic; by Berkeley—but God becomes the Absolute; by Kant—but the thing-in-itself is banished; by Schelling—but nature is God's thoughts imperfectly perceived by man; and by Hegel—but he, says Boström, wrongly brings God into time.[23]

In sum, central to Boström's doctrine was a conception of ultimate reality was different from Söderblom's. Söderblom thought of ultimate reality as life and activity; Boström thought of ultimate reality in the form of the supreme self-consciousness.[24] But they shared the idea of two planes—the plane of sensual reality and the plane of ultimate reality. This idea was radically contested by Axel Hägerström.

THE DEMISE OF SUBJECTIVISM AND THE PHILOSOPHY OF AXEL HÄGERSTRÖM

The Swedish philosopher *Axel Hägerström* is perhaps most famous for his doctrine of value, known under the name of "ethical nihilism," according to which value-sentences do not refer to any kind of property, neither natural nor non-natural. They are merely used to express certain emotions.[25] In the present context, I will focus on another aspect of his philosophy, namely, his ontology, which is in clear contrast to Boström's.

Axel Hägerström argued that Boström and other idealists or subjectivists made a basic mistake. They presupposed that the only thing given is the content of our own consciousness. But this is wrong and Hägerström became convinced about this early in his youth. He recollects a memory during a walk near the mansion in Södermanland, where he acted as a private tutor. It came to him, suddenly, as a revelation.

---

23. Boström, Review of *Philosophy of Religion*, 381.
24. See Söderblom, *Religionsproblemet*, 410–11.
25. See Mindus, *A Real Mind*, chap. 3. The book is especially valuable when it comes to the practical philosophy of Hägerström.

The external world cannot be a part of our consciousness, because I and my representations are but parts of the world and the whole cannot be contained by the part. In every representation of reality, we apprehend something other than the representation itself.

Hägerström regarded this idea as basic to his thought.[26]

Konrad Marc-Wogau (1902–1991)—professor in theoretical philosophy in Uppsala from 1946 until 1968—is one of the most important and reliable interpreters of Hägerström's works, the most important of which were published in German.[27] Marc-Wogau writes that many of them are "extremely difficult to understand." This notwithstanding, Marc-Wogau made some heroic efforts to interpret and present Hägerström's thought to a larger public. Central is Hägerström's critique of metaphysics, but his own position has been subject to different interpretations. One reason for this is the complicated relationship between Hägerström's different works on central ontological and epistemological questions (especially the possible difference between *Prinzip der Wissenschaft*, 1908, and *Die Philosophie der Gegenwart in Selbstdarstellung*, 1929). I shall bypass these problems in the present context. Marc-Wogau considers three different ideas central to Hägerström's theory of reality.[28]

The *first* is that the concept of the real means the same as determinate and self-identical. In a non-technical way it means that the real cannot be in one way and at the same time not.[29] That which is real is also non-contradictory. This seems to imply a certain understanding the relation between logic and reality. Is logic (a) derived from reality or is logic (b) a human construction which we bring apply to reality? Hägerström opts for (a). Logical principles such as the law of self-identity and the law of non-contradiction

---

26. Hägerström, *Das Prinzip der Wissenschaft*. See Marc-Wogau, *Studier*, 18, referring to a biography of Hägerström by his sister Margit Waller. Waller, *Axel Hägerström*.

27. But it must be pointed out that Marc-Wogau does not mention the opposition of Hägerström and Phalén against Einstein and quantum physics. Quantum physics discovered that subatomic particles may not be determinate in the sense supposed by Hägerström Patricia Mindus has highlighted this in the closing chapter of her *A Real Mind*.

28. Marc-Wogau, *Studier*, 85–112.

29. This does not seem to accord with quantum physics according to which subatomic electrons may appear as both particles as waves. Hägerström never resolved this problem and it may have contributed to the demise of his theory.

are not human constructions, but discovered through reflection on reality.[30] Hägerström is a conceptual realist (in the Aristotelian sense).[31]

The *second* idea is that there is but *one* determinate comprehensive context of reality (*das 'in toto' gegebene Wirklichkeit*).[32] Knowledge comes through experience, but the basis is this context of reality. In this sense, Hägerström is an epistemological rationalist. The existence of one determinate context of reality is a presupposition of science.[33] It encompasses everything that is determinate either as a content of representation only (such as dreams or phantasies) or as object alongside of the content of representations (such as humans, flowers, or galaxies). Nothing beyond this determinate context of totality is conceptually possible.[34] God as a being beyond time and space is a self-contradiction—so are values and subjects as spiritual substances.

The *third* is that this *in toto* given reality is the spatial-temporal context. This context is presupposed in all judgments. It is the only comprehensible context of reality and it is determined by perceptions and inductions from such perceptions. A spiritual reality is a *contradictio in adiecto*.[35]

Each of these three points provides the basis for Hägerström's critique Boström's idealism and metaphysics more generally, and have been extensively elaborated by Hägerström himself and in secondary literature.[36] The main point in the present context is to explain how his theory of reality is inconsistent with belief in a transcendent God.[37] But—and this is impor-

---

30. Early in his career the American philosopher Ernest Nagel had a similar thought in explicit opoosition to his teacher John Dewey. "[E]xistence is determinately what it is; determinate existence excludes certain characters, and a determinate subject must either possess or not possess a given attribute; the characters which things possess fix the relations between propositions about them. It is true that propositions, and not things, contradict each other. But the ground of contradiction is to be found not in the propositions, but in the nature of the things which the propositions say." See Nagel, "Can Logic Be Divorced from Ontology?," 709. Nagel later came to change his mind about this. See above, chap. 14.

31. Hägerström, *Filosofi och vetenskap*, 84–86.

32. Hägerström, *Filosofi och vetenskap*, 141.

33. Nordin, *Den Boströmska skolan*, 178–79.

34. Hägerström, *Filosofi och vetenskap*, 106–7.

35. Hägerström, *Filosofi och vetenskap*, 144. This refers to a chapter of Hägerström's *Die Philosophie der Gegenwart in Selbstdarsellungen* (1929). As already indicated, is a matter of debate how far this work is consistent with *Das Prinzip der Wissenschaft* (1908). See Marc-Wogau, *Studier*, 53–64.

36. Another Uppsala philosopher, Adolf Phalén contributed significantly to Hägerström's critique of idealism. See Marc-Wogau, *Studier*, 76–81.

37. At least a transcendent God having causal effects in the spacetime world (see further, chap. 12).

tant—Hägerström also had a specific theory of religion, which can be described as a more radical version of the model of separation between science and religion. He distinguished between the metaphysical religiosity and the positive religions. The essence of both is an objectification of a feeling, more precisely, a feeling of divine presence. Dogmas and fanaticism threaten to suffocate this feeling and there is only one way to save religion from destruction, and that is to tear it from the question of truth. Applying the question of truth to religious experience is as meaningless as a person with a toothache asking herself: is this pain that I feel in reality, i.e., independent of my feeling of it?[38] Religion can be saved, but only by severing its ties with metaphysics. Hägerström found religion in the form of feeling of utmost importance, but he was hostile to religion in the form of truth-claims. Feelings do not involve an apprehension of a reality, which exists independently of the feeling. A sentence, which expresses a feeling, cannot be expressed in the form of a statement, which can be either true or false. In consequence, no conflict between science and religion can occur.[39]

Söderblom never engaged Hägerström's philosophy, nor his theory of religion. But a number of other twentieth century Swedish theologians did so, namely, Emilia Fogelklou, Torsten Bohlin, Ragnar Bring, Anders Nygren, Hjalmar Lindroth, John Cullberg, and Jarl Hemberg.[40] When it comes to the relation between science and religion, all of these theologians advocate a model of separation, but *it must be strongly underlined that they understand the separation in different ways*. To understand the way this is done, we need to take a closer look at the different ways they relate to the philosophy of Axel Hägerström.

---

38. Nevertheless, the feeling of toothache could be described as cognitive; is indicates a certain state of affairs in my tooth, indirectly a certain state of affairs outside my body (such as an object that broke my tooth).

39. Hägerström, *Religionsfilosofi*, 168; Marc-Wogau, *Studier i Axel Hägerströms filosofi*, 21–22. For a critique of Hägerström's theory of feeling and religious experience, see Hemberg, "Axel Hägerströms skäl att vara ateist."

40. Jarl Hemberg belongs to another generation of theologians and will be discussed in chap. 10.

7

Theological Responses to Axel Hägerström

Axel Hägerström had a lasting impact on Swedish twentieth century theology. Some elements were taken up in efforts to provide theology with a sharp line of demarcation against metaphysics. His atheistic theory of reality required a more constructive theological response. Emilia Fogelklou is an illustrative example of the latter.

WITH SÖDERBLOM AGAINST HÄGERSTRÖM —EMILIA FOGELKLOU

Emila Fogelklou is one of the significant persons in Swedish twentieth century intellectual life. She became the first Swedish bachelor of theology in 1909. Interestingly, Fogelklou had both Söderblom and Hägerström as teachers. She made important contributions to religious studies, was in the foreground of women's emancipation movement, and contributed to educational thought.[1]

The focus in the present context is her philosophical heritage from Söderblom and Hägerström. According to Fogelklou, we have access to reality not only through science and logic, but also through an experience of value.[2] Science captures that part of reality, which can be objectified and

---

1. Jeffner, "Sjuk av jungfrufödelsen."

2. This is the theme in a short essay by Fogelklou from 1911 later incorporated in her biography *Barhuvad*, 1950. An anlysis of Fogelklou's theological and philosophical ideas is found in Jeffner, "Sjuk av jungfrufödelsen.

incorporated in a coherent view of reality. But certain aspects of reality cannot be described with the concepts of science. We are acquainted with these aspects in the experience of value, which is something more than an awareness of certain values in a restricted sense. It is an encounter with reality and cannot be expressed in in the same logical form as our knowledge of material reality. She argues that there is not only one, but two complementary ways to approach reality.[3] The following quotation summarizes her view:

> ... all *things* are subjected to the jurisdiction of inference. But when it comes to life and the extraordinary logic of the invisible, it nearly always is put to shame, because it surely is capable of combining and explaining, but not imagine, surely clarify, yet never really understand, surely take the tissue of grass asunder, but never hear the grass grow.[4]

Fogelklou acknowledges the scientific worldview, but supplements it with another source of insight which is in important for theology. Her ideas have a certain affinity with Nathan Söderblom's theory of the two planes of reality, but her understanding of the experience of value brings a dynamic element into the relationship between religion and science. What can be incorporated into an objectified and coherent view of reality is not given once and for all. New areas of knowledge appear and these areas may be more or less open to the theological ideas with the help of which experiences of value have been interpreted.

One principal question remains unanswered by Fogelklou, namely, how do we know that the experience of value is an experience of something real? Are the methods of natural science in any way relevant in answering this question? If not, the two planes between sensual reality and ultimate reality are not only distinguished ontologically, but also epistemologically. Torsten Bohlin developed a model of the relation between science and religion, which follows this path—with certain important reservations.

THE INFLUENCE OF KIERKEGAARD
—TORSTEN BOHLIN

Torsten Bohlin studied in Uppsala and published his dissertation on Søren Kierkegaard in 1925. He became professor in dogmatics at the Faculty of Theology in Uppsala in 1929 and was bishop in the diocese of Härnösand 1936–1950. Anders Jeffner characterizes Bohlin as a member of the tradition

---

3. On complementarity bewteen science and religion, see below, chap. 10.

4. See Fogelklou, *Medan gräset gror*, 7. The influence of Henri Bergson is evident and also acknowledged by Fogelklou.

from Söderblom. The emphasis is on religious experience in the form of a personally acquired faith. In his dissertation, Mikael Lindfelt underlines that Bohlin understood concept of faith was described in sharp contrast to scientific knowledge.[5]

The overarching source of inspiration was Kierkegaard, but not the Kierkegaard of Karl Barth. Bohlin makes at distinction between two lines in Kierkegaard's thought, namely, the line of paradox and the line of religious experience. The line of paradox is characterized by abstract absolutism, while the line of religious experience in the form of devotion to God conceived as "holy will of love."[6] This experience is connected to a certain from of certitude characterized as impressedness and insight in analogy to the impressedness and insight we have in relation to human persons. He contrasts this certitude against scientific knowledge, but more importantly, and following Kierkegaard, he distinguishes between objective truth and subjective truth. This suggests a model of separation with a truth in the form of correspondence on the one side and personal or emotional truth on the other. But Lindfelt argues that this is not a correct interpretation of Bohlin. Both objective and subjective truth are concerned with truth in the form of correspondence with reality, but they are so concerned in different ways. In explicit contrast to Hägerström, Bohlin affirms that there is a religious question of truth. Is this religious experience an experience of something real—or is an illusion? Bohlin writes:

> Christianity rests on the conviction that the innermost of reality is a creating and good will, that the Father of Jesus Christ is love. Is this conviction truth, i.e., does this certitude correspond to a supernatural entity, or is it just a reflection of the wishes and dreams of human heart under duress?[7]

In emphasizing that there is a religious question of truth, Bohlin seems to come into conflict not only with Hägerström, but also with two other contemporary Swedish theologians, namely, Anders Nygren and Ragnar Bring (which I shall present in a moment). In this he is in agreement with the Swedish philosopher Ingemar Hedenius, but was nevertheless an object of Hedenius' forceful attack against Christianity and Christian theology in 1949. The ground they share is never exploited, because Hedenius chooses to attack Bohlin's theory of religious language and his conception

---

5. Lindfelt, *Teologi och kristen humanism*, 137.

6. Bohlin, *Tro och uppenbarelse*, 118, quoted by Jeffner in "Teologin inför vetenskapens utmaningar," 154.

7. Bohlin, *Den ofrånkomlige*, 60. Quoted by Lindfelt, *Teologi och kristen humanism*, 133.

of theology, areas on which Bohlin is philosophically and rhetorically weaker than Hedenius.[8]

In the present context, the focus is on the relationship between science and religion. It is then of vital importance to understand how Bohlin responds to the question how we arrive at religious truth. How do we know that it is something real and not an illusion or wishful thinking? Well, not in the same way that we seek to verify or falsify empirical statements by experiment. Bohlin suggests a *religious pragmatic way of responding to the question of religious truth*. His remarks are, however, rather fragmentary with scattered references to William James. In general terms, the criterion for a religious truth is that a religious belief is true if it strengthens the realization of a real human life or a conviction to have found "a vision of that the personality has found his idea, his originally given determination."[9] But Bohlin has no theory about pragmatic arguments in religious discourse and their relation to epistemic arguments in scientific discourse. For example, Bohlin is far from the American philosopher Jeff Jordan and his elaborated argument in his work on Pascal's wager from 2006.[10] Interestingly, Jordan argues that pragmatic arguments for the existence of God have rational force only in the case of an evidential "tie" between arguments for and against God's existence.[11] Such a claim would be closer to a model of interplay than a model of separation, and foreign to Bohlin, who underlines the difference between religious faith and mere theoretical speculation and abstract metaphysical statements "which have nothing to do with personal confidence and religious trust."[12]

I-THOU-PHILOSOPHY—JOHN CULLBERG

In his classic taxonomy of ways of relating science and religion (conflict, independence, dialogue, and integration), Ian Barbour explains three different ways in which the independence of science and religion can be justified. The first is through protestant neo-orthodoxy, the second through existentialism and the third with the help of linguistic analysis (for example,

---

8. See Jeffner, "Teologin inför vetenskapens utmaningar," 153–57. See also Hedenius, *Tro och vetande*, 146–63; Hedenius, *Att välja livsåskådning*, 299–312, Lindfelt, *Teologi och kristen humanism,* 261–70; Lundborg, *När ateismen erövrade Sverige*, 96–97; 143–52. Lundborg gives a more positive description of Bohlin's argument against Hedenius than Jeffner.

9. The latter quote is taken from an article by Bohlin published in 1931 and quoted by in Lindfelt, *Teologi och kristen humanism*, 137.

10. See Jordan, *Pascal's Wager*.

11. Jordan, *Pascal's Wager*, 110.

12. Bohlin, quoted in Lundborg, *När ateismen erövrade Sverige*, 144.

Ludwig Wittgenstein's later philosophy).[13] (Independence through the interpretation of religion as an autonomous *a priori* falls outside this scheme, but Nygren's approach in his later phase comes close to this additional line of reasoning.) In Swedish twentieth century theology there is one clear example of sharp separation through the existentialist contrast between the realm of personal selfhood and the realm of impersonal objects. It is found in the thought of John Cullberg. He was assistant professor in Uppsala and later bishop of the Diocese of Västerås between 1940 and 1962. His thought developed during the years and at the last page in his book from 1968, Cullberg acknowledges that God exists out of his hiddenness in the resurrection of Jesus Christ, which as a historical fact belongs to the realm of I-it-relationships.[14]

Cullberg's main objections against Axel Hägerström were delivered already in 1930.[15] They have been thoroughly analyzed by Jarl Hemberg and will not be discussed in detail in the present context. Suffice to say, Cullberg's main point is that Hägerström's theory of reality suffers from internal inconsistencies, particularly when it comes to the concept of reality. Cullberg claims that Hägerström's theory—contrary to Hägerström's own intentions—presupposes an I, a subject, who has certain conceptions about reality. Hägerström's main concern is to challenge subjectivism, but he lapses in this trap himself! According to Cullberg's own opinion, the concept of objective reality should rather presuppose not only an I, but a community of persons, which has this intersubjective reality in common. This is also the main point in his constructive philosophy.[16]

Cullberg distinguishes between first-person-knowledge (I-knowledge), second person-knowledge (you-knowledge) and third-person-knowledge (it-knowledge). Natural science consists of it-knowledge of material reality. But this kind of knowledge is established by cooperation between scientists, when the results of a scientific experiment are verified (or falsified) by other scientists. Intersubjective verification or falsification presupposes mutual trust, i.e., mutual recognition that other scholars are persons and not only objects. A person can be reduced to an object and studied as a thing, a "mere" organism. But such studies do not give us knowledge about the other person in the form of you-knowledge. You-knowledge is an essentially different kind of knowledge than it-knowledge. It is a kind of knowledge of qualities that are non-measurable by behavioral

13. Barbour, *Religion and Science*, 84–87.
14. Cullberg, *Den fördolde Guden*, 199.
15. Cullberg, *Religion och vetenskap*.
16. Cullberg, *Tro och verklighet*.

or biological science. Nevertheless, such a kind of non-measurable relationship is a basic presupposition for science.[17]

Now, Cullberg applies this philosophical analysis to religion and the knowledge of God. He argues that knowledge of God cannot be analyzed as a kind of I-knowledge. This results in a mysticism of boundlessness, where God and I merge into an indistinguishable unity. Nor can knowledge of God be described as it-knowledge. Religion is reduced to metaphysics. Rather knowledge of God is a kind of you-knowledge.[18]

From this follows a sharp division between science and religion. Human persons known in a radically different way than material things are known in natural science—and God is similarly known in a unique way, which cannot be expressed in scientific propositions and cannot be supported with reasons similar to those used in science. This difference between science and religion is a difference in principle.

This version of the model of independence between science and religion suffers from a serious weakness. Human you-knowledge is clearly connected with—albeit not reducible to—it-knowledge. We know other persons through their bodies and behavior and through our knowledge about the relationship between the physical organism and mental life.[19] But according to Cullberg, God is known *only* in the form of an Eternal Thou and not as an object.

Cullberg achieves a sharp separation between science and religion by his sharp distinction between you-knowledge and it-knowledge. In contrast to Martin Buber, he does not allow for an intertwined mixture between the two kinds of knowledge.[20] When he interprets the relationship to the eternal Thou, God, as a revelation, he is obliged to conclude that this revelation is without any objective content! Not only science stands in contrast to revelation, but any kind of objective knowledge. Hampus Lyttkens concludes that the distinction between you-knowledge and it-knowledge is unclear. There is an awareness of an object also in second

---

17. See Cullberg, *Das Du und die Wirklichkeit*, 197, 211.

18. In his earlier writings Cullberg boldly affirmed that recognition of the God as the eternal You is implicitly or explicitly presupposed by the recognition of human you-knowledge. See ibid., 217–18; Cullberg, *Religion och vetenskap*, 265, and critically commented by Hemberg, *Religion och metafysik*, 178. Cullberg retreats from this kind of moral argument for the existence of God in Cullberg, *Tro och verklighet*, 76n1.

19. See Jeffner, *Filosofisk religionsdebatt*, 50.

20. See Cullberg, *Das Du und die Wirklichkeit*, 45. See also Lemberg, *Jag, du och verkligheten*, 77.

person-knowledge.²¹ Such an analysis opens for the possibility of an interrelationship between science and religion.

It should be noted that Cullberg himself modifies the model of independence in important ways. His main idea concerns the relationship between science, the humanities and religion (in the form of Christian faith in God). There is a progression from an analysis of science to the analysis of religion. In a Kantian fashion Cullberg makes a kind of transcendental deduction of science. What makes science possible? Cullberg's response is: science contains an irreducible subjective element (without I—no object).²² Furthermore, the humanities contain an irreducible social element (without you, no I). Ultimately, both science and the humanities presuppose faith in God (without God—no you) for their possibility. "The given conception of God in revelation has real significance, because it is a presupposition for the consciousness of reality as a whole.²³

Cullberg's transcendental deduction of science was questioned by Hjalmar Lindroth, professor of dogmatics in Uppsala (1937–1960). I will discuss Hjalmar Lindroth's approach to the relationship between science and religion later in this chapter, but it is important to mention his critique of Cullberg in the present context. It took place in the theological journal *Svensk teologisk kvartalsskrift* in 1931.²⁴ Both Lindroth and Cullberg were pupils of Axel Hägerström.

Lindroth presupposes the truth of a model of independence in a much stronger form than Cullberg and is critical of Cullberg's transcendental deduction. Lindroth argues, for example, that Cullberg lapses into subjectivism in defending his formula "without I—no object." The humanities fall short of providing us with a you in the sense of an *absolute* you (not only in the sense of a limited psycho-physical you). Furthermore, Lindroth has misgivings about Cullberg's concept of religion. Moreover, "the *emphasizing* of religious categories of reality is foreign to true religion."²⁵ This is not a fair critique of Cullberg; his argument does not concern the existence of God, but that science implicitly presupposes the existence of God by affirming

---

21. Cullberg, Review of *Den fördolda Guden*, 131.

22. This is Cullberg's basic critique of Hägerström's philosophy. See Cullberg, *Religion och vetenskap*, 229–31. See also Lemberg, *Jag, du och verkligheten*, 133–40.

23. See Cullberg, *Religion och vetenskap*, 268–69; Cullberg, "Religion och vetenskap än en gång," 405; Cullberg, *Das Du und die Wirklichkeit*, 235; Cullberg, *Tro och verklighet*, 78–80. (In Cullberg, *Den fördolda Guden*, 76, Cullberg retreats from his earlier position in Cullberg, *Das Du und die Wirklichkeit*, 217.)

24. Lindroth, "Religion och vetenskap." Cullberg, "Religion och vetenskap ännu en gång." Lindroth, "Religion och vetenskap ännu en gång,"

25. Lindroth, "Religion och vetenskap," 280.

objective reality, which cannot be presupposed without belief in God. To be sure, Cullberg may be wrong in this, but the mistake lies somewhere else than Lindroth thinks it does.

Unfortunately, Lindroth does not highlight the central problem with Cullberg's transcendental deduction, namely, the step from the I-you-relationship (especially important in the humanities) to faith in God. Cullberg writes: "Die existentielle Du-beziehung setzt die ebenfalls existentielle Beziehung zu Gott voraus" ("The existential You-relation presupposes the existential relation to God).[26] As far as I can see, Cullberg gives no specific reason for this daring claim.

In short, Lindroth's main concern is driven by the insistence on a sharp interpretation of the separation thesis. I shall return for his argument for this interpretation shortly.

In his last work, Cullberg himself shows almost no sympathy for "muddling the waters" between science and religion. He rejects Rudolf Bultmann and his program of demythologization and Bultmann's claim that we must take leave of the mythological worldview of the New Testament because our contemporary thinking is irrevocably determined by science. Cullberg argues (1) that not only is the message of the New Testament totally distinct from the worldview of antiquity and, for that matter, from any kind of worldview. Furthermore, (2) science is forever changing and an adaption to the "modern" worldview would result in an endless series of adaption to the ever-changing process of scientific progress.[27]

Cullberg concludes his book on the hidden God with a chapter on miracles with special attention to the New Testament witness about the resurrection of Jesus Christ. In contrast to Söderblom, Cullberg is critical to the idea of the *causal closure of the physical deterministic universe*.[28] Miracles in the sense of violations of natural laws are possible and do not go against science. For example, Cullberg argues that the resurrection of Jesus is a historical fact. It is an act by the hidden God based on the existential encounter between risen Lord and the first disciples as well as Christian believers (through the sacraments). One would expect Cullberg's sharp distinction between existential you-knowledge and factual it-knowledge should lead him to dismiss the historical and factual accounts in the New Testament

---

26. Cullberg, *Das Du und die Wirklichkeit*, 237. See also Cullberg, *Den fördolde Guden*, 79: "The existential community would completely break down if it was not sustained and inspired by divine love."

27. Cullberg, *Den fördolde Guden*, 126–27. (2) is a common argument for distinguishing science and religion, relativizing science and absolutizing religion. It is not convincing, but I shall not argue against it in the present context.

28. See above, chap. 7.

stories about the resurrection. But, as already noted, this is not the case. For example, Cullberg argues that the empty tomb of Jesus is not irrelevant for the existential belief in the risen Lord. It is relevant in the subsequent reflection on the encounter, because the existential encounter involves the conviction that something "really happened."[29] And the empty tomb is evidence for that something "really happened." At this point Cullberg goes against his sharp separation between you-knowledge and it-knowledge. Science and religion are in interplay and historical evaluation of the narratives of the empty tomb seems relevant for the credibility of the existential encounter between the disciples and the risen Lord. Cullberg argues that the relevance is "essentially ambiguous" and requires a "personal decision."[30] At the same time he admits that "miracle" is a boundary concept between science and faith. Cullberg's basic point of departure is modified and the result is a departure from the model of independence to a plea for the model of connection, i.e., that religion and science comes into interplay with each other and can mutually affect each other. Possibly, he would argue that there is a conflict, but a conflict that may be resolved by a distinction between science proper and science combined with the idea of a closed universe.[31]

THE PHILOSOPHICAL THEOLOGY OF HJALMAR LINDROTH

Hjalmar Lindroth succeeded Torsten Bohlin as professor in dogmatics at Uppsala University in 1937.[32] His dissertation was a study in Schleiermacher's concept of religion (1926, with a second volume in 1930), but his interest soon turned to his philosophical teacher Axel Hägerström. He admired Hägerström's critique of subjectivism, but argued that Hägerström himself fell prey to metaphysics, when he affirmed that the *only* reality is the spatial and temporal context of experience. Such a universal view is untenable for many reasons and Lindroth summarized these reason in various publications throughout his academic career.[33] One reason is that unique self-consciousness is real, even if the self does not belong to the spatial-temporal context of experience.[34] Another reason corresponds to an idea,

29. Cullberg, *Den fördolde Guden*, 192–93.
30. Cullberg, *Den fördolde Guden*, 197, 199.
31. See further, Plantinga, *Where the Conflict Really Lies*, 84–90.
32. Wrede, "A Hjalmar J Lindroth."
33. See Lindroth, *Verkligheten och vetenskapen*; Lindroth, *Tron och vetenskapens gräns*; Lindroth, *Filosofiska och teologiska essayer*; Lindroth, *Kyrklig dogmatik*.
34. Lindroth, *Tron och vetenskapens gräns*, 34–50; and *Filosofiska och teologiska essayer*, 27n3.

which we already discussed in connection with Emilia Fogelklou: there is an experience of value, which is something more than an awareness of certain values in a restricted sense. It is an encounter with reality and cannot be expressed in in the same logical form as our knowledge of material reality. A third reason has to do with Hägerström's idea that reality is determinate. But reality *as a whole* can never be determinate. Determination implies that it is determined by something and if reality as a whole is determinate it is determined by something, which (to be real) must be determined by something *ad infinitum*. Total reality escapes our conceptual grasp. It is transcendent.[35]

There are certainly weak points in Hägerström's ontology, but Lindroth's critique is not entirely convincing. The last argument, that Hägerström's ontology leads to an infinite regress, rests on misinterpretation. When Hägerström claims that the spatial and temporal context of experience (*das 'in toto' gegebene Wirklichkeit*) is determinate, he does not mean that it is determined in the sense of casually dependent. He argues that it is non-contradictory, i.e., in one way and not at same time not in that way. For example, the spatial and temporal context of experience is such that it came into existence more than five hours ago. This property of *das 'in toto' gegebene Wirklichkeit* does not lead to an infinite regress.

The claim that total reality escapes our conceptual grasp may nevertheless be correct, but is it not possible that a so-called "theory of everything" someday might be verified? If that is the case, total reality could be ambiguous in the sense that several different interpretations of total reality might still remain open, for example, an interpretation that suggests something beyond the spatial and temporal context of experience (and even all forms of enduring contexts).[36] In this sense the point of entrance for Lindroth's constructive theological proposal might still be open. Faith has to do with that which is beyond the border of knowledge and science.

> [Faith is] the opposite of knowledge, the opposite of capturing and determination. It is a question of . . . being seized and determined by an ultimate power, it may be called reality in itself or in the living language of faith, God. It is a question of faith in the key of transcendence, faith, with the agape-direction, i.e., faith as the reception of the eternal reality and God's fullness.[37]

Lindroth's language is closer to a Sunday sermon, than academic theology. This is also evident in the final lines of his discussion with John Cullberg.

35. Lindroth, *Tron och vetandets gräns*, 100–102.

36. The parentheses is inserted to cover other universes according to theories of the multiverse.

37. Lindroth, *Tron och vetandets gräns*, 127.

His underscores his critique that Cullberg *in reality* wavers when it comes to the separation between religion and science, but welcomes Cullberg's intention *in principle* to uphold the model of independence. This would require Cullberg's abandonment of the idea that science requires faith in God for its claim to reach objective reality.

Lindroth affirms that God's love in Jesus Christ is an incomprehensible and impenetrable mystery. God is the Irrational Unknown and the theologian must reckon with this against all logic and all the comments from reason. It is nothing less than a paradox that the author of these words is also the author of three volumes of *Church Dogmatics* with over 900 pages on the Christian doctrine with special attention to eschatology and salvation history (published in 1975). His thought in these volumes reflects a paradigmatic change in his theological outlook, which took place in the beginning of 1940s when Lindroth came under the influence of Uppsala studies in New Testament eschatology and Anton Friedrichsen, professor in New Testament exegesis.[38] On the one hand, Lindroth came to give prominence to Biblical revelation as the ultimate source for Christian theology. Moreover, he came to affirm the possibility of an objective interpretation of divine revelation. On the other hand, the Church received a more prominent place in Lindroth's view of history. It is telling that the Church occupies the second and main part of his dogmatic. Lindroth identified the Church as the *Regnum Christi* between the resurrection of Christ and the general resurrection at Christ's second coming. *Regnum Christi* includes society and nature, but the "inner circle" is the Church. The Church is the proclamation of the word of God, the sacraments and the ministry by which they are administered. In this day and age, the Church is the place where the spiritual life of the age to come can be realized. In this sense, the Church is something exclusively unique, incommensurable with everything else.[39]

Two other Swedish theologians stand in sharp opposition to Hjalmar Lindroth. The first one is Harald Eklund, the son of J. A. Eklund, a well-known Bishop in the diocese of Karlstad (1907–1938). Harald Eklund was professor in Lund from 1949 until his death in 1960. During his whole career as an academic theologian, Eklund was critical against all efforts to separate religion from science and knowledge. This is evident in different ways from his time as a professor in Åbo (1945–1949). He distanced himself from the idea that religion was independent from science, an autonomous "form of life" with other laws than the theoretical life, "atheoretical," "existential," "practical," or

---

38. Three reprinted lectures from 1948 in Lindroth, *Filosofiska och teologiska essayer*, 195–235, explain this shift.

39. Lindroth, *Kyrklig dogmatik*, 2:61–62.

"personal."[40] In a short article from 1945 he specifically addresses the relation between science and Christianity and argues that the relation is insufficiently analyzed. He makes the important point that already to engage in these questions is to involve oneself in a scientific discussion. And this analysis shows that there are possibilities of conflict between science and Christian theology. One example is "the modern eschatological faith."[41] Is it anything more than an experiment in mythology? Eklund is especially critical of the new church-theology "with its dissolution of all logic, its peculiarly overblown self-confidence, sectarian zeal and demand for belief in authority."[42] Religious belief is considered as something other than knowledge and theology as a separate science with its own territory. Eklund claims that this leads sectarianism and one expression of this is the sheer impossibility for external observers to understand what it's all about.[43]

In a much-discussed article in *Svensk exegetisk årsbok* in 1948, Eklund voiced heavy criticism against the eschatological theology, which was advocated by Lindroth and Friedrichsen in Uppsala. Eklund made a strong plea for an everyday view of reality, which could be described as a form of naïve realism. This view of reality was the criterion against which stories in the New Testament had to be judged—including the Biblical stories of the end of the world expressing a "mayhem-theology" accompanied by an exclusivist ecclesiology totally impossible in our time and day.[44] I will come back to Eklund's theology in the chapter 9.

Another critique of Lindroth and Friedrichsen comes to expression in the theology of another Swedish theologian, namely, Gustaf Wingren. Wingren focuses upon God's activity in society and nature. In the proclamation of the word of God and the sacraments the Church makes the sustaining, emancipating, and universal power of God visible. Wingren describes the Kingdom of God as the unhindered flow of the spontaneous expressions of life.

---

40. Eklund, *Tro, erfarenhet, verklighet*, 149, in an article already published in 1944.
41. Wrede, "A Hjalmar J Lindroth."
42. Eklund, "Kristendom och vetenskap," 10.
43. Eklund, "Kristendom och vetenskap," 10.
44. See Eklund, "Nystestamentlig och modern verklighetssyn," 1–8. See also Anton Friedrichsen's response: "For him [Harald Eklund] the problem is this: Can modern human beings discern anything real in the proclamation of the Church? For me it is like this: can the radical revolution of the ordinary view of reality that the Bible represents come to expression in such a way that the 'theologically unprepared' human being feels the ground shake under her feet and begin to look for a safe spot on which she can stand?" Friedrichsen, "Epilog," 120.

These controversies appear to be irrelevant to the main theme of this book. But I will return to Eklund's view on religious language in chapter 13 and the disagreement between Lindroth and Wingren in chapter 14 of this book and show that it is relevant to the interpretation of science in a theistic perspective.

8

A Model of Independence and Its Critics

Most of the Swedish theologians whom I have presented so far are in different ways proponents of a model of independence or separation between science and religion. Nevertheless, most of them find this model uncomfortable at various points. Söderblom underlines that "the two planes do not lack connection the one with the other"[1] and few of them go as far as Steven Jay Gould went with his NOMA-thesis denying that religion should be understood in contrast to science by not making any kind of truth-claims. But there are two Swedish theologians who come close to Gould's NOMA-thesis, namely, *Anders Nygren* and *Ragnar Bring*.[2]

THE THEORY OF ATHEORETICAL RELIGION — ANDERS NYGREN

Nygren was professor in systematic theology at Lund University 1924 to 1948 and after that bishop in Lund until his retirement in 1959. He conceived his main theological approach early in his life and elaborated his main idea already in 1921.[3] After his retirement, he published *Meaning and*

---

1. See above, 67.
2. On a closer analysis, there is a signficant difference. See further below, chap. 14.
3. Nygren, *Det religionsfilosofiska grundproblemet*; Nygren; *Religiöst Apriori*. English translation: Nygren, *Anders Nygren's Religious Apriori*.

Method: Prolegomena to a Scientific Philosophy of Religion and a Scientific Theology (1972).[4]

The theology of Anders Nygren is of special significance in the present context. Unfortunately, there is no scholarly agreement about its interpretation. I shall come back to this, but initially depart from the conventional interpretation of his basic theological idea, which also undergirds a model of separation or independence. Nygren emphasizes that science and religion are two separate fields without contact or overlap. This may come as a reminder of Stephen Jay Gould's principle of Nonoverlapping Magisteria (NOMA). But there are at least two important differences between Gould and Nygren. First, Nygren does not use the concept of magisteria. Instead he speaks about categories or contexts of experience or meaning. Secondly, Nygren writes that there are (at least) *four* different contexts of experience: the theoretical, the ethical, the aesthetical, and the religious.[5] Each of these is determined by specific presuppositions. The theoretical field (primarily science) is linked to the question of truth, the ethical to the question of goodness, the aesthetical to the question of beauty, and the religious to the question of the eternal.

Nygren emphasizes that confusion and unnecessary conflicts arise if the borders between these categories are not respected. If science oversteps the boundary to religion and claims to be the ultimate norm of truth and rationality, we end up with scientism. And if religion adopts the methods of science, the result is metaphysics. Scientism transforms science into an unjustified absolute and metaphysics distorts religion.

Nygren's theory provides an elaborated justification of the model of separation and independence, i.e., that science and religion are two autonomous contexts of meaning and experience without overlap or contact. This is at least how the theory is transmitted to students of theology in articles and textbooks.[6] But how does the theory appear on closer scrutiny and is the theory reasonable? An answer to this question requires a more detailed look into Nygren's main idea with focus on his theory of religion.

---

4. Nygren, *Meaning and Method*. In the present context the original Swedish publication will be cited: Nygren, *Mening och metod*; Hall, *Anders Nygren*. A scholarly survey of Nygren's thought is Kegley, *The Philosophy and Theology of Anders Nygren*. Nygren's major works are summarized in Hall, "The Nygren Corpus."

5. Nygren, *Anders Nygren's Religious Apriori*, 88–96; Nygren, *Meaning and Method*, 331–42. For the significance of Bring for further systematization of Nygren's theory, see Månsson, *Vetenskap, evighet och religion*, 91–95.

6. See, for example, Rasmusson, "A Century of Swedish Theology," 133; Capps, *Religious Studies*, 25–27.

Nygren's point of departure was heavily influenced by Immanuel Kant. Central to Nygren is the Kantian concept of *a priori*, i.e., non-empirical conditions for the possibility of experience. Kant used the concept of a priori in more limited sense of non-empirical conditions for the possibility of scientific judgments. Nygren extends the use of the concept by seeking not only the ethical and aesthetical *a priori*, but also posing the question: is there a religious *a priori*? Through a transcendental deduction of the basic category of religion, Nygren arrives at a positive answer:

> If the religious apriori is a given, the necessity of religion can be demonstrated. Certainly this does not mean that religion has a factual necessity, but, rather, that it possesses a formal necessity: *religion is an inherent, essential ingredient of human consciousness*. Without the modality of religion, consciousness cannot be regarded as being complete. Without religion, consciousness must be portrayed as being, in a certain sense, underdeveloped.[7]

What is the substantial result of the transcendental deduction of religion? Nygren claims that "religion elevates human life beyond the temporal, finite sphere." And he continues:

> ... there is an eternal world that does not stand in exclusion if the temporal within which we live, such that we would have to leave temporality behind in order to participate in that which is eternal. On the contrary, it is the nature of piety to seek the traces of the eternal in everything that occurs.[8]

A central claim of Nygren is that religion, ethics and aesthetics are atheoretical in contrast to theoretical science. But the differentiation between, say, religion and theoretical knowledge is not the same as isolation. Nygren is clear about this:

> Starting from religion, the same principle applies: while the distinctiveness is clear, the differentiation is not absolute, for *there are significant overlaps between religion and the other spheres*. We can even say that there is no religion that is not, at the same time, knowledge. This is an important point, for there is a strong tendency today to contest this view; the opponents wish to claim that "religion is not theory, but life." Of course, on its positive

---

7. Nygren, *Anders Nygren's Religious Apriori*, 48; my italics.

8. Ibid., 95. This quote seems to undermine the model of independence. If there are "traces of the eternal in everything that occurs," how is it possible to separate the theoretical and scientific context of experience from the religious?

side, this is a true statement. But it is also true that the religious life penetrates theoretical considerations. For example, religious life expresses itself in conceptions of faith. It would be incorrect to separate these spheres absolutely.[9]

The overlap between the theoretical and religious context notwithstanding, Nygren does not speak about truth in religion. Looking for truth in religion results in metaphysics. The proper question is to look validity. When it comes to religion, a distinction must be made between validity of a particular religion and validity of religion in general. Validity of a particular religion has to do with the "fundamental motif" of a certain religion. "There are after all so many religions Christianity, Buddhism, Taoism, etc. each characterized by its own fundamental motif."[10] Walter Capps has suggested that this is in fact a basic program for religious studies. Nygren himself conducted such a study of the basic motif in Christianity—love, *Agape*—in contrast to the basic motif of Platonism, i.e. *Eros,* and of Judaism (*Nomos*).[11] Capps hints that similar studies could be accomplished with other religious traditions in mind. A theory of the essence of religion like Nygren's need not stand in the way of a scientific study of religion.[12]

Validity of religion in general is closely related to the formal category of religion, i.e., the eternal. It is the category upon which all the other three categories (or cultural forms) depend.[13] This feature of Nygren's theology has interesting parallels to Nathan Söderblom's enigmatic phrase that the two planes of science and religion "do not lack connection the one with the other" as well as John Cullberg's idea that the it-knowledge of science presupposes an existential you-relation and "the existential You-relation presupposes the existential relation to God." At the same time, these suppositions are problematic because they compromise the science and religion-model of separation and independence.[14]

9. Ibid., 48–49; my italics.
10. Nygren, *Mening och metod*, 361.
11. Nygren, *Eros and Agape*.
12. See Capps, *Religious Studies*, 49–52.

13. Nygren, *Anders Nygren's Religious Apriori*, 95: "[R]eligion is deeply rooted within the life of consciousness since all forms of culture are built upon the axioms, principles, and presuppositions that find their realization in religion. That is, what the other categories need, if they are not to be left floating in the air, is provided distinctively in religion."

14. See also Nygren, Review of *Meaning and Method*. Paul R. Clifford notes the incoherence between arguing for the autonomy of the different contexts of meaning and at the same time claim that it is a proper task of philosophy to search for the "presupposition of presuppositions," i.e., metaphysics! See Nygren, Review, 496.

This problem becomes even more pressing in Nygren's later thought.[15] Spurred by the reading of the later Ludwig Wittgenstein's idea about language-games, Nygren seems to leave the idea of overlapping contexts of meaning behind and sharpens the autonomy of these contexts.[16] Paul R. Clifford's verdict seems justified: "Nygren invokes philosophy as the handmaid of isolationism, not of dialogue."[17] Furthermore, Nygren in *Meaning and Method* has substituted his theory in *Religious Apriori* that religion is a basic presupposition for all the other three contexts of meaning with a theory that the different contexts of meaning are *mutually* dependent on each other. As Clifford notes this later theory of mutual dependence is difficult to square with the idea of the complete autonomy of the different contexts of meaning.[18]

Anders Nygren's philosophical theology and theory of religion has been widely discussed in Swedish theology and throughout the international academic world.[19] Nygren's view on the atheoretical religion was, for example, strongly criticized by the Swedish philosopher Ingemar Hedenius. But Nygren can—as Jarl Hemberg and Johan Lundborg (2002) have shown—be interpreted in different ways. This becomes clear if one links Nygren's theory to one of its chief interpreters, namely, Ragnar Bring. In Hedenius'critique, the works of Nygren and Bring are often mentioned together. But Ragnar Bring deserves a separate presentation.

MAKING THEOLOGICAL SENSE OF HÄGERSTRÖM'S THEORY OF REALITY—RAGNAR BRING

Ragnar Bring received his theological training in Lund and defended his doctoral dissertation on Luther in 1929. He was colleague with Nygren as professor in systematic theology from 1934, but stayed on at the academy after Nygren became bishop (in 1948) until 1962. Luther became a focus in Bring's later research, but earlier in his career he was more interested in philosophical issues in part through studies of Axel Hägerström. Bring agrees with Nygren that religion belong to another context of meaning and

15. Nygren, *Mening och metod*.
16. See Nygren, *Mening och metod*, especially chap. 10.
17. Nygren, Review of *Meaning and Method*, 496.
18. Nygren makes a reference to the Calcedonian formula of Christology, but, argues Clifford, it is not very helpful. "Still less does it help us understand how Nygren reconciles the presuppositions of presuppositions with the strict autonomous contexts of meaning. Has he not opened the door to metaphysics at least in the sense of admitting that philosophy is somehow concerned with a universal perspective?" Ibid, p. 497.
19. See Hemberg, *Religion och metafysik*, chap. 7 for a thorough, but not entirely clear, analysis of critics of Nygren and Bring.

it is a fundamental mistake to understand them as truths about reality in the scientific sense of truth. They are valid as far as they are connected to a fundamental motif. And the fundamental motif of Christianity is the love of God, Agape.[20] But Bring made an important supplement to Nygren's theology by identifying Hägerström's *temporal and spatial context of experience* with the theoretical-scientific context of meaning.

Bring made significant efforts to clarify Nygren's thought—especially in the wake of the debate on faith and knowledge initiated by Ingemar Hedenius in 1949. I shall return to this below, but anticipate it at this point because of the relevance it has for the relationship between religion and science.

Johan Lundborg discusses various interpretations of Nygren's theory of atheoretical religion.[21] Hedenius makes a stronger interpretation and assumes that Nygren affirms that religious belief is strictly and radically separate from the theoretical context of meaning. Religious belief does not contain any propositions that can be true or false. Ragnar Bring makes a weaker interpretation. Bring argues that the intention of Nygren by providing religion with a specific context of meaning was *not* to deny that religious beliefs could refer to something real, but to deny that they refer to something real in the sense of referring to something which belongs to the spatio-temporal context. God cannot be objectified in the way that was common in classical theism, but this is not to deny that God is real at all.[22] But in what sense is God real if God is not real in the usual sense? This question was raised by Hedenius in the debate on faith and knowledge, but receives no answer by Nygren or Bring.[23]

Presupposing the stronger interpretation of Hedenius, what conception of the relationship between science and religion is the result of Nygren's and Bring's theological outlook? Science and religion belong to two radically different contexts of meaning ultimately anchored in different regions of human consciousness. It goes without saying that there can be no way in which they can or should affect each other. The sphere of science is essentially separated from the sphere of religion—and vice versa. The relationship between science and religion is not open to any kind of mutual interplay.

If we—on the other hand—presuppose a weaker interpretation of Nygren's and Bring's theory of atheoretical religion, then the relationship between science and religion appears in a different light. Religious claims

---

20. See Bring, *Gud och människan*, esp. chap. 4.

21. Lundborg, *När ateismen erövrade Sverige*, 122–31.

22. Lundborg, *När ateismen erövrade Sverige*, 126; Bring, "Kristen tro och vetenskaplig forskning," 211.

23. Hedenius, *Att välja livsåskådning*, 332–35.

cannot be verified in the same way as claims are verified (or falsified) in science. Upon the weaker interpretation it is still possible to ask how a supposed religious reality is related to the reality revealed by science. I will return to this important point later in this book. Nygren himself does not open for such questions. On the contrary, he underlines that "[i]t is of importance to let faith remain faith and knowledge knowledge."[24] This speaks in favor of the stronger interpretation by Hedenius.

It may appear that Bring's and Nygren's approach to science and religion belongs to the past has no relevance in contemporary Scandinavian theology. But it has and especially in the thinking of the Finnish theologian Tage Kurtén. He has made efforts to revive Nygren's approach and link up with Nygrens's later efforts to clarify his theory with the help of Ludwig Wittgenstein's later philosophy.[25]

Beside the very abstract argument for the separation thesis, Nygren and Bring had some other arguments for drawing a firm borderline between science and religion. Nygren argues that faith is separate from ideologies and science because it is a *total relationship to life.*[26] God is not only a doctrinal conviction that there exists a reality which we name God, "but it is the wholly concrete and everywhere in life intervening, that the center of my life does not lie within myself but over me, in Him, who is the Lord of my life."[27] However, it remains unclear in what way this justifies the claim that Christian belief is *radically* separated from scientific theories and results. Secondly, Nygren argued that science is relative and changing.[28] Faith has to do with that which is basically human, which remains in principle similar during the ages. If science is made the norm of Christian faith, we may lose something valuable, which may or may not be retrieved by later generations. Thirdly, Bring argues that not only faith is perverted if science is made the norm and arbiter of religious truth; science is also transformed into something absolute thereby usurping the place of faith. Bring does not consider a weaker quest, namely, seeking "consonance between science and religion."[29]

24. Nygren, "Tro och vetande," 173.
25. See, for example, Kurtén, "Trust—the Hidden Presence of God."
26. Nygren, "Tro och vetande," 168.
27. Nygren, "Tro och vetande," 168.
28. Nygren, "Tro och vetande," 172. Bring has a similar argument in Bring, "Kristen tro och vetenskaplig forskning," 221: "Science is, as noted, not a series of irrevocable theses, a collected knowledge, but an ongoing activity, a work, which according to its nature must always move forward. Each fixing of a specific knowledge such *can* definitely be a prejudice that must be demolished if the scientific success is not to be prevented."
29. See below, chap. 14.

A NEW CHALLENGE TO THEOLOGY
—INGEMAR HEDENIUS

In 1949 and onwards the theologies of Nathan Söderblom, Torsten Bohlin, Anders Nygren, Ragnar Bring, and John Cullberg received heavy critique from the Swedish philosopher Ingemar Hedenius. During his earlier years at Uppsala University he received strong impulses from Axel Hägerström and Adolf Phalén, the two leading philosophers of "Uppsalafilosofin." Later he was more influenced by logical empiricism and the Cambridge school (G.E. Moore, Bertrand Russell, C.D. Broad, and others). He also took an interest in "speculative analysis of more general problems, which cannot be solved by exact science."[30] His critique of religion—and especially Christianity in its Lutheran form—belongs to this area of interest.

In the present context, I will limit myself to the philosophical aspects of Hedenius' critique. The debate he initiated by his book *Tro och vetande* and several articles in the national newspaper *Dagens Nyheter* was largely focused on other intellectual problems of Christianity, but the question of science and religion entered the debate in various ways. Hedenius was advocating a model of conflict, but it was not of the same kind as the one presupposed by Hägerström. Hedenius argued that religious beliefs contained truth-claiming—but false—propositions and were not merely expressions of feelings. The most extensive part of his critique of religion, was devoted to Nygren's and Bring's doctrine of religion as atheoretical. I shall come back to this in the next chapter, but first give an idea about his main points of departure.

In his analysis of Christianity, Hedenius proceeds from Leibniz' distinction between eternal (analytic) truths and factual (synthetic) truths. Science are those truths which at a given moment available eternal or factual truths. Hedenius acknowledges that this definition might be too wide in that it includes trivial everyday truths and—possibly—such truths, which are revealed by God, but he suspends his judgment on this until further analysis. Now, reason is that kind of activity by which we believe in or adjusts our action after scientific truths and rejects those truths, which contradict scientific truths.[31] Hedenius introduces what he calls *the principle of intellectual morality*: you should not believe anything, for which there is no rational reason to suppose it to be true.[32]

---

30. See his self-characterization in Hedenius, "Hedenius, Ingemar," 64.

31. Hedenius, *Tro och vetande*, 83.

32. Hedenius, *Tro och vetande*, 35. Further explained in Hedenius, *Att välja livsåskådning*, 74–78.

Hedenius combines this form of rationalistic scientism with three so-called postulates. The first is the *postulate of psychology of religion*. It says that religious faith to an essential degree contains truth-claims of certain conceptions and suppositions, which science cannot verify.[33] The second is called the *language-theoretical postulate*, which affirms the concepts of religious language are possible to communicate in a way that is understandable to non-believers. (In a later work, Hedenius modifies this postulate and says that the explanation of religious concepts cannot be "exhaustive."[34]) The third postulate is the *logical*, i.e., that two true statements cannot contradict each other and that any view of the world that contains contradictions is false.

The logical postulate is important and Hedenius argues that it proves (in the strong sense of the word) that the God of classical Christianity does not exist, because it is a contradiction to affirm the goodness of God and at the same time affirm that God sends most of humanity to eternal punishment.[35] Nygren, Cullberg, Bohlin, and many other Swedish theologians questioned this argument and did so by questioning one or all the postulates of Hedenius. The general conclusion of subsequent analyses of the debate is that Hedenius was victorious—although not always fair to his theological critics.

The principle of intellectual morality is only briefly explained by Hedenius. For example, it is not clear which reasons are rational. One interpretation is that the principle of intellectual morality comes close to what Mikael Stenmark terms *rationalistic scientism, i.e.,* "the view that we are entitled to believe only what can be scientifically justified or what is scientifically knowable."[36] Rationalistic scientism together with the postulate of psychology of religion leads to the conclusion that we are not rationally entitled to affirm religious beliefs.

> *Premise 1*: we are rationally entitled to believe only what can be scientifically justified or what is scientifically knowable (rationalistic scientism)
>
> *Premise 2*: religious faith to an essential degree contains truth-claims of certain conceptions and suppositions, which science cannot verify (Hedenius' postulate of psychology of religion)

---

33. Hedenius, *Tro och vetande*, 65.

34. Hedenius, *Att välja livsåskådning*, 233. See Lundborg, *När ateismen erövrade Sverige*, 68.

35. Hedenius, *Tro och vetande* 89–102.

36. See Stenmark, *Scientism*, 6.

Conclusion: we are not rationally entitled so believe religious truth-claims.

Nygren and Bring denied both premise (1) and (presupposing that Hedenius' stronger interpretation of their model of independence is correct) premise (2). Söderblom, Cullberg, and Bohlin denied only premise (1). Each in their own way they argued for different versions of the model of independence, i.e., that science and religions are different spheres of activities with separate questions, aims and/or methods. But in contrast to Nygren and Bring, they did not deny that religious belief contains truth-claims.

It is easy to understand the dilemma into which Nygren is placed by his theory of religion. If he accepts no overlap between the fields of experience, Nygren must deny the obvious, i.e., that religious life involves theoretical considerations. If he, on the other hand accepts an overlap, then religion and Christian belief is exposed to the philosophical critique of Hedenius. There is no easy way out of this dilemma, but one possible solution might be to distinguish between propositional and non-propositional experience. When a person has a propositional experience, that person knows what she sees. But it is also possible to experience something without knowledge in the sense of knowledge by description. It is possible that it was this kind of experience that Nygren found at the source of religion.

There is another voice in the discussion, which remained in the background in the 1940s but was more visible in the 1950s. It is *Harald Eklund*, philosopher of religion at the University of Lund (already mentioned in the previous chapter). Eklund also denies premise 2, but not because he wants to deny that religious beliefs contain truth-claims, but *because they do not lie outside science*—at least not entirely. Eklund argued the contact view on science and religion, i.e., that science and religion are different activities but they might nevertheless come in interplay with each other and affect the way they are—or should be—presented. This view signals a new era in Swedish twentieth century theology. During the years he was active, he appears to be something of an academic outsider. But in hindsight he comes forward as a theologian with profile and integrity. I shall come back to him in in just a moment.

9

Lundensian Models of Contact

The advance and prestige of science places Christian theologians in a serious dilemma. How is it possible to acknowledge the significance of science and at the same time develop a credible interpretation of Christian belief? All of the Swedish theologians I have discussed share this dilemma—but they resolve it in different ways. Anders Nygren and Ragnar Bring represent the most radical solution: religious beliefs are atheoretical. I have discussed the problem of interpreting this claim, but it may represent one possible resolution of the dilemma. Nevertheless, it did not win broader acceptance. The fierce criticism of Ingemar Hedenius made Nygren's and Bring's theologies less credible. The polemics from other Swedish theologians was similarly uncompromising. One example is the critique from the internationally well-known theologian Gustaf Wingren. A more philosophical argument comes from Harald Eklund, who in direct opposition to the theory of atheoretical religion pleads for making the question for truth in religion central.

THE EMPIRICAL ELEMENT IN RELIGIOUS BELIEF—HARALD EKLUND

Harald Eklund studied in Uppsala and defended a dissertation on the ethics of Max Scheler in 1932. After a few years in Finland, he became professor in Lund in 1949. He published several books on ethics and theology, important in the present context.[1] As already underlined, he represents an independent voice in Swedish theology.

Eklund opposes what he calls the "faith-theological claims of the century," namely that religious belief refers to something that is inaccessible to theoretical analysis. His main argument is empirical. Religious belief as we meet it in concrete life, includes elements of theoretical nature. It does not hover in an abstract sphere separated from evidence and argument.[2] One of the main points in Eklund's philosophy of religion was that there was such a theoretical element in religion. His main thesis is that the very concept of belief implies knowledge in the same theoretical meaning as all other knowledge, and that it is the task of doctrinal analysis to assess these claims without dogmatic prejudice.[3] Eklund takes many examples of the object of this assessment from the New Testament, but also from later Christian tradition. It is an undisputable fact, he argues, that religious thinkers throughout history have tried to give reasons for their beliefs. Religious beliefs were never considered as being something other than truth-claims.[4] In the New Testament "one believes because one has seen and experienced new and great things and draws conclusion from these."[5] Moreover, theological efforts to interpret religious belief as a form of interplay with realities beside or in contrast to ordinary knowledge seems to be misconceived. Eklund writes:

> If the entire complex of claims in a religious context is to be declared totally independent of others one must, however, bear clearly in mind what it is one claims and what possibilities there are for asserting it. The general negation of all proof and counter-proof is a difficult operation, and the question is whether such a general exclusion is under any circumstances reasonable and possible. Who can say that some proof, for or against, will not have some significance for a creed?

Eklund adds a footnote to this paragraph: "It should also be borne in mind that the religions, in fact have theirs proofs" and refers to Acts 1:3 where Luke reminds Theophilos that Jesus after suffering death presented himself

---

1. Eklund, *Troslärans perspektiv*; Eklund, *Tro, erfarenhet, verklighet*.

2. See Eklund, *Troslärans perspektiv*; Bejerholm, *Harald Eklunds religionsfilosofi*, 13–15.

3. Eklund, *Troslärans perspektiv*, 68; Eklund, Review of *Troslärans perspektiv*, 293–94.

4. See, for example, Eklund, *Tro, erfarenhet, verklighet*, 45–46; Bejerholm, *Harald Eklunds religionsfilosofi*, 35–44.

5. Bejerholm, *Harald Eklunds religionsfilosofi*, 48.

alive "by many proofs, appearing to them during forty days, and speaking of the kingdom of God."[6]

During his whole career as an academic theologian, Harald Eklund was critical against all efforts to separate religion from science and knowledge. He distanced himself from the idea that religion was independent from science, an autonomous "form of life" with other laws than the theoretical life, "atheoretical," "existential," "practical," or "personal."[7] In a short article from 1945 he specifically addresses the relation between science and Christianity and argues that the relation is insufficiently analyzed. He makes the important point that already to engage in these questions is to involve oneself in a rational discussion. And this analysis shows that there are possibilities of conflict between science and Christian theology. One example is "the modern eschatological faith" (reference to Hjalmar Lindroth?[8]). Is it anything more than an experiment in mythology? Eklund is especially critical of the new church-theology "with its dissolution of all logic, its peculiarly overblown self-confidence, sectarian zeal and demand for belief in authority."[9] Religious belief is considered as something other than knowledge and theology as a separate science with its own territory. Eklund claims that this leads to sectarianism and one expression of this is the sheer impossibility for external observers to understand what it's all about.[10]

SWEDISH THEOLOGICAL RESPONSES TO NINETEENTH CENTURY DARWINISM—TORD SIMONSSON

Eklund's view opens for an interesting analysis of the interplay between science and religion, even if such a detailed analysis is nowhere to be found in his writings. Nevertheless, he inspired one of his doctoral students to such an analysis, namely, Tord Simonsson. He was active at the theological faculty in Lund after which he pursued an ecclesial career and became bishop of Strängnäs between 1982–1989. His dissertation represents a significant step forward in the discussion on science and religion in Sweden. By examples from theological authors from later nineteenth century, Simonsson is able to verify that the contact view had strong support in late nineteenth century Swedish theology. Simonsson concludes:

---

6. Eklund, "On the Logic of Creeds," 82. Needless to say, there is a difference between a *claim* to proof and a *justified* proof.

7. Eklund, *Tro, erfarenhet, verklighet*, 149 (in an article already published in 1944).

8. See Wrede, "A Hjalmar J Lindroth."

9. Eklund, "Kristendom och vetenskap," 10.

10. Eklund, "Kristendom och vetenskap," 10.

> All statements presented here have either directly asserted or in some question demonstrated a common field of interest of Christianity and science. In this common field were certain of those matters, concerning which Darwinism has given utterance. However, some occasional statements contained, at the same time and despite this, words constituting a denial of the existence of such a field.[11]

It is interesting to compare this result with the basic outlook of the theologians, which we presented earlier in this book. They all argue some version of the model of independence, i.e., that science and religion deal with different issues, methods or realms of reality. In contrast, the model of contact—that science and religion overlap at various points—is in the foreground of the material analyzed by Simonsson. This notwithstanding, it is not entirely clear how representative the voices cited in Simonsson's work is for Swedish theology in the later nineteenth century. A large part is published in Church periodicals, papers at meetings of the clergy and works published by religiously disposed publishing houses. But the material is not sufficiently contextualized, nor clearly delimited. Moreover, another scholar—Kjell Jonsson—gives a different picture in his dissertation. Jonsson emphasizes the influence of Boström's idealism, but more generally the polarity between idea of restrictionism (another name for the model of independence) and expansionism (that science should replace religion[12]). He argues that restrictionism was the general strategy of argumentation among theologians and cites many examples.[13] Following Jonsson's dissertation, the model of independence in the first half of the twentieth century is in continuity with the restrictionism in later twentieth century Swedish theology.

It is interesting to take part of Simonsson's analysis of the different versions of the contact view. First, he distinguishes between (1) demonstration of irreconcilability, and (2) demonstration of consistency between individual Christian statements and various Darwinian theories. His analysis is subdivided into four different parts: (a) the when and how of creation, (b) the state of perfection (in the Garden of Eden), (c) the exclusive position of human beings in creation, and (d) the creator. I shall take a closer look at these different parts in turn.

The when and how of creation (a) concerns primarily the difference in the time scales of the Bible and the time scales of evolution. One author

---

11. Simonsson, *Face to Face with Darwinism*, 43.

12. This idea is often combined with the idea of irreconcilability between science and religion, but these views are in principle different.

13. Jonsson, *Vid vetandets gräns*, 104–5, 122–24.

writes that new discoveries are alleged to have been made, "through which an existence far beyond the boundaries of all legends and history should be assigned to man. In what way this can be reconciled with the statements of the Holy Scripture, we do not know."[14] When it comes to (b), *the state of perfection*, another author in a theological periodical writes that "Darwinian views of man's gradual evolution from the lowest animal organism to what he now is cannot possibly be reconciled with the doctrine of the human state of perfection . . . and man's fall from this and the subsequent need of redemption."[15] *The exclusive position of human beings* (c) is also challenged by Darwinism. "How can one acknowledge man's differentiation from the animal, if the whole of the human spiritual life is potentially to be found in the soul of the animal?"[16] Finally, the same author (systematic theologian C.G. Rosenqvist) writes concerning (d) the *relationship between Darwinism and the Christian doctrine of God, the creator*: "If one now seeks this principle of pure outer occasional causes, which exclude all teleology and all creative activity, it is then clear that an irreconcilable opposite exists between this theory and every theistic religious theory of life."[17]

All these authors exemplify a model of contact—in the form of conflict—between science and religion. They recognize a conflict between Darwinism and Christian belief at various points, but this does not amount to the definitive incompatibility. And the reason is, of course, that these points of conflict can be resolved in different ways. Thereby, the relation between science and religion can be harmonized.

In chapter 2, the study proceeds to such an analysis of the ways in which these contradictions between Christian belief and Darwinism can be handled. The irreconcilable statements can be reinterpreted so that the conflicts are resolved or at least diminished. Such an elimination of conflict between science and religion can be achieved through (1) correction of the scientific statements or through (2) correction of Christian (Biblical, doctrinal) statements. It is, for example, claimed that "science presupposes belief" and that these beliefs will ultimately stand corrected by Christian revelation. Or that science—in the form of Darwinism—has overstepped its rea of competence or is only a hypothesis. But some also suggest that Christian statements should be corrected and that they—as far as they contradict reliable science—do not belong to the true Christian revelation.

---

14. Quote from Simonsson, *Face to Face with Darwinism*, 28.

15. Simonsson, *Face to Face with Darwinism*, 30. The same author, J. A. Englund, minister in the Church of Sweden, is quoted by Jonsson, *Vid vetandets gräns*, 122.

16. Quote from Simonsson, *Face to Face with Darwinism*, 31.

17. Simonsson, *Face to Face with Darwinism*, 33.

This, of course, raises the larger question how to address the question of the "true" significance of the Bible and, particularly, the question of the literal and symbolic aspects of the Bible.[18]

Furthermore, Simonsson engages the problem of authority in Christian belief and, particularly the problem of Biblical authority. This part of his work is the least developed and not as rewarding as the previous chapters. Nevertheless, Simonsson's work is an important reorientation in Swedish theology when it comes to reflection on the relation between science and religion. The emphasis shifts from normative theology to a more descriptively or analytically oriented analysis.[19]

Simonsson published some other monographs in philosophy of religion. Of special importance in the present context is his contribution to an anthology by Jarl Hemberg and Anders Jeffner (published in 1963). Here Simonsson is reflecting on the wider relationship between worldviews and science and makes the following general remark:

> Science determines and moves the boundary to the unknown. Worldviews thrives on the nourishment among other things from insufficiencies of knowledge.[20]

Science may dissolve worldviews deeply cherished. And the reason for this is that worldviews often contain explanations of mysterious events in nature or history. In religious worldviews, these events may be understood as a divine "master act" in creation of the world, but also as a number of "special acts" in relationship to individual persons. When such "special acts" are challenged by science, believers may retreat referring to the divine "master act." When the challenge has calmed down, many earlier conceptions abandoned by science may be retrieved. In this way elements in Christian doctrine may live a kind of "amoebic existence."[21] A special form of contradictions with science is thereby preserved. Simonsson does not give any example, but one is reminded of the way the presuppositions of Newtonian physics are preserved in contemporary theology. The

---

18. Ibid., 119. This section could have benefitted from attention to the long discussion in Christian theology concerning the interpretation of the Bible. See, for example, Thistelton, *Hermeneutics,* esp. chap. 5.

19. In Uppsala, this shift was emphasized by Axel Gyllenkrok (professor in dogmatics, 1960–1973) in Gyllenkrok, *Systematisk teologi.* He is, among others, heavily criticized by Gustaf Wingren in Wingren, *Tolken som tiger* (1981). Even if Wingren was professor in systematic theology in Lund from 1951 to 1977, neither Harald Eklund nor Tord Simonsson are mentioned in this book.

20. Simonsson, "Livsåskådning och vetenskap," 67.

21. Simonsson, "Livsåskådning och vetenskap," 70.

definition of miracles as violations of natural law, thereby sustains a conflict between science and religion.

Simonsson also considers other devices by which worldviews can be preserved in face of efforts by science to falsify them. One example is modification of the religious worldview, another is use of developments in science, for example in physics. Modification of religious worldviews is reasonable even if the authority of religious worldviews might be questioned, when an external authority modifies them. Simonsson is explicitly critical of the second device. To interpret the worldviews of yesterday with the concept of modern science does not raise the credibility among serious scientists. And he adds an argument, which we have met earlier. Those who marry science today, may find themselves a widower tomorrow.[22] Simonsson does not consider the opposite risk, i.e., that a religion which is separated from the science of today will not be able to produce any offspring for tomorrow.

Simonsson's remarks notwithstanding, he seems clearly within the paradigm of Eklund's model of contact between science and religion. He acknowledges the difference between religion and worldviews on the one hand and science on the other, but at the same he allows for mutual corrections (at least the correction of theology with scientific arguments). He explicitly affirms that declarations about a total separation between the two domains seem "empirically and analytically leaky."[23]

STUDIES INTO THE EXPERIENTIAL VERIFICATION OF RELIGIOUS BELIEFS—HAMPUS LYTTKENS

Harald Eklund died in 1960 and was succeeded the following year by Hampus Lyttkens. His dissertation on Thomas Aquinas is a good example of the solid historical-analytical research characteristic of the many doctoral dissertations under the guidance of Hjalmar Lindroth.[24] After his dissertation Hampus Lyttkens gradually changed focus to problems in analytical philosophy of religion.[25]

One example is his interest in the meaning and cognitive significance or religious experience. This can be described as a continuation of Harald Eklund's philosophy of religion, even if Lyttkens went further than Eklund and studied the possible parallels between scientific and religious hypotheses.

22. Simonsson, "Livsåskådning och vetenskap," 73–75.
23. Simonsson, "Livsåskådning och vetenskap," 74.
24. Lyttkens, *The Analogy between God and the World*. Lyttkens describes the significance of Lindroth's leadership and his strong belief in the objectivity of science and scholarship in his farewell lecture 1982. See Lyttkens, *Religiös samhällsanalys*, 10–12.
25. See, for example, Lyttkens, "Kan kristna trosutsagor verifieras?"

He argued that such parallels exist, but also significant differences.[26] Departing from the Catholic philosopher Jósef Maria Bóchenski, he argued that there are at least two such differences.[27] First, religious hypotheses refer to a much larger class of experiences than hypotheses in natural science. Secondly—and more importantly—religious hypotheses refer to other forms of experiences than sensual experiences, i.e., moral, aesthetical and religious experiences. But this parallel is complicated by the general difficulty of determining whether there are *specific* religious experiences.[28]

Lyttkens observed that there is a more obvious possible parallel between science and religion in the frequent use of auxiliary hypotheses. Auxiliary hypotheses are often introduced in science to "save" a theory from unexpected falsification. For example, the main hypothesis that there is organic life on the planet Mars is falsified by the research of the Mars rover Curiosity. But the main hypothesis can be left standing if we presuppose that all possible areas on Mars have not been researched or that the organic detection methods on the Mars rover are insufficient to detect life.[29] Similarly, the Christian belief that God answers prayer can be constructed as a religious hypothesis in parallel with hypotheses in science. It is often objected that if the Christian belief in the efficacy of prayer is constructed according to scientific hypotheses, they are clearly falsified. The Christian theologian might respond with an auxiliary hypothesis. Prayer needs to be conducted with intensity and/or the right frame of mind. If the hypothesis that God answers prayer is not verified, then the reason might be that some or all the conditions of this auxiliary hypothesis are not fulfilled.[30]

Another example is the doctrine of creation. It can be described in the form of a hypothesis that God created the world, Adam and Eve and the Garden of Eden. The theory of evolution seems to falsify this hypothesis, but an auxiliary hypothesis might be formulated stating that God did indeed create the world, but not Adam and Eve, nor the Garden of Eden. This auxiliary hypothesis must be distinguished from the main religious hypothesis that the world is God's creation. It is a difficult, but important, theological work to distinguish between main hypotheses and auxiliary hypotheses in doctrinal systems.

---

26. Lyttkens, "Kan kristna trosutsagor verifieras?"

27. Bóchenski, *The Logic of Religion*.

28. Lyttkens, "Kan kristna utsagor verifieras?," 57–58.

29. These auxiliary hypotheses can, but need not, be rejected as ad hoc-hypotheses, i.e., hypotheses devised merely to save the original hypothesis.

30. Lyttkens, "Kan kristna utsagor verifieras?," 58–59.

A more general difficulty for the argument concerning parallels between science and religion is that in science we know what observations would constitute evidence for the existence of, say, life on the planet Mars (even if there is a twilight-zone between no evidence and conclusive evidence). What, if any, experiences would constitute conclusive evidence for the existence of the eternal God of theism or eternal life?[31]

Swedish theologians in the first half of the twentieth century often argued that it is a total and serious misunderstanding to describe Christian belief in the form of a scientific hypothesis. Lyttkens' detailed analysis puts this general judgment in dispute, even if it does not give a definitive reason to reject the separation thesis. His colleague Urban Forell adds some further reasons for the model of separation in his analysis of miracles.

AN ANALYTIC THEORY OF MIRACLES —URBAN FORELL

Urban Forell defended his dissertation 1967 in Göttingen, Germany.[32] He became assistant professor at the Faculty of Theology in Lund and after that professor in systematic theology at the Faculty of Theology in Copenhagen. Of interest in the present context, is Forell's exploration of the relationship between an explanation of an event as an act of God and a scientific explanation.[33]

There are—according to Forell—two alternative interpretations of the relation between a religious and a scientific explanation. The *first* interpretation is that reference to God's action is a form of scientific explanation. God's action lies on the same plane as natural causes and powers. God's action is conceived as a factor alongside of other factors in the empirical world. Against this background, there are two ways to understand an act of God. The first is *type a*, i.e., such acts, by which God supports a chain of events, without determining any specific event. The second is *type b*, i.e., those events, which occurs only if God acts. Only events of *type b* can be

---

31. In his theory of eschatological verification, John Hick has argued that there are such definitive experiences. See Hick, *Philosophy of Religion*, 94–95; Bråkenhielm, *How Philosophy Shapes Theories of Religion*, 99–100.

32. Forell, *Wunderbegriffe und logische Analyse*.

33. Forell, "Begreppet 'Guds verksamhet' i ljuset av naturvetenskapliga förklaringstyper," 33–51. One other important contribution of Urban Forell to the issue of science and religion deserves attention, namely, Forell, "Gud och rummet." For a more extensive exposition of Heim's thought, see Holmstrand, *Karl Heim on Philosophy, Science and the Transcendence of God*.

regarded as genuine acts of God. They would amount to miracles in the strong sense of being a violation of a natural law.[34] Forell writes:

> Every inexplicableness of even a single empirical event from known, purely immanent causes, can according to this conception be used as *evidence*—not *proof*—that God intervened at precisely this point. On the contrary, every successful explanation from purely immanent factors can be used as evidence for that God did *not* intervene here precisely at this this moment.[35]

The inexplicableness of an event by immanent causes is—in other words—a necessary but *not* a sufficient condition for an event being an act of God.

This is denied by the *second* interpretation of the relationship between an explanation of an event as an act of God and a scientific explanation.[36] The inexplicableness of an event by immanent causes is *not* an argument for God being active in this event any more than in other events. Neither is a successful explanation by immanent causes for an event evidence for that God did *not* act in that particular situation. Acts of God lie on another plane than the plane studied by natural science. In short, scientific explanations are not relevant in religion.[37]

Forell has a more general argument, which claims that religious beliefs cannot be understood as scientific explanations in the form of the so-called Hempelian model of explanation. This model ultimately presupposes an abstract theory about the natural laws determining, for example, the movements of atoms and molcules. But no act of God can be explained in terms of natural laws. Nor is there any known natural law that connects an act of God with certain empirical events. Hence, the first alternative (in the sense of *type b*) should be rejected. Forell argues that this result is in line a basic attitude in religion, namely, the attitude of submission and dependence.

It should be added that Forell rejects what he calls the concept of "singular causality," i.e. a singular factor (such as the intention of a human person or the will of God), which cannot be subsumed under a natural law. Such a concept of explanation has been proposed by—for example—Max

---

34. This is not the case if there are events which are not determined by natural laws (such as events described in quantum physics).

35. Holmstrand, *Karl Heim on Philosophy, Science and the Transcendence of God*, 36.

36. Holmstrand, *Karl Heim on Philosophy, Science and the Transcendence of God*, 50–51.

37. Holmstrand, *Karl Heim on Philosophy, Science and the Transcendence of God*, 36–37. It can be argued that the inexplicableness of an event by immanent causes can provide an inroad for a special religious interpretation of the event as an act of God. See further MacIntyre, *Difficulties in Christian Belief*, 49–50. Needless to say, such interpretations always run the risk of falling prey to the fallacy of "God of the gaps."

Weber. But Forell claims that such a concept is an impossible concept in science, because there is no logical-deductive connection between *explanans* and *explanandum*.[38]

It is an historical irony that George Henrik von Wright the following year presented an argument against a causal theory of action and assembled evidence for the significance of teleological explanations of human actions. To be sure, this kind of teleological explanation cannot be described as a model of singular *causality*; to explain an action is according to von Wright to make a *logical* connection.[39] But von Wright's "Logical Connection Argument" can nevertheless be said to have shown to be an intelligible model of singular (teleological) explanation of human action. In principle, it does not seem impossible to elucidate the concept of divine action in relation to such a concept.[40]

von Wright's work sparked a discussion, which contributed to a renewal of interest in personal explanations. It had an impact in the philosophy of religion through—for example—Richard Swinburne. Swinburne writes:

> For scientific explanations by their very nature terminate with some ultimate natural law and ultimate arrangement of physical things, and the questions I am raising are why there are natural laws and physical things at all. — However, there is another kind of explanation of phenomena which we use all the time and which we see as a proper way of explaining phenomena. This is what I shall call personal explanation. We often explain some phenomenon E as brought about by a person P in order to achieve some purpose or goal G. The present motion of my lips is explained as brought about by me for the purpose of delivering a lecture. The cup being on the table is explained by a person having put it there for the purpose of drinking out of it. Yet this is a different way of explaining things from the scientific. Scientific explanation involves laws of nature and previous states of affairs. Personal explanation involves persons and purposes.[41]

Swinburne argues that explaining an event as an act of God can be interpreted as a personal explanation. Whether this is possible (let alone, justified) is a matter of continuing discussion.

38. Forell, "Begreppet 'Guds verksamhet' i ljuset av naturvetenskapliga förklaringstyper," 41.

39. von Wright, *Explanation and Understanding*, 96–97, and, for a "final formulation of the inference schema," ibid., 107.

40. One effort to do this can be found in Emmet, *The Effectiveness of Causes*, chap. 11.

41. Swinburne, "Arguments to God from the Observable Universe," 124.

Hampus Lyttkens and Urban Forell display an effort to promote a more direct dialogue between scientific method and Christian beliefs in the activity of God. This represents a significant advance in the Swedish discussion. This dialogue is further developed by some other theologians, which represent a movement out of the shadow of the model of separation.

SCIENCE AND WORLDVIEWS —LARS HAIKOLA

Lars Haikola submitted his dissertation on the Wittgensteinian notion of language game in 1977.[42] After that he became assistant professor at the university of Lund and concluded his career as University Chancellor at the Swedish Higher Education Authority.

Haikola made at least two important contributions to the discussion on science and religion. The first article originated from the *Eight European Conference on Philosophy of Religion* in 1990 and was a response to a lecture by the British theologian Maurice Wiles.[43] The main thesis of Wiles' lecture was that an interventionist understanding of God's action in the world (in contrast to God's action in an act of creation and sustenance of the world) is metaphysically and morally untenable.[44] He concludes with a plea for a non-interventionist reinterpretation of four concepts in Christian theology: miracle, revelation, virgin birth, and physical resurrection of Jesus Christ.

The main purpose of Haikola's article is to clarify the significance of modern science for the possibility of God's action in the world (understood as an intervention in the world). In short, Haikola's conclusion is that whereas divine actions are problematic presupposing the mechanistic and closed worldview, they are nevertheless possible presupposing a "modern" worldview of a "fine-tuned" and causally "open" universe.[45] Haikola rejects John Polkinghorne's idea of the hidden work of God on the subatomic level of indeterminacy, but is positive to Arthur Peacocke's theory of God's continuous action in "new emergent forms of matter." So—in principle— there are new possibilities for resolving the traditional conflict between science and religion. But this does not resolve the more general conflict highlighted by Maurice Wiles, namely, between radical transcendence and interventionistic action.

At this point in his argument, Haikola suggests a model of contact, according to which science and religion are two separate areas of human

---

42. Haikola, *Religion as Language-game*.
43. Haikola, "Skapelse och försyn i ljuset av modern vetenskap," 122–29.
44. This position seems to be similar to the conclusions of Urban Forell.
45. Compare this to Alvin Plantinga's analysis mentioned in chap. 7.

thinking, but nevertheless overlap at certain points. This notwithstanding and surprisingly, it is not necessary to strive for integration. The tension between science and religion is relatively unimportant and will probably become obsolete. The reason for this is that there is an increasing awareness that science is incomplete and not a firm and self-evident basis of our understanding of reality. "The optimism which was present in the early natural science that all the enigmas of the world will be solved has been dissolved."[46] Furthermore, the negative consequences of science in the form of nuclear weapons and environmental destruction have led to the loss of the positive evaluation of science. The problem of harmonizing science and religion might be regarded as irrelevant.

Another significant article is interesting because of Haikola's analysis of the model of separation between science and religion as closely associated with a particular worldview. His argument can be summarized in the following way.[47]

Science has increasingly shaped our everyday life, but also gained influence by forming a specific worldview. Such a worldview contains not only a system of morality but also a simplified description and explanation of reality, an "ideological superstructure." Classical myths and cosmologies were precursors to the Aristotelian worldview, which was based on teleological explanations. These were discarded in the mechanistic worldview, which relied on causal explanation. The universe was described as a watch and was closely related to a specific idea of religion as reduced to the individual person's relation to God and ethics. In short, *the mechanical worldview and the model of independence are closely entwined*. Haikola cites a statement from the National Academy of Sciences in 1972:

> . . . religion and science are . . . separate and mutually exclusive realms of human thought whose presentation in the same context leads to misunderstanding of both scientific theory and religious belief.[48]

But the mechanical worldview has been seriously questioned by scientific developments during the twentieth century. The generality of causal laws does not hold any more. Chance has a more central position. We cannot describe reality without at the same time affecting it. Reality does not consist of small material entities. Time and space are not the secure basis in the way that was presupposed in the physics of Newton.

46. Haikola, "Skapelse och försyn i ljuset av modern vetenskap," 128.
47. Haikola, "Naturvetenskapen—en ifrågasatt världsbildsskapare."
48. Haikola, "Naturvetenskapen—en ifrågasatt världsbildsskapare," 299.

A new worldview is needed. But—Haikola argues—no such is forthcoming, at least not from science. There are many reasons for this. One is that chance, chaos, maximal entropy, relativity and so on are difficult to comprehend by common sense. They cannot be the basis of a new worldview. Science has more humility and acknowledges that a worldview cannot be built on science alone. It is also relativized and speaks of models instead of definitive theories. It does not provide us with secure knowledge, even if such knowledge is in increasing demand in society.

Furthermore, science has a backside, which has become more and more evident. Weapons technology, acid rain, climate change and ecological disasters are just a few of the more evident examples. Ethical problems in connection with application of new medical techniques arise repeatedly. The original intention of science and technology was to make life easier and more secure. But today science is creating as many problems as it solves.

In sum, modern science will provide us with no new worldview. Science will continue to play an important role, but science will never achieve the same self-evident and unquestioned role as it did in association with the mechanical worldview.[49]

At present—over twenty years later—we may have an answer to Haikola's forecast. Empirical studies show that science functions as an important factor in the world worldviews of men and women—at least in Sweden.[50] Haikola might respond that the mechanical worldview has not been succeeded by a worldview with anything like a similar dominance. More importantly, this raises the question of the consequences of the contemporary cultural situation in Western Europe and USA and rising preferences of model of contact between science and religion.[51] Haikola predicted that the problem of harmonizing science and religion might be regarded as irrelevant. There is, however, evidence for the view that a *model of incompatibility* between science and religion is prevalent among the contemporary public. The reader might recall the earlier mentioned survey by the *Swedish Association for Science & the Public* made in 2008, about 47 percent of a

---

49. Haikola, "Naturvetenskapen—en ifrågasatt världsbildsskapare," 302.

50. See above, chaps. 1 and 2.

51. Stenmark notes that as science and religion are evolving practices, there is no *a priori* or once-and-for-all answer to the question how they should be related. Stenmark, *How to Relate Science and Religion*, 268. In line with this, one might argue that the independence view is natural in a situation where a mechanical worldview dominates, but not natural in a similar way when such a worldview is superseded by a more holistic worldview.

representative sample of the adult Swedish population agrees that science and religion cannot be reconciled.[52]

Finally, it should be added that members of the Faculty of theology in Lund made important institutional contributions to the dialogue between science and religion. In 1996 Ulf Görman was elected the president of *European Society for the Study of Science and Theology* (ESSSAT) and contributed to a number of volumes by this society. One example of significance in relation to Lars Haikola's thought is the volume *Design and Disorder: Perspectives from Science and Theology,* containing several of the contributions to the ESSSAT Conference in 2002. In the introduction of the anthology, Görman notes that "[c]haos theory shows that our received view of determinism may be far too limited" and raises the question if the traditional image of God, God's providence, pre-knowledge and predestination must be remade in the light of chaos theory.[53]

---

52. Hermansson, *Vetenskap att tro på?*
53. See Gregersen and Görman, eds., *Design and Disorder,* 2.

10

New Philosophical Perspectives on Science and Religion

There are several sources to the contemporary theological attention to the relationship between science and religion. One is the development in physics such as quantum physics, chaos research, and theories of dissipative structures and self-organization highlighted by Lars Haikola and Ulf Görman.[1] Another source can be found in philosophy of science. One example is Thomas Kuhn's theory of scientific paradigms and a second Georg Henrik von Wright's misgivings about the mechanical worldview and the limits of "de-teleologization."[2] Still another source is the emergence of claims that scientific method and/or theories have atheistic implications and that atheism and science are inseparable. One expression of this is Richard Dawkins assertion that "Darwin made it possible to be an intellectually fulfilled atheist."[3] This claim spurred many theologians to dig deeper into the whole issue of the relationship between science and religion.

Another factor for the emergence of interest in science and religion is certain developments in philosophy. There are many examples of existentialist thought encouraging a critical stance towards science. Other ideas such as Wittgenstein's theory of language game have assisted theologians (such

---

1. Another example of this is the Danish theologian Viggo Mortensen in his dissertation *Teologi og naturvidenskab,* reviewed by Ulf Görman in Görman, "Vänlig växelverkan mellan naturvetenskap och religion?"

2. von Wright, *Vetenskapen och förnuftet,* 113. See also the work by Urban Forell in the preceding chapter.

3. Dawkins, *The Blind Watchmaker,* 6.

as Anders Nygren) to sharpen the demarcation line between religion and science. A number of theologians at Uppsala University have found a source of inspiration in other parts of English-speaking analytic philosophy. Three examples are Anders Jeffner, Jarl Hemberg, and Mikael Stenmark. Stenmark is internationally renowned for his contributions to the field of science and religion from the middle of the 1990s and onwards. Jeffner's publications since the middle of the 1960s are not as focused as Stenmark's on issues in science and religion. But his philosophical theory of religious belief has provided a theoretical framework for the interpretation of the relationship. In later publications, he has explored these implications. I shall begin with a presentation of Jeffner and some of his works, continue with Jarl Hemberg (already mentioned in connection with the presentation of Anders Nygren in Chapter 8) and then come back to Stenmark's contributions.

THE THEORY OF FUNDAMENTAL PATTERNS —ANDERS JEFFNER

Anders Jeffner was professor in Studies in faiths and ideologies in Uppsala from 1975 to 2000. His dissertation on Butler and Hume was published already in 1966 and was highly acclaimed. The philosophical theory of religion presented in the last chapter of this work is of special significance in the present context. First, I will present some significant conceptual remarks made by Jeffner in a work from 1976. Then I will go on and give the outlines of his general theory of religion presented in his dissertation from 1966 and in another book from 1992. Finally, I will spell out the implications of his theory for the interpretation of the relationship between science and religion.

Anders Jeffner outlined the concepts and methods for a new area of empirical as well as philosophical research, namely, the study of worldviews.[4] In the present context, there are two distinctions that deserve attention. These distinctions concern the general properties of worldviews as they appear not only in an academic context, but also in more widely disseminated worldviews among the general public. Worldviews are defined as "a certain complex whole, consisting of a person's central values and norms, the basic feeling which can be said to characterize his experience of life and those cognitive elements which influence, or are dependent on the ethical and attitudinal components."[5] The conceptual distinctions I have in mind concern the cognitive component and have to do with the "belief-norms," i.e. norms regulating the cognitive claims allowed in a particular worldview.

4. Jeffner, *Livsåskådningsforskning*, 41–47.
5. Gyllenkrok and Jeffner, "Theological and Ideological Studies," 129.

The *first* distinction is between scientific *results and theories* on the one hand and scientific *worldviews* on the other. Scientific results and theories are built on established facts, i.e., individual occurrences in the world of experience common to all. Scientific worldviews on the other hand are constituted by a belief-norm according to which the reasons accepted for statements about reality are (a) evidences of the kind accepted in empirical science and (b) only such evidence. *Secondly*, scientific worldviews can be distinguished from experiential worldviews containing belief-norm (a), but excluding belief-norm (b), and adding another belief-norm (c), according to which certain cognitive experiences *not* accepted in science are accepted as reasons.[6] In experiential worldviews not only sense experiences are taken as reasons, but also experiences of transcendence, ethical experiences, I-thou-experiences—and fundamental patterns.

Jeffner's distinction between (1) scientific results/ theories and scientific worldviews and (2) between scientific worldviews and experiential worldviews opens for an appreciation of science as well as for a kind of remaking of Christian doctrine in accordance with scientific results/theories (in distinction from scientific worldviews). These conceptual innovations might have been another factor for developments from the model of independence to a model of contact, accepting not only coherence, but also certain conflicts with traditional theology. It is a central task of philosophical theology to—if possible—solve these conflicts.

Jeffner is of special significance in that he approaches the science and religion-issue from the point of a more general theory of religious belief (Jeffner's Theory of Religion, JTR). In his dissertation, he advanced a theory of fundamental patterns primarily designed to understand the disagreement between Bishop Joseph Butler and the empiricist philosopher David Hume.[7]

JTR works within the perspective of a representational understanding of religion acknowledging the significance of cognitive truth-claims. His theory is designed to justify the possibility of such an understanding against the critique of Antony Flew and the non-cognitive interpretation of religion by R.B. Braithwaite and others. Jeffner departs from a significant observation (derived from—among others—the later Ludwig Wittgenstein). Two persons study a certain composite object and arrive at a disagreement about some property of this object. Person A says, "this is f" and person B says, "this is g." Their disagreement notwithstanding, they agree on all empirical

6. Jeffner, *Livsåskådningsforskning,* 46–47.

7. Jeffner, *Butler and Hume on Religion.* He reinterpreted this theory in another terminology in Jeffner, *The Study of Religious Language.* In the present context I will stick to his original version.

facts in the field. Jeffner gives the following definition of "fact." "*By a "fact" we mean here an individual occurrence in the world of experience that is common to all.*" [8] Whether a statement of fact is true can always—at least in principle—be decided by observation.

Now, an important question is the following: what kind of disagreement is the disagreement between *A* and *B*?

Drawing on the later Wittgenstein, Jeffner distinguishes between three kinds of disagreements, i.e., disagreements in attitude (i.e. a propensity to act and feel in a certain way), disagreement of gestalt (i.e. the composition of facts to a whole), and disagreement of attitude *and* gestalt. In the last form of disagreement, the same attitude is bound up with the same gestalt experience, and a change in gestalt experience always brings a change in attitude, or vice versa. When two persons disagree in this way (while agreeing on all facts in the field), they perceive different *patterns*.[9]

At this point Jeffner introduces the idea of a *fundamental pattern* (FP). FPs involve (a) the adoption of a certain attitude or complex of attitudes "to the totality of what I experience in life."[10] This attitude/complex of attitudes are (b) bound up with a certain organization of my experiences, a gestalt. Disagreement concerning FPs is possible even if there is agreement concerning the facts. Facts delimit the fundamental patterns, which in some sense are possible. New facts may make old fundamental patterns obsolete. Fundamental patterns can be falsified by science, but *they can never be verified by science, because by the very nature of FPs, they go beyond science.*

JTR was primarily developed to explain the kind of disagreement that obtained between Bishop Joseph Butler and David Hume. But secondarily, it was also suggested as a general theory of religious language, i.e., that certain sentences in religious contexts could be interpreted as expressions of FP.

A crucial point in the present context is the distinction between *existence-implying fundamental patterns* (EIFP) and *fundamental patterns not implying existence* (NIFP). An EIFP implies the existence of something, for which there is no reason to assume if one has a NIFP. For example, David Hume works with an empirical knowledge-restricting principle. The only thing we know about the world is what is "present to our senses" and "those lively images with which memory presents us."[11] This principle results in a radical NIFP (later developed in different sense data-theories by twentieth

---

8. Jeffner, *Butler and Hume on Religion*, 240; Jeffner's italics.

9. Jeffner, *Butler and Hume on Religion*, 242.

10. Jeffner, *Butler and Hume on Religion*, 243.

11. Hume, *A Treatise on Human Nature*, 265, quoted in Jeffner, *Butler and Hume on Religion*, 247.

century British empiricists). But Hume's theory of natural beliefs (assuming—among other things—that sensual impressions are experiences of an objectively existing material reality) goes beyond this radical NIFP and assumes a simple EIFP, even if it does not imply the existence of any new entities over and above material things. Such naturalistic or physicalistic NIFP differ from the religious EIFPs assumed by Butler and other physico-theologians. They regarded the world as designed, thereby implying the existence of a designer. Some of them had a pretty crude idea of the nature of the implication in their EIFPs, constructing it in analogy to an ordinary empirical method. Jeffner, however, argues that the nature of the inference from a fundamental pattern to something beyond that pattern deviates from common sense and scientific reasoning. Words whose function ordinarily is used to describe empirical facts are used in some transferred or symbolic sense.[12] It is not possible to decide between a physicalistic NIFP and a religious EIFPs with the help of scientific arguments. The main issue has to be decided by a subjective experience of what "suits" the things we know and experience.

Nevertheless, there might be one thing that speaks against transcending FPs—even if it "suits" the things we know and experience. And it is Occam's razor, i.e., to be economical when it comes to new forms of truth claims in our explanations of events and processes. The application of Occam's razor when it comes to FPs is philosophically controversial. It has been a useful principle in science, but this does not imply that it is applicable in the same way outside the scientific realm.[13]

After this presentation of JTR, we shall make some observations concerning its relevance for the science and religion-relationship. Two points are especially important. The first concerns the model of the relationship between science and religion suggested by JTR. The second concerns science as a provider of fundamental patterns (such as biologism).

At first glance, it may appear that JTR supports a model of independence and separation between science and religion. The EIFP of, for example, theism, lies on a more general level than scientific explanations and theories. Fundamental patterns cannot be scientifically supported in the normal sense, because several such patterns are possible on the basis of a certain level of scientific knowledge. *But it should be noted that a number of*

---

12. Jeffner, *Butler and Hume on Religion*, 254; see further Jeffner, *The Study of Religious Language*—especially his theory on untranslatable metaphors on 52–54. See further below, chap. 14.

13. Jeffner, *Filosofisk religionsdebatt*, 77. It should be noted that Sören Halldén is also quite undecided how the principle of parsimony is to be applied. See Halldén, *Universum, döden och den logiska analysen*, 36–38.

FPs can—and has been—falsified. One example is the fundamental pattern of the medieval worldview with seven spheres. It was a possible interpretation of the universe before the Copernican revolution. But it has now been falsified. Jeffner writes:

> . . . *only a certain number of fundamental patterns are possible at each stage of factual information*, and . . . a fundamental pattern which is possible for an ignorant person can be impossible for one who is more informed.[14]

With this Jeffner rejects what he in another context has termed a *positive criterion of science*.[15] In other words: fundamental patterns are ultimately underdetermined by scientific evidence. If JTR is correct, then it is not possible to support a religious belief with a scientific theory or result. And the reason for this is that at a certain level of scientific knowledge a number of different fundamental patterns can still be possible. Which fundamental pattern should be accepted is a question beyond science. On the other hand, if JTR is correct, it is possible that a scientific result or theory can falsify a certain religious belief. Hence, JTR allows for a *negative criterion of science*. This model of the relationship between science and religion falls between the strong independence thesis and the contact model. I shall call it the *asymmetric model of independence*.[16]

A second application of JTR concerns science as a provider of FPs. Scientific theories and results are not only the object of interpretation of fundamental patterns. It's not like we are scanning the landscape of science from a distance and try to fit what we see in the sciences into a preconceived picture of reality. Rather scientific results themselves contribute to the organization of knowledge into a particular gestalt. Lars Haikola has already given some examples in one of his highlighted article.[17] One was the Aristotelian worldview, which influenced the medieval picture of the world, which in turn was challenged by Galileo and Copernicus. Copernicus did not only make some interesting astronomical discoveries. His new picture of the solar system falsified the traditional FP of the world. By the time of Newton order was restored through a new transcending FP. "This world had its origin in God's design and it was the task of science to find out by empirical methods what God had thought when planning the wonderful order of

---

14. Jeffner, *Butler and Hume on Religion*, 243; my italics.
15. See Jeffner, *Kriterien christlicher Glaubenslehre*, 87.
16. I shall return to this asymmetric model in chap. 14.
17. See above, chap. 9.

the universe."[18] Jeffner argues that something similar is now happening with new scientific findings particularly in biology, which can be characterized as a full-blown scientific worldview such as biologism. But whereas the former shift driven by the discoveries of Copernicus affected only a limited part—albeit important—of our world, the new biology threatens not only the idea of teleological order, moral insights and inherited ideas of mental life; it is "a total take-over of our cognitive life and thereby demonstrates that the empirical knowledge is our only link to reality."[19]

A radical alternative to biologism is to start our reflection about the world form the unique capacities of humans being (and interpret science as an exercise of this capacity). There is—according to Jeffner—a dividing line in philosophy between those who try to understand the world from the perspective of empirical science and those who start from human beings as *Dasein*. The paradigmatic example of the latter approach is—of course—Martin Heidegger. Jeffner accepts this philosophy as a complementary perspective, but emphasizes what he calls the *primacy of empirically based knowledge*, i.e., that no other data than sense data are relevant for the propositions and theories concerning that part of reality which in principle accessible to our senses. Mystical and moral experiences are simply irrelevant in biology and quantum physics has nothing to learn from Eastern mysticism.[20]

Scientific knowledge and existential experiences are two important sources based on the common sense understanding of the world. The experience of transcendence is the third. A synthesis between these three sources of knowledge gives rise to a FP, which Jeffner calls *religious evolutionary materialism*. "Behind all and in everything that happens is a totally different reality which for the believers of three great world religions is interpreted as similar to personal love. This is the basis of being—God."[21]

Which interpretative pattern should be chosen? Biologism, subjectivism or religious evolutionary materialism?

> When it comes to general interpretations of our situations as humans in the world we must, according to my view, abandon the hope for security. . . . What we have to do is to face the risk and make a choice on the basis of our total experience of all the different accepts of life.[22]

---

18. Jeffner, *Biology and Religion*, 7.
19. Jeffner, *Biology and Religion*, 7. More on this in chap. 14.
20. Jeffner, *Biology and Religion*, 22. See further below in chap. 14.
21. Jeffner, *Biology and Religion*, 25.
22. Jeffner, *Biology and Religion*, 26.

Even if certain FPs are rationally excluded, different fundamental patterns are nevertheless possible on the basis of a given amount of information.

In sum, Jeffner's theory of religion has extensive application both in the field of empirical studies and on issues in philosophy of religion. In the present context, the most interesting application concerns the relationship between science and religion as well as the rise of science (especially biology) as a provider of naturalistic FPs. Interestingly, Jeffner has recently returned to the theory he developed half a century ago. He does this in the context of art. Art provides experiences which must be acknowledges in our quest for a comprehensive gestalt of reality. Moreover, he points out an important complication, which his theory of FPs might lead to, and that is relativism. Many different FPs are possible and it is tempting to be content with a repertoire of FPs without any efforts to make a responsible decision. Such a decision presupposes that we accept FPs not as changing without truth-value, but as uncertain with a real but hidden truth-value.[23]

IN DIALOGUE WITH SCIENCE—JARL HEMBERG

During the 1960s and 70s Swedish theology experienced a time of theological reorientation from more narrowly defined ecclesial issues to broader issues of social ethics and international justice. To a large extent this reorientation was inspired by the General assembly of the World Council of Churches in Uppsala 1968. This left an imprint on the themes chosen for theological dissertations and was spurred by established theologians such as Gustaf Wingren, Lars Thunberg, and Per Erik Persson. The Christian journal *Vår lösen* (with Anne-Marie Thunberg as editor) provided theological reflections on these issues as well as contributions from social scientists and scholars in the humanities. But Anne-Marie Thunberg also opened for questions on science and religion already in the 1950s—even if contributions from academic theologians were relatively spars.[24] Another publication frequently opened up for questions on science and religion, namely, *Årsbok för kristen humanism*. One example is an article on science and religion by the theologian Jarl Hemberg.

Jarl Hemberg submitted his dissertation in 1966. It had the title *Religion och metafysik* and was focused on Axel Hägerström and Anders Nygren and their influence in Swedish theology. After his dissertation Jarl Hemberg became assistant professor in Uppsala and in 1977 full professor

---

23. Jeffner, *I vetandets gränsmarker*, 168–71. More on the distinction between changing and uncertain gestalts is found in Jeffner, *The Study*, 45–48.

24. Thunberg, *Tradition i rörelse*. See, especially, the articles of Anders Jeffner and Gunnar Tibell.

in ethics at Lund University until his untimely death in 1987. I shall highlight three articles on science and religion by Jarl Hemberg from 1963, 1967, and 1980.

The first article was sparked by an important book by the Swedish physicist Tor Ragnar Gerholm.[25] This book was an important contribution to the public understanding of modern physics, and occasionally also made references to science and religion. But it is not entirely clear what model about the relationship between science and religion that Gerholm endorsed. According to a stronger interpretation, Gerholm was an adherent of the incompatibility theses, i.e., that there is an irreconcilable conflict between science and religion. But I would suggest that his position is more accurately described as a contact-model with emphasis on the tensions or oppositions between science and religion. Jarl Hemberg main point is that "rightly understood" there is no such tension. Two questions emerge: (1) How does the appearance of a tension or opposition occur? and (2) how is this tension or opposition reduced?

Hemberg gives two answers to the first question: religious persons are simply ignorant of scientific facts, and scientists make false interpretations of their results. Basically, Hemberg insists that conflicts between science and Christian belief are impossible. This is so, because a clear line of demarcation must be drawn between scientific facts and genuine religious claims (such as God loves mankind). Another reason is that science is in principle agnostic, i.e., the scientific method excludes in principle the reference to non-scientific authorities such as "God or Marx or any kind of ethical value system."

This line of reasoning raises several questions, but I want to focus on the basic model that Hemberg implicitly presupposes, and it is a separation in principle between science and religion. This position is also presupposed in another article by Hemberg from 1967. This article is an analysis of another much-discussed work of a Swedish philosopher, Sören Halldén.[26] Halldén makes a sensitive analysis of classical philosophical problems and Hemberg focuses on Halldén's analysis of the mystery of existence: why is there something rather than nothing?[27] Is this a genuine problem? If so, how do we go about answering it? In short, Halldén argues that the mystery of existence is a genuine problem and should be interpreted as a quest for an explanation. But this explanation must sought in empirical science. Hem-

---

25. Hemberg, "Naturvetenskaplig världsbild och kristen tro." Gerholm, *Fysiken och människan.*

26. Halldén, *Universum, döden och den logiska analysen.*

27. Hemberg, "Världens ursprung och den logiska analysen," 104–16.

berg has misgivings about this response. It is, indeed, a genuine problem, but a problem requiring a teleological rather than a casual explanation. The mystery of existence may have a metaphysical answer in the form of claims that about reality as a whole. Hemberg argues that "God exists" is such a metaphysical statement.[28] His argument presupposes a model of separation between science and religion.

In 1980 Jarl Hemberg returns to the question of the relationship between science and religion in an article in the scholarly journal *Tro och liv* (1980). At the outset, it seems that Hemberg would allow for a genuine conflict between science and religion. The strict separations thesis is left behind; there is now a possibility of an overlap between religion and science. But it is an overlap in a specific sense. Hemberg argues that a religious belief must always be consistent with (i.e. not contradict) well-established scientific theories and results. His primary example is the theory of evolution. Theologians upholding a literal interpretation of the creation stories in Genesis are simply wrong. But this conflict is not a *pure* conflict. At a deeper level a distinction between scientific and existential explanation is called for. Scientific explanations are on another logical level than belief in God.[29] Scientific explanations have no positive or negative relevance for existential explanation. Take, for example, chance as a factor in the emergence of life. Some scientists have taken chance as an insufficient principle of explanation in these and other key events in biological evolution. God must be introduced as a factor in these events. Doesn't this show that model of separation breaks down? Hemberg answers this in the negative. It's just "bad apologetics."[30] Chance is just another name for problems for which science has not yet found a solution to. Much could be said in favor of this response, but Hemberg does not explain why there could not be objective chance in nature.

What about miracle, prayer and divine providence? Hemberg argues that science is open to certain theological interpretations of particular events. For example, a reasonable scientist cannot a priori deny that no violations of natural law can occur.[31] Furthermore, a religious believer can interpret blessings (explainable by science) as a response from God to a prayer. But how is it possible that God can answer prayer from afar, when nothing in the universe can exceed the speed of light? The praying believer must be dead long before

---

28. Hemberg, "Världens ursprung och den logiska analysen," 114–16.
29. Hemberg, "Om naturvetenskap och kristen tro," 6.
30. Hemberg, "Om naturvetenskap och kristen tro," 7
31. Compare this to Nathan Söderblom's interpretation of miracles in chap. 6.

his prayer reaches God! Hemberg's response is that spiritual communication occurs under other laws than physics. God is omnipresent.

One further contribution from Jarl Hemberg must be acknowledged. In 1973 he took the initiative of arranging a symposium on Pierre Teilhard de Chardin under the theme "Humans in Cosmos." The Symposium was realized in the beginning of 1974 and contributions were published in 1975.[32]

In sum, Hemberg came to modify his strict interpretation of the model of independence. There are no valid scientific arguments supporting or debunking religions belief in the form of explanations of life as a whole. These explanations are metaphysical. In other areas conflicts may arise. Some of them are "impure," but others require an accommodation between science and religion. If science and religion are properly interpreted, these conflicts may be resolved.

SCIENCE AND RELIGION AS COMPLEMENTARY ASPECTS OF REALITY—RAGNAR HOLTE

In Chapter 2, I presented a survey made by Ulf Görman in 1999. The survey concerned the understanding of the relation between science and religion among the Swedish public. The respondents were presented with three alternatives: science and religion (1) have no point of contact, (2) . . . complement each other, express different aspects of one and the same reality, and (3) . . . are in conflict with each other. Alternative (2) was the most popular alternative, especially among religious believers in God.

One Swedish theologian—Ragnar Holte, professor in ethics in Uppsala— has contributed to a philosophical clarification of the second alternative. Holte's main contributions in theology can be found in the field of systematic theology and ethics, but he has one book on theories and methods in studies of faith and ideologies from 1984.[33] One chapter in this book is a critique of ontological reductionism in theories on human nature (particularily, Skinner's form of behaviorism). This critique is based on Kantian epistemology and a contemporary interpretation of science and quantum theory.

Holte interprets Kant's epistemology as a middle way between the rationalism of Plato and the empiricism of John Locke. While Plato relies on

32. See Hemberg, *Teilhard de Chardin*. Of special interest is the contribution by Kerstin Anér (1920–1991), who had an important public impact as a scholar and politician, but also made several contributions to the discussion on science and religion. See, for example, Anér, "Teilhard de Chardin." She was among the first Swedish theologians to highlight Ilya Prigogine and his relevance for theological issues. Anér, *Framtidens insida*, 186–88.

33. Holte, *Människa*.

the ultimate source of knowledge in our intellectual acquaintance with the world of ideas, Locke departs from the idea that all knowledge is ultimately derived from the senses. Aristotle combines the rational and empirical elements in our knowledge in a receptionist theory (reality has a rational structure which is accessible through sense experience), but Kant assumes that reality itself ("noumenal" reality) is inaccessible; what we experience is the phenomenal world, structured through the categories of understanding and the forms of the senses. Consciousness has a "synthetic unity" and functions and has basically the same properties in all human beings, which guarantees the intersubjective validity of knowledge.[34]

Holte uses Kant's activistic combination-theory in an interpretation of an article by a Swedish pioneer in quantum physics, namely, professor Per-Olov Löwdin at the Royal Institute of Technology in Stockholm.[35] Löwdin is also critical of principle of ontological reductionism in the sense of reduction from biology to physics. This reductionism is closely associated with the mechanistic worldview, i.e., that everything—including human consciousness and actions—is determined by natural laws. He argues that modern quantum theory effectively refuted this worldview and with it ontological reductionism. Quantum theory implies that the observer influences the reality he/she observes and Löwdin questions how it is possible to arrive at "physical reality" and scientific objectivity from this point of departure.

Holte interprets Löwdin's solution in terms of the Kantian combination-theory. The existence of a physical universe, which can be described only in interplay with our senses, is "an implicit point of departure." Furthermore, Holte supplements this theory with a distinction between models and reality.[36] Models are simplified description of an infinitely complicated reality. Maps are an example of such models, but so are physical descriptions and poetry. The same object can be described by different models, such as a tree may be described in physical, as well as agronomic, and poetic and mythical terms. These different models do not compete, but complement each other. Needless to say, poetic models are not as exact as physical descriptions, but they may nevertheless allow for a broader access to reality. Later in his book Holte argues that models in science may be complemented by models, for example, in religion.[37]

---

34. Holte, *Människa*, 53–54.

35. Löwdin, "Människan och hennes psyke."

36. This distinction is taken from an article by Lars Kristiansson (from Chalmers Univerity in Gothenburg). See Holte, *Människa*, 57–62.

37. Holte, *Människa*, 104.

In Chapter 1, I made reference to a more elaborated theory about the role of models in human thinking, namely, Steven Horst and cognitive pluralism (CP). Interstingly, Horst is equally critical of reductionism, but he does not rely as heavily as Holte does on quantum theory.[38] I will return to Horst and cognitive pluralism in the introduction to Part 3 of this book.

The idea of science and religion as complementary models can be interpreted in different ways. It seems to be one way to eliminate apparent conflicts between science and religion. If science and religion are complementary, they cannot be in conflict. But this is not evident. Science and religion may be complementary models and still stand in conflict in the way that different models are claimed to be conflict in quantum physics. It is important to notice that Holte—drawing on another Swedish physicist, Lars Kristiansson—does not allow for models that contradict each other.[39] In this way, one may understand science and religion as complementary models and still advocate a model of conflict between science and religion. The alternative idea of science and religion as complementary might also inform a model of independence as well as a model of contact. A model of independence would equal the position that religious beliefs are independent and separate from science. This would be denied by those who advocate a model of contact between science and religion. I will further elaborate these issues in Part 3.

RATIONALITY, SCIENCE AND RELIGION
—MIKAEL STENMARK

Mikael Stenmark defended his dissertation on *Rationality in Science, Religion and Everyday Life: A Critical Evaluation of Four Models of Rationality* in 1995 and has henceforth contributed with several other monographs, articles and anthologies in the area of science and religion. He has made significant analyses of different aspects of the study of science and religion, such as (1) the general problem how to relate science and religion,[40] (2) general theoretical challenges from science to religious belief such as the challenge from scientism, evidentialism and relativism,[41] (3) particular theoretical chal-

---

38. Horst, *Cognitive Pluralism*, loc. 4196.

39. Holte, *Människa*, 60.

40. For example, in his monograph *How to Relate Science and Religion: A Multidimensional Model*.

41. See, for example, Stenmark, *Scientism*; and Stenmark, *How to Relate Science and Religion*, chap. 5. See also Stenmark, "Rationality and Different Conceptions of Science," 47–72; and Stenmark, "Relativism—Pervasive Feature of the Contemporary Western World?," 31–43.

lenges from science to religion such as conceptions of God, human nature and ethics, and[42] (4) the value/ ideology challenge of religion and worldviews to science.[43] Needless to say, much of the "fine-print" of Stenmark's analyses have to be bypassed. Instead, I shall focus on the main tenets of his most important works and begin by summarizing his main contribution to the general problem how science and religion should be related.

Of special importance is Stenmark's more elaborated typology of different relationships between science and religion. His argument proceeds in a stepwise manner. In a *first* step, he begins by noticing the distinction between *scientific restrictionism* (roughly equivalent to the model of separation[44]) and *scientific expansionism* (roughly equivalent to the model of substitution[45]). Stephen Jay Gould with his theory of non-overlapping magisterial, NOMA, is taken as an example of scientific restrictionism and Richard Dawkins as an example of scientific expansionism.[46]

In his *second* step Stenmark recognizes three science-religion views: the *independence view* (exemplified by Gould), the *monist view* (comes in two versions: total harmony or—according to Dawkins—irreconcilable conflict), and a third intermediary model, the *contact view*, which stand for more or less partial harmony or—possibly—reconcilable conflicts between science and religion. The contact view is also Stenmark's own position.[47]

In a *third* step, Stenmark departs from the observation that former typologies have the weakness of only taking the theoretical aspect of this relationship into consideration. The relationship between science and religion is a multilevel or multidimensional relationship. Besides the epistemological and theoretical dimension, there is also a social and teleological dimension. Science and religion can be described as practices with certain functions. Moreover, persons practicing science and/or religion have certain, possibly different but not necessarily opposing, aims with these practices. Stenmark summarizes:

---

42. See Stenmark, *Scientism;* and Stenmark, "Competing Conceptions of God," 31–43.

43. See Fuller, Stenmark, and Zackariasson, eds., *The Customization of Science.*

44. See below, chap. 13.

45. See above chap. 2 and below in the Introduction to Part 3.

46. It should be observed that there is also a version of religious expansionism represented by, for example, Alvin Plantinga and his idea of Augustinian science (see Stenmark, *How to Relate,* chap. 8).

47. Stenmark, *How to Relate,* 267. Note, however, that Stenmark endorses "restrictionism in respect to scientific theory justification" and disallows "an overlap between science and religion on this point." See ibid., 230–34 and 258.

The basic idea, then is, that *somebody who wants to successfully understand how to relate science and religion needs to take into account at least the social structure of science and religions, the aims of these practices, the kind of epistemology they exhibit and their theoretical content (their theories, beliefs, or stories).*[48]

Furthermore, these dimensions of science and religions can be studied (1) from a historical or contemporary perspective, or (2) descriptively (telling us how they in fact are related) or normatively (how they *should* be related). The following matrix emerges:[49]

	Social dimension	Teleological dimension	Epistemological dimension	Theoretical dimension
I. Contemporary studies				
1. universal				
a. descriptive				
b. normative				
2. contextual				
a. descriptive				
b. normative				
II. Historical studies				
1. universal				
2. contextual				

Table 13: Dimensions of science and religion

Stenmark underlines that these different kinds of studies could also focus on secular alternatives to religion, i.e., worldviews, ideologies and views of life.

In the last chapter of *How to Relate Science and Religion,* Stenmark brings in Ian Barbour's well-known typology of science and religion. But more importantly, Stenmark shows how the multidimensional approach widens the area of application for the three models (the monist view, the independence view and the contact view). For example, those who claim that not only the theoretical content of science and religion is completely

48. Stenmark, *How to Relate,* 13.
49. Stenmark, *How to Relate,* 14.

different, but also the goals and methods, could be counted as exponents of the independence view. Similarly, if there is an overlap between the goals, methods and theoretical outputs of science and religion, we have an example of the contact view. Lastly, if the goals, methods and theories of science and religion are or should be the same, we have the monist view.[50]

The next area of Stenmark's philosophy of religion is his study of general theoretical challenges from science to religious belief such as the challenge from scientism, evidentialism, and relativism.

His monography on *Scientism* (2001) has been the subject of wide attention and significantly contributed the development in the study of science and religion. Of special importance is his distinction between different forms of scientism and his critical scrutiny of these claims.[51] The main distinction is between academic-internal and academic-external scientism. Methodological scientism—the attempt to extend the use of methods of natural science to other academic disciplines—is one example of academic-internal scientism. Academic-external scientism includes epistemic, rationalistic, ontological, axiological, existential (or redemptive), and comprehensive scientism. In Chapter 8, I have already made use of the concept of rationalistic scientism. Epistemic and ontological scientism is of special importance in the contemporary discussion between naturalism and theism and one important focus of Stenmark's discussion. One conclusion of his analysis is that epistemic scientism—the view that the only reality that we can know anything about is the one science has access to—is self-refuting. Epistemic scientism "seems to be something that science cannot tell us."[52] And if epistemic scientism cannot be true, neither can ontological scientism.

Another focus is Stenmark's scrutiny of (1) arguments from the scientific explanation of morality, and (2) arguments from the scientific explanation of religion. The aim of these arguments is to show that science alone can explain and replace traditional ethics (axiological scientism) and religion (existential scientism). Stenmark concludes that while it is an important scientific project to explain ethics and religion, "there is a clear limit to the exten evolutionary theory can explain religious belief and behavior."[53] Furthermore, new scientific knowledge about life may undermine some religious beliefs and ethical conviction, but not necessarily all.[54]

50. Stenmark, *How to Relate*, 260–61.
51. Stenmark, *Scientism*, chap. 1.
52. Stenmark, *Scientism*, 135.
53. Stenmark, *Scientism*, 138.
54. Stenmark, *Scientism*, 138–39.

The basic thesis of his dissertation (1995) is that our model of *rationality* is basically to be informed by everyday belief formation rather than belief formation in science— "everyday believing" is the "control case of rationality."[55]

Stenmark enters into dialogue with Wentzel van Huyssteen's plea for a post-foundational model of rationality.[56] Together with Thomas Kuhn and, particularly, Harold I. Brown, van Huyssteens maintains that rationality is (1) not ruled-governed, but (2) ultimately about making informed or responsible judgments. In contrast to modernist models, where emphasis is laid on the logical relation between evidence and belief, the post-foundational model takes rational agents to be the basic notion and rational beliefs to be the derivative. But the modernist and post-foundational model of rationality have one thing in common, namely, evidentialism, i.e., that it is rational to accept a belief only if there are good evidence to think that it is true.[57] In sum, judgment-based evidentialism is common to all forms of human rationality.

Stenmark is critical of this post-foundational model of reality and particularly the component of evidentialism. He mentions several reasons for this, among which one is that everyday beliefs are more fundamental in our belief system and the second is what Stenmark calls the axiom of reasonable demand, i.e., we should not demand more of rationality than what our limited human resources allow. Instead Stenmark argues for what he calls *presumptionism*, i.e., that the most rational thing to do is to continue to believe what we already believe until we find good reasons to believe something else.[58]

When it comes to religious practice, Stenmark maintains that it resembles everyday rationality more that scientific rationality or theological rationality. Hence presumptionism should be a basic principle when ordinary religious believers consider their faith in—for example—relation to evolutionary theory. But when they reconsider their belief in this way another principle kicks in, i.e., the "cautious principle of belief revision." "*[W]hen revising something we already belief we should pick, among those alternatives that are available to us, the one that is nearest to our original conviction.*"[59] For example, evolutionary theory does not require the ordi-

55. Stenmark, *Rationality in Science*, 200.
56. Stenmark, *How to Relate*, chap. 5.
57. See further, Stenmark, *Rationality in Science*, 134–35.
58. Stenmark, *How to Relate*, 90. Needless to say, science and scholarly activities— such as theology—should be guided by more stringent demand for rationality. See further, Stenmark, *How to Relate*, 97–104.
59. Stenmark, *How to Relate*, 104.

nary believer to abandon their belief in God as personal providing there are less radical alternatives such as abandoning the belief in God's direct intervention and substituting that belief for the belief that God created life and humans through a natural evolutionary process.

In addition, Stenmark proposes two other principles guiding the transformation of religious beliefs in face of contrary reasons. One is the "principle of concern." *"we should require stronger reasons for giving up something we believe which has greater depth of concern in our belief-world than for giving up something which plays a more peripheral role."*[60] This is a natural principle in everyday life, but also in religious life. Furthermore, religious everyday rationality should be guided by a "strength of belief-principle."

"In a situation where I meet many other apparently reasonable and honest people who happen to believe something other than I do, this should affect the strength with which I hold on to my belief at least in such a way that I realize that there is a real chance that I actually could be wrong."[61] In short, a pluralistic cultural situation makes a dogmatic attitude irrational.

One further thing is of significance in the context of Stenmark's theory of religious rationality. It affects our understanding of science and religious practice, namely, the distinction between "agent-rationality" and "spectator-rationality." Religious practice has to do with the way we live our lives. Religious practice is about coming to grips with sin, evil and meaninglessness. Certain choices have to be made and cannot be postponed (as Kierkegaard emphasizes in his writings). Stenmark writes:

> Boldly stated, my claim is that religious rationality is a different species of rationality than both scientific and theological rationality. It is a kind of "agent-rationality," whereas rationality in science and theology is a kind of "spectator rationality" . . .[62]

Here is point at which I find myself at variance with Stenmark's analysis. Stenmark argues that (1) scientific rationality is the same as spectator-rationality and (2) religious rationality is not identical with agent-rationality. Let me call this Stenmark's *bold statement*. I would argue that there is a possible overlap between the two forms of rationality in both scientific and religious practice. This overlap is not sufficiently reflected upon and which basically requires some form of empirical studies. Obviously, scientific practice requires participation between scientists themselves and, in certain instances, between scientists and their objects of research. Moreover, religious practices

---

60. Stenmark, *How to Relate*, 106.
61. Stenmark, *How to Relate*, 107.
62. Stenmark, *How to Relate*, 111.

include significant moments of theological reflection, which requires a kind of "spectator rationality." The Sunday sermon is an important part of the Christian practice and so is religious education.

In fact, Stenmark at other times acknowledges the overlap between rationality in "science, theology and religion." [63] Indeed, Stenmark wants to argue *against* "the view that non-epistemic, pragmatic, moral, or practical reasons are the only reasons we really have to take into account within the epistemology of religion."[64] This seems to contradict Stenmark's *bold statement*. It is possible, however, that Stenmark can evade a contradiction by a sharp distinction between theology and religion. Pragmatism is an untenable position in theology, while his *bold statement* about the two opposite forms of rationality (spectator-rationality and agent-rationality) concern the relation between scientific and religious practices.

Stenmark has also made some significant contributions to the discussion about particular theoretical challenges from science to religion such as conceptions of God, human nature and ethics. This has to do with the relationship between the theoretical content of science and the theoretical content of religion. Stenmark has—for example—engaged Gordon D. Kaufman on the question whether belief in a personal creator is compatible with the modern physics, cosmology and biology.[65] Furthermore, Stenmark has challenged biologists and philosophers who argue that evolutionary theory shows that life has no purpose or meaning. I will present the essentials of these discussions in turn.

The first argument departs from Gordon Kaufman's thesis about the incompatibility between science and religion in the form of belief in a personal God the theory of evolution. Evolutionary thinking shows that persons with intentions, will, love, morality etc. emerge only at the end of a long evolutionary process. This is contrary to belief in a God with humanlike qualities such as will, intentions, love etc. at the beginning of the whole process.

Stenmark is critical of this argument for at least two different reasons. First, there might be limits to scientific knowledge. Kaufman disregards the possibility that religious believers are entitled to believe certain things about the universe, which scientists qua scientists cannot find reason to believe. And one such thing—common to the great religious traditions of the Western world—is belief in a personal creator of the world. Secondly, Kaufman

---

63. Stenmark, *How to Relate*, 121.

64. Stenmark, *How to Relate*, 122.

65. Stenmark, *How to Relate*, 138–56. See also Stenmark, "Science and a Personal Conception of God."

originally suggested an additional premise presupposing scientism, i.e., that we are rationally entitled to believe only that which can be justified through the sciences. But scientism in this form is not beyond criticism and in a rejoinder Kaufman claims that theology can and should go beyond what can be scientifically established.[66] If so, why is it not possible for religious believers to do the same and affirm a belief in a personal creator? In sum, Kaufman has not provided religious believers with any good scientific reasons for abandoning their personal conception of God.

The second argument—advanced by several philosophers and biologists—is that evolutionary theory shows that our species, *Homo sapiens*, cannot be intended by God.[67] And the reason for this is that *Homo sapiens* "ranks as a 'thing so small' in a vast universe, a wildly improbable evolutionary event, and not the nub of universal purpose." Therefore, there is no meaning in the form a purpose for human life. The theory of evolution shows that we are not planned by God or anything like God.

To evaluate this argument, Stenmark considers the conditions that must be satisfied in order for something to exists for a reason or be planned by someone. In short, this something has to be (a) intended by a person and (b) come into existence for that reason. If so, it seems that God could not create *Homo sapiens* through evolution, because of the haphazard character of evolutions. It is impossible for the same reason that humans—using normal means of reproduction—cannot realize an intention to have a child with a specific genetic make-up. But—extending the metaphor—humans can intend to have a child *simpliciter*. Similarly, God could plan to have universe with living, self-conscious creatures by realizing certain initial conditions of the universe.

This response to the "no purpose argument" requires a certain modification of classical theism. Stenmark writes:

> Instead of believing that God had a *particular species* in mind, they should (or at least could) believe that what God had in mind was the emergence of a generation of communities of free, self-aware, self-directing sentient being. The benefit of thus revising our beliefs is that the likelihood that such a life-form would appear in evolutionary history is much higher than that a particular instance of this type of life, Homo sapiens, would emerge.[68]

---

66. Stenmark, *How to Relate*, 153–55.
67. See *How to Relate*, 157ff., for a number of examples.
68. See *How to Relate*, 167.

This argument needs some further elaborations in view of possible objections, but I would concur to its general structure. Needless, there is a similar issue—and much more difficult—concerning the meaning and purpose of our individual lives. It seems to be a part of the classical doctrine of divine providence, that God has a purpose for each and every person of the human species. If *Homo sapiens* qua a specific species was not part of God's plan (even if the creation of some intelligent species somewhere in the universe was), how is it then possible that each and every individual of our species is intended by God? Philosophers of religion have engaged this question in other contexts, but it goes beyond the scope of the present book.[69]

Another important part of Stenmark's work concerns the value/ ideology challenge of religion and worldviews to science.[70] The phenomenon in question concerns different efforts to "customize" science, be it designing science to fit economic interests, political ideologies, religious or antireligious convictions of different institutions or groups. Whether or not this phenomenon really exists or whether it is of significance is a matter of debate. If it exists, it might be evaluated in different ways—welcomed by some or criticized by others.

One focus on the customization of science could be upon efforts to shape science according to the interests of a particular religion. Stenmark terms this tendency *religious expansionism* and gives two examples: Augustinian science (proposed by Alvin Plantinga) and Islamic science (proposed by Seyyed Hossein Nasr).[71] The main idea is to put science in the perspective of Christianity or Islam. For example, the Christian academic and scientific community "ought to pursue science in its own way, *starting from* and taking for granted what they know as Christians."[72]

Stenmark argues that proponents of a religiously customized science have given convincing examples of a naturalist bias within contemporary science. From this Stenmark concludes:

> We need a scientific education which contains a study of examples of worldview influences on past and present scientific research, so that scientists can develop a better and less naïve understanding of how their own and other people's ideological

---

69. See Clark, "God, Chance, and Purpose."

70. See Stenmark, *How to Relate*, chap. 8; and Fuller, Stenmark & Zackariasson, *The Customization of Science*.

71. For references, see Stenmark, *How to Relate*, 184–94.

72 Stenmark, *How to Relate*, 187.

or religious commitments interact with scientific practice at different levels.[73]

I would concur with this conclusion and base this concurrence on earlier studies of the interaction between views of human nature and knowledge about genetics and its applications.[74] More controversial is the answer to the question whether a religion (such as Christianity) or worldview (such as naturalism) should shape the practice of science. Stenmark's response to this question is guarded and nuanced. He distinguishes between different phases of the scientific process and argues that a worldview-partisan science is legitimate in the problem-stating phase, the development phase and application phase, *but for a worldview-neutral science in the justification phase.*[75] The problem—as I see it—is that it will in reality be difficult to make an exception from the worldview-partisan paradigm in the justification phase if the scientific problem is shaped by concepts, methods and hypotheses in the development phase.[76]

---

73. Stenmark, *How to Relate*, 249.
74. Bråkenhielm and Hansson, *Livets grundmönster*, chap. 7.
75. Stenmark, *How to Relate*, 249.
76. See further below, chap. 14.

11

From Generalities to Specifics

In the first half of the 1990s, Swedish theology takes leave of a longer period of dominance of the model of independence. The weight tilts more definitely to the model of contact with varying accent on harmony, tension, conflict, or integration. One explanation is the growing influence of American theology and contributions to the relation between science and religion. Kantian influences were left behind already in the 1970s and British philosophy of a Wittgensteinian flavor is becoming less important during the 1990s.

This shift opened for more specific analyses of different problems in the interplay between scientific theories and Christian doctrines. Tensions and conflicts are clarified and solutions suggested. This development is visible in a number of dissertations from the middle of the 1990s up until the present. The present chapter consists of a presentation of seven dissertations and in some cases subsequent publications by the authors in the area of science and religion.

THE CHALLENGE OF EVOLUTIONARY THINKING TO CHRISTIAN THEOLOGY—ANDERS NORDGREN

Anders Nordgren defended his dissertation in Uppsala in 1994.[1] His stated aim was to investigate the relevance of biological theory of evolution for the philosophical treatment of problems related to mind, rationality, morality and religion. Important to his approach is that Nordgren found strong

1. Nordgren, *Evolutionary Thinking*.

reasons to supplement the traditional idea of natural selection with the idea of self-organization. Nordgren argues that theistic interpretations of the theory of evolution are logically possible, but that they are undermined by different arguments and that they appear less coherent than naturalistic interpretations.[2] In conclusion, Nordgren defends a proposal based on nature-oriented religious premises.[3]

At the outset Nordgren makes a significant distinction a *general approach to the relation between science and religion* and a more *specific approach* concerning the relation between evolutionary theory and belief in God. He describes and defends the significance of the specific approach and proceeds with a normative comparison between naturalistic versions of the theory of evolution, person-oriented religious versions (such as theism) and nature-oriented religious versions. But it should be noted that he presupposes a version of the contact model with a strong emphasis on the necessity to integrate science and religion. This is in contrast to Haikola's earlier mentioned argument.[4]

Nordgren claims that common for theistic interpretations of the theory of evolution is their teleological character. "God has determined the course of evolution. The natural mechanisms of evolution are the instruments of God in the act of creation."[5] After rejecting different arguments to relate the theory of evolution to belief in God in the first place, Nordgren advances five different problems with the theistic versions as well as efforts to solve these problems. The problems are the following:

- The theory of evolution makes the problem of evil deeper and more difficult
- Biological non-functionality and maladaptation contradicts God's design
- Human being is the result of random mutation, which contradicts God's design
- Theistic interpretations of evolution radically transforms the theory of evolution
- No traces of God in the world (as described in the theory of evolution)

All these problems are no problems for the naturalistic versions of the theory of evolution. And even if these objections do not make the theistic

2. Nordgren, *Evolutionary Thinking*, 219.
3. Nordgren, *Evolutionary Thinking*, 232–33.
4. See above, chap. 9.
5. Nordgren, *Evolutionary Thinking*, 213.

versions of the theory of evolution logically impossible, they "undermine the information that the arguments for the theistic version usually build upon."[6] The theistic conception must be modified or even radically reinterpreted.

Nature-oriented religious versions of the theory of evolution can be understood as such radically reinterpreted forms of theism. They are based upon an immanent view of God imagined with the help of space-oriented, time-oriented or mind-oriented models.

> In the *space-oriented models*, God is identified with nature (cosmos) as a whole, with our planet (the Gaia hypothesis), or with the deepest dimension or ground of nature. In the *time-oriented models*, God is perceived as the starting-point of the universe or as its terminal point. In the *mind-oriented models*, God is regarded as the mind of the universe or of the nature on Earth.[7]

The general problem with these models concerns the view of God and which view of God is to be preferred. Nordgren surveys different views of God suggested by the three different types of models and concludes with his proposal of a view of God as identified with the dynamic universe as a whole, but viewed form different aspects: spatial, temporal and mental. In the (1) spatial aspect, God is the highest level of self-creating, self-organizing universe as whole, in the (2) temporal aspect, the original singularity of Big Bang as well as the moral terminal point of evolution, and in the (3) mental aspect, God is the mind of the universe and the mind of the earthly nature.

Anders Nordgren makes a bold synthesis of science and religion, but he leaves some important questions unanswered. His conclusions are stimulating, but not sufficiently developed. After his dissertation Nordgren concentrated his efforts on ethical issues and several questions concerning his "constructive proposal" have remained unanswered. For example, does it survive the criterion of simplicity, which he raises against several other interpretations of the theory of evolution? And how can this impersonal view of God be "a living God with direct practical relevance."[8]

SCIENCE, TIME, AND ETERNITY—ANTJE JACKELÉN

Antje Jackelén submitted her dissertation to the Faculty of Theology, Lund University, and defended it in 1999. Since then it has been edited and

---

6. Nordgren, *Evolutionary Thinking*, 219.
7. Nordgren, *Evolutionary Thinking*, 220; my italics.
8. Nordgren, *Evolutionary Thinking*, 233.

republished in three different editions.⁹ Antje Jackelén has been a productive scholar since the middle of the 1990s with 8 monographies, over 70 articles in scholarly journals, and the editor of over 20 books.¹⁰

Central in her production is her dissertation from 1999. One reason for her choice of subject is that the question of time has wide-ranging consequences not only for eschatology, but also for concepts of God, the understanding of incarnation and numerous other fields in theology. It also provides a key to a general understanding of the relationship between science and religion. She takes the model of contact seriously, but instead of asking for the contribution of science to theology (which is widely acknowledged), she also asks for the contribution of theology to science. The answer to this question concerns her methodological profile evident in the outset of her dissertation.

One of the most important things that theology can do for science comes from its experience in the field of *theological hermeneutics*.¹¹ She explains this concept in the following way:

> Theological hermeneutics, as I understand it, is a way of expressing the dimension of transcendence. It is always pointing to the "more" and to the "beyond." It is a way of living faithfully with limitations and recognizing that there is more to the world than theologians, philosophers or scientists ever dreamt of. Expressed differently: where philosophical hermeneutics can be said—in Hans-Georg Ghadames's terminology—to be occupied with "the fusion of the horizons" of understanding, theological hermeneutics claims that there always remains a horizon that can never come to nothing in any fusion of horizons. The ultimate beyond, transcendence, cannot be merged into immanence.¹²

---

9. Jackelén, *Zeit und Ewigkeit*; Jackelén, *Tidsinställningar*; Jackelén, *Time and Eternity*. From 1999 to 2001, Antje Jackelén was regional director for Europe of the Science and Religion Course Program of the Center for Theology and the Natural Sciences, Berkeley, CA. From 2001–2007, Jackelén taught systematic theology and religion and science at the Lutheran School of Theology at Chicago and was director of the Zygon Center for Religion and Science. She is a founding member of the International Society for Science and Religion (ISSR) and served as president of the European Society for the Study of Science and Theology (ESSSAT) from 2009–2013. She was the bishop of Lund 2007–2014. From 2014, she is the archbishop of the Lutheran Church of Sweden.

10. An exhaustive bibliography can be found at https://www.svenskakyrkan.se/omoss/om-arkebiskop-antje-jackelen.

11. Jackelén, "What Theology Can Do for Science," 291.

12. Jackelén, "What Theology Can Do for Science," 299.

Against this background, the main thesis of her dissertation can be described in the following way: Christian eschatology contributes to a full and appropriate understanding of the relationality of time as it is suggested in a narrative interpretation of human understanding and in contemporary science.

A "narrative interpretation of human understanding" begins not with philosophy nor with the "language of metaphysics," but with "concrete symbol systems."[13] The subject of time makes this approach particularly natural. With reference to Paul Ricœur and his work *Time and Narrative*[14] she argues that "[n]arrative understanding deserves precedence over narratological rationality."[15] But narrative understanding is not only necessary in the analysis of time, but more broadly. In a later article on the theories of emergence, she makes the same point in no mistaken terms:

> In philosophy and theology, emergence contributes to the critique of ontological metaphysical statements. Narrative understanding is always necessary: metaphysics cannot do without myth.[16]

Jackelén's narrative understanding of time is closely associated with two hermeneutical principles: (1) "the self-evidence of the discussion" and (2) "the desire for contact." The second principle of desire for contact is of special importance. This means that the dialogue between science and religion "must be brought into contact with the lives of those for whom these ideas and experiment were (supposedly) made and who are affected by their consequences."[17] This can be interpreted as a plea for ethical criteria in the dialogue between science and religion, but it remains unclear how such criteria are to be applied.

In the present context I shall bypass chapter 1 and 2 and pass directly to the last and pass to the conclusions of chapter 3 and the final chapter 4, which has the title "Aspects of a Theology of Time." Here theological insights from the previous chapters are brought into contact with the Christian doctrines of Trinity and eschatology. Basic is her conclusion towards the end of chapter 3 that time in contemporary quantum physics and theory of relativity is a relational phenomenon. Jackelén writes:

---

13. Jackelén, *Time and Eternity*.
14. Ricœur, *Time and Narrative*, 2, 158. The distinction between narrative understanding and narratological rationality would require further analysis, which I will bypass in the present context.
15. Jackelén, *Time and Eternity*, 11.
16. Jackelén, "Creativity through Emergence," 88.
17. Jackelén, "Creativity through Emergence," 6.

> An adequate understanding of time cannot be satisfied with the analysis of individual elements; it must include structures and relations, being and becoming. It must also reckon with the fact that genuinely new things are possible. This openness encourages interpretations that go far beyond physics.[18]

Jackelén underlines a *relational understanding* of time in contrast to substance-thinking. "Whereas substance-thinking proceeds from a clear subject-object distinction, relationality attempts to describe processes and dynamic interactions." Furthermore, "while substance-thinking, as orientation to that which exists, is more oriented toward the past, relational thinking, as orientation toward that which is possible, is more strongly oriented toward the future."[19] In conclusion, Jackelén notes that physics suffers from an "eschatological deficit" and that it must do so if it wishes to remain true to its nature. Nevertheless "quantum physics and chaos theory add greater meaning to the concepts of relation, dynamics, and openness without providing us with clear definitions."

Furthermore, Jackelén considers "the extent to which scientific development can be received theologically."[20] Jackelén summarizes her findings in the earlier parts of her dissertation. After reexamining the central concepts of statics, dynamics and relation, she returns to the doctrines of the Trinity and eschatology. As many other twentieth century Christian theologians, Jackelén finds relational thinking at variance with an abstract theistic understanding of God. But it is more consonant with a trinitarian concept of God. Many contemporary theologians have found a bridge between trinitarian theology and modern science, but Jackelén finds that but this concept invites further problems.[21] Moreover, "[t]rinitarian models include a relatively large portion of arbitrary speculation."[22] Therefore, Jackelén in the final section of her dissertation turns towards the doctrine of eschatology.[23] She understands "eschatology as the key to a relational understanding of time."[24] She is, however, critical to Moltmann's theological eschatology as well the scientific eschatologies of Frank Tippler and Freeman Dyson.[25]

---

18. Jackelén, "Creativity through Emergence," 180.

19. Jackelén, "Creativity through Emergence," 189. This analysis raises the question in what sense relationality can dispense with some kind subjects or objects between which there is a relation.

20. Jackelén, "Creativity through Emergence," 180.

21. Jackelén, "Creativity through Emergence," 109, 189.

22. Jackelén, "Creativity through Emergence," 197.

23. Jackelén, "Creativity through Emergence," 198–232.

24. Jackelén, "Creativity through Emergence," 198.

25. Jackelén, "Creativity through Emergence," 204–6. See also, Jackelén,

On the physical eschatologies of Dyson and Tipler, she cites John Polkinghorne's verdict that they represent "a cosmic Tower of Babel" and "the most extreme *reduction in absurdum* of an exclusively evolutionary optimism."[26] Jackelén's own interpretation focuses on eschatology as a recreation of existence in continuity and discontinuity with the past and writes:

> The dynamic nature of the "already" and the "not-yet" also holds together ethics and expectation, struggle and hope. This dynamic relation is the quintessence of Christian eschatology. It can be reconciled with findings from the field of natural science, but it cannot be derived from them.[27]

In later articles Jackelén has turned towards the biological science and emphasized the significance of concepts of relationality as well as complexity, emergence and self-organization.[28]

Jackelén also underlines that theological hermeneutics is the main contribution of theology to science. It is not a question of intruding the realm of scientific research or what Mikael Stenmark calls "the justification phase of science."[29] But because of "the potential of scientific concepts to build ideology,"[30] there is a role for theological hermeneutics. Jackelén gives several examples of such building of ideologies. "Einstein, for example, was quite unhappy with the ideological use that was made of relativity. Quantum physicists have had reason to complain about the ideological exploitation of the uncertainty principle and the measurement problem." She continues:

> it seems to me that in our days, "complexity" and "emergence" constitute yet another example of concepts that, like a sponge, absorb meaning from different contexts of inquiry, transfer such meaning from one context to another, and pour it into conceptual bowls that may be very different from the original source of the concept. Thus increases their capacity for building ideology.

This "migration of concept" need not be negative. But a critical stance is nevertheless necessary and justified as such a migration often brings false claims of scientific warrants with them.

"Eskatologi—ett naturvetenskapligt ämne?," 43–47.

26. Jackelén, *Time and Eternity*, 207; Polkinghorne, *The Faith of a Physicist*, 165.

27. Jackelén, *Time and Eternity*, 221. I will return to the question of this dynamic in chap. 15.

28. See Jackelén, "Mellan ett reduktionistiskt och teistiskt förflutet och framtidens dynamiska komplexitet," 146–52; and Jackelén, "Creativity through emergence," 69–95.

29. Stenmark, *How to Relate Science and Religion*, 215–16.

30. Jackelén, "What Theology Can Do for Science," 292.

One example could be one of the main results in Part 1, namely, the impact of "biologism" among the Swedish public. Biologism is an interpreting pattern of the human life as a whole. Theological hermeneutics could play an important role in a critical assessment of this and other interpreting patterns. Another example of such theological hermeneutics in action is Anders Jeffner's previously mentioned essay.[31]

Jackelén's methodological approach could be brought into an interesting interplay with Anders Jeffner's theory of fundamental patterns. The basic theories in physics and biology are often interpreted in the form of a non-transcending materialistic pattern, but Jackelén's theological work could be described as a search for an alternative transcending pattern. Jackelén is critical of biologism and suggests that physics and biology are open to an alternative transcending fundamental pattern. Needless to say, it is much more challenging task to show that such a transcending fundamental pattern could find interest among physicists, biologists and astronomers (in their professional capacity). This notwithstanding, it is an important task. Science impacts not only our understanding of the empirical world, but also our understanding of the world as a whole—even if this understanding lies beyond science.

SCIENCE AND THE THEOLOGY OF CREATION —ÅKE JONSSON

One further example of the rising interest within Swedish theology on issues in science and religion is the dissertation by Åke Jonsson studying Gunnar Edman, was a Christian writer and a poet (1915–1995) with broad knowledge in science and strong conviction that a dialogue between science and religion leads to mutual enrichment.[32] The method of Jonsson's dissertation is to clarify Edman's texts with the help of professional theologians engaged in the dialogue between science and religion. These theologians are mainly Arthur Peacocke, John Polkinghorne, Jürgen Moltmann but Jonsson also refers to John Macquarrie and Janet Martin Soskice. The focus is upon the world as the creation of God and human beings as God's co-creator.

THE THEOLOGICAL RELEVANCE OF PHYSICS—ÅSA NORDÉN

A dissertation of a different kind was defended by Åsa Nordén. This philosophical study is right at the center of studies on science and religion, more

---

31. Jeffner, *Biology and Religion*. See further below, chap. 14.
32. Jonsson, *Skapelseteologi*.

precisely on the relevance of physics to theology. Of special significance is the basic thesis of her dissertation, namely, (1) that science is not directly, but indirectly relevant for religious belief. Nordén rejects (2) that science and religion are in conflict, but it is also misleading to say (3) that scientific theories and religious beliefs are directly in agreement. Needless to say, (1), (2) and (3) are vague and general statements. Scientific theories are "indirectly relevant" for theistic—or atheistic—ideas in the sense that they are "based upon metaphysical suppositions which are in some way connected with scientific theories."[33] Nordén illustrates this indirect relevance with reference to three different (allegedly) scientific theories, namely, the theory of Big Bang, the theory of the fine-tuned universe and Quantum-theory.

One of Nordén's main points is that this explains why different philosophers come to different conclusion as to the relevance of the Big Bang-theory for the religious belief that God exists. As an illustration, Nordén takes the disagreement between William Lane-Craig and Quentin Smith. Quentin Smith argues for the strong claim "that big bang cosmology is actually *inconsistent* with theism."[34] The argument is very complicated, but the basic steps are simple. Smith departs from Stephen Hawking's *principle of ignorance*, which states that the Big-Bang-singularities are inherently chaotic and unpredictable. Presuming that God is omniscient, omnipotent, and loving, God wants and can and knows how to create a life-containing universe. But this contradicts Hawking's principle of ignorance. The singularity is unpredictable in principle. Not even God can know what emerges out of it. Hence, God, if God exists, cannot be omniscient.

William Lane Craig has various theological and philosophical misgivings about this argument. They coincide with various counter arguments mentioned by Åsa Nordén. One of the most important is that singularity (1) cannot be described as an empirical event, which is part of a scientific theory and (2) does not belong to the standard theory of Big Bang.[35] Nordén suggests that the standard Big Bang-theory can be framed by an alternative metaphysical assumption that God exists and the universe is dependent upon God. But the cogency of this argument presupposes that (1) the standard Big Bang-theory is more convincing than quantum theories of the Big Bang, and (2) that there are stronger arguments for a theistic metaphysics than for an atheistic metaphysics of singularity. Such an argument is not included in her dissertation.[36] This notwithstanding, her observation that

33. Nordén, *Har nutida fysik religiös betydelse?*, 28, 158.
34. Craig and Smith, *Theism, Atheism, and Big Bang*.
35. Nordén, *Har nutida fysik religiös betydelse?*, 72–74
36. See Nordén, Review of *Har nutida fysik religiös betydelse?*, 53–62.

arguments about science and religion often are framed by metaphysical assumption is of significance.

ALTRUISM AND ORIGINAL SIN IN AN EVOLUTIONARY PERSPECTIVE—EVA-LOTTA GRANTÉN

Eva-Lotta Granténs presented her doctoral thesis in 2003.[37] Granténs takes her starting point is the science and religion dialogue and makes a critical comparison between sociobiological theory and the theological ethics of agapeistic ethics. Sociobiology is a heated subject of debate, but it is not the ideological component of sociobiology that is the object of Granténs research; it is sociobiological science. Furthermore, agape in the form of selfless love is an important part of Christian ethics—even if there is more to Christian ethics than agape. There is also more to Christian ethics than Anders Nygren's thoughts on eros and agape.

The main question of the thesis is formulated in the following way: Does the biologist's description have any relevance at all for a Christian ethical discussion of human altruism, taken as a special quality and directed to everyone?[38] Granténs answers this question in the positive and her dissertation gives the reason for this answer. It has to do with both descriptive (i.e. what is altruism, how is it possible given the basic presuppositions of evolutionary theory) and normative issues (is altruism a good thing?). The method is analytical and comparative.

Granténs suggests that there are three different ways to relate sociobiology to Christian theology. The *first* (conflict) accepts only one of the two theories as valid. Presuming that the two cannot med reconciled, it's a question of *either* sociobiology *or* Christian theology. The *second* (independence) makes a difference between human and non-human altruism: non-human altruism can be explained by sociobiology, but human altruism is a matter of cultural developments. The *third* approach (integration) argues that human altruism has emerged through an interplay between cultural and biological factors. Sociobiology is a necessary, but not sufficient condition to understand human actions.[39] This model is the basis of the so-called integrative model, which Granténs presents in the final chapter of her dissertation. The chapter has the telling title "Beyond the In-Group Border."

The main purpose of the integrative model is to suggest a synthesis of different "normative strategies to move beyond the in-group border."[40]

37. Granténs, *Patterns of Care*.
38. Granténs, *Patterns of Care*, 1.
39. Granténs, *Patterns of Care*, 14–26.
40. Granténs, *Patterns of Care*, 217.

There are three such strategies. The first seeks to move the borderlines, the second to transcend them and the third to transform and be liberated from them. All these three strategies presuppose that sociobiological theories of altruism—or other-directedness—can be related to the Christian tradition of agape. Grantén suggests that the relationship between the different strategies is one of complementarity rather than conflict.

In 2013, Grantén made an important contribution to the study of the Christian doctrine of original sin. She formulated her main question in the following way: "is it possible to develop a theology and an ethic about original sin—with a point of departure in Lutheran tradition—which appears reasonable today, and how would it look like?"[41] She formulates three criteria for such an acceptable theology of original sin:

1. It should be consistent (not contradict) the present scientific view of the world (the criterion of correlation)

2. It should be consistent with and be meaningful to human experience today (the criterion of relevance).

3. It should have reasonable consequences—particularly for children (the pragmatic criterion).[42]

In the present context, I will concentrate on Granténs response as regarding the first criterion of correlation. It is examined in a chapter entitled "Origin," where Grantén enters into dialogue with several contemporary scholars such as Wolfhart Pannenberg, Wentzel van Huyssteen, Philip Hefner, Patricia Williams, and—especially—Ted Peters.[43] Grantén affirms the significance of the traditional doctrine and its emphasis on the ambivalence of human existence, which she describes as "an existential insight."[44] Human existence is filled with happiness, joy, and possibilities as well as suffering, guilt, and death. This insight comes to expression in Lutheran tradition as well as in a universal human experience. It comes to clear expression in the doctrine of original sin.

At the same time, there are elements of the doctrine that must be left behind. The main reason is that they are not in accordance with the criterion of correlation. In agreement with Pannenberg, Granté denies an historical interpretation of paradise and the fall.[45] The stories in the first chapters of

---

41. Granté, *Utanför paradiset*, 29.
42. Granté, *Utanför paradiset*, 27–29.
43. Granté, *Utanför paradiset*, 67–98.
44. Granté, *Utanför paradiset*, 28, 191.
45. Granté, *Utanför paradiset*, 72.

Genesis should be regarded as myths. Death and suffering must be understood with help of the theory of biological and human evolution. The doctrine of humanity as created in the image of God can be reformulated in line with Wentzel van Huyssteen (with reference to Steven Mithen) has suggested and as the outcome of the development of human consciousness when different parts of the brain were connected to a "cognitive flow." It is more difficult to understand natural selection as an expression of God's creation. The problem of evil is sharpened. Death and evil seems to belong to the very fabric of life itself. Granténagrees with Robert John Russell that death and suffering cannot be solved within the context of the doctrine of creation. A solution must be found in Christology and eschatology.[46]

Of special significance for Granténis the Lutheran scholar Ted Peters. His doctrine of universal grace receives support, but Grantén's attention is more directed to Peters' development of Luther's emphasis on the theology of the cross. "An interpretation of creation without understanding it from the cross leads astray."[47] God's presence in the suffering and evil of the evolutionary process is hidden. The message of the cross is not that God relinquishes God's power (kenosis), but rather that God participates in the suffering of his creatures.

Other lines of thought in contemporary theology could be linked to Grantén's reflections on evolution and the problem of evil. One is Christopher Southgate. As Peters and Grantén, he takes Christology and the co-suffering of God with his creatures seriously, but explores further possibilities in the realm of the doctrine of creation. He suggests what has been known as the "only way-argument," i.e. that death and evil is the inevitable outcome of the beauty, diversity and sentience that the biosphere contains.[48]

NEUROSCIENCE AND CHRISTIAN THEOLOGY—ANNE RUNEHOV

Anne Runehov defended her thesis in 2004.[49] As the title reveals, the purpose of the dissertation is to investigate the potential of contemporary neuroscientists to explain religious experiences. Runehov analyses and evaluates the research performed on religious experiences of the Canadian neuropsychologist Michael Persinger and the American neurologist

46. Grantén, *Utanför paradiset*, 79–80.
47. Grantén, *Utanför paradiset*, 89 and 94.
48. Southgate, *The Groaning of Creation*, 16. See further below, chap. 12 and chap. 14.
49. Runehov, *Sacred or Neural?* Republished in a revised version with the subtitle "The Potential of Neuroscience to Explain Religious Experience." The citations refer to the original thesis defended in 2004.

Andrew Newberg and his fellow researcher, the late Eugene d'Aquili. The main questions are the following: in what way and to what extent can neuroscientists explain religious experience? And what are the philosophical conclusions?[50]

In the final chapter, Runehov provides an answer the core questions of her research project. She makes a distinction between two types of explanations, namely, accurate reducing and erroneously reducing explanation. Further, she distinguishes between ontological reductionism and methodological reductionism. "Ontological reductionism has the purpose of an exhaustive explanation of some phenomenon, process or event (the *explanandum*), in terms of another phenomenon, process or event (the *explanans*). Accordingly, something x is explained in terms of nothing but y."[51] Reductive materialism is a form of *ontological* reductionism. Runehov argues that this form of explanation is presupposed by Michael Persinger in his research, but his explanation is erroneous. In contrast, the explanation provided by Andrew Newberg and Eugene d'Aquili is a kind of *methodological* explanation, i.e., reducing "a whole to a specific set of parts for a scientific purpose without claiming that this is all there is to that whole."[52]

Michael Persinger explains experiences of the ultimate Reality or God in a way that *exhaustively* reduces them to a neural activity in the brain (or the underlying biology of a personal crises). This is a form of ontological reductionism. An experience of God is nothing but certain neural processes in the brain. This is an insufficient explanation because "it fails to identify the experiencer's own descriptions of what he or she experiences."[53] Andrew Newberg and Eugene d'Aquili explain religious experience in a non-reductive and methodological way, and suggest that there are other factors besides the neural ones which explain such experiences, for example, cultural factors and, moreover, the actual existence of an Absolute Unitary Being. Expert information (emphasized by Wesley Wildman and Leslie Brothers) would further strengthen this conclusion.[54]

I think that Runehov is correct in underlining that an experience of God or an Absolute Unitary Being cannot be explained as an experience of something real on the subjective authority of the experiencing subject *alone*. The reason for this is that there are many experiences with an extraordinary content, which we know, are illusory. Therefore, we need some

---

50. Runehov, *Sacred or neural?* Abstract.
51. Runehov, *Sacred or neural?*, 232.
52. Runehov, *Sacred or neural?*, 233.
53. Runehov, *Sacred or neural?*, 244.
54. Runehov, *Sacred or neural?*, 253.

method to discriminate between experiences of something real and illusions. Runehov suggests that there is expertise among sociologists, theologians, philosophers of religion, ethicists, and psychologists, who have such a method. But, in fact, it is far from clear that such a method exists and, furthermore, Runehov has not presented it. Moreover, "we cannot base any knowledge-claims concerning the external world exclusively upon extraordinary experiences."[55] The philosophical core question of Runehov's dissertation remains unanswered. But she is right in her conclusion that there are no generally agreed scientific explanations that show that religious experiences have *purely* natural causes. It is possible that God partly explains certain religious experiences and reveals God's self to us humans by these experiences.[56] But showing that this possibility is a reasonable assumption requires an additional argument.

COGNITIVE SCIENCE AND THE RATIONALITY OF RELIGIOUS BELIEFS—LOTTA KNUTSSON BRÅKENHIELM

In December 2016 Lotta Knutsson Bråkenhielm defended her thesis in philosophy of religion on cognitive science of religion (CSR) and its relevance for religious beliefs.[57] Research in CSR is philosophically interesting because the explanations given are used both to explain away and to support the rationality of religious beliefs. The question examined is this: *what relevance should research findings and theories in CSR have on whether it is rational to hold religious beliefs as true?* The concept of "relevance" as used here is taken from Mikael Stenmark: "Science (or some part of science) is *religiously* or *ideologically relevant* if what it does or contains could either support or undermine ideologies, religions, or worldviews."[58]

Two different positions are examined: 1) *negative relevance*: the findings and theories in CSR undermines the rationality of religious beliefs; and 2) *positive relevance*: religious beliefs need not be irrational in the light of CSR. In fact, CSR may actually support the rationality of religious beliefs.

In order to assess whether it is rational for a person to hold religious beliefs as true, Knutsson Bråkenhielm develops a model of rationality starting from the one assumed in the CSR, namely, a *reliabilistic* epistemic theory: a person is rational to hold a certain belief, (1) if this belief can be assumed to have been generated by one or more reliable cognitive mechanisms; (2)

55. See Jeffner, *The Study of Religious Language*, 109.
56. See Jonsson, *Med tanke på Gud*, 328.
57. *Religion—evolutionens missfoster eller kärleksbarn? Kognitionsvetenskaplig religionsforskning och dess relevans för religiösa trosföreställningars rationalitet* [Religion—Evolutions' Freak, or Love Child? Cognitive Science of Religion and Its Relevance for the Rationality of Religious Beliefs].
58. Stenmark, *How to Relate*, 180.

applies whether or not he is aware of what these mechanisms are; but (3) only as long as it does not exist or arise some reasons (defeaters) to question the belief; if they occur, she must (4) reflect on it and find other reasons or grounds to hold the belief in question.

Knutsson Bråkenhielm distinguishes two lines of argument among those who argue for negative relevance. According to the first, the natural causal explanations given by CSR constitute a better option than that people's religious beliefs are caused by God or reflect a religious reality. *Natural explanations are therefore preferable.*

According to the second line of argument, religious beliefs are irrational because they are *caused by unreliable cognitive mechanisms,* such as HADD (hypersensitive agency detection device), i.e., those mechanisms, which have a tendency to greatly cause false beliefs.

Two arguments can be distinguished among those philosophers that argue for the positive relevance of CSR. The first argument argues that *religious beliefs seem to come naturally* to us humans, that is, belief in God is something that, thanks to our cognitive equipment, appears spontaneously and intuitively. This suggests that such beliefs are also probably true or at least *prima facie* rational to hold as true.

The second argument is based on the idea that God provided each human being with a *sensus divinitatis,* a special ability to form knowledge of God. That we are equipped with such a "divine mechanism" is now confirmed empirically by CSR research. There are also reasons to believe that the *mechanisms that generate religious beliefs are reliable* in terms of truth.

The dissertation provides a comprehensive overview of the discussions taking place on the basis of theories and results of the CSR. Knutsson Bråkenhielm arrives at the general conclusion that CSR has negative relevance to particular types of religious beliefs, but not for all. As long as we restrict ourselves to the unreflective beliefs about finite supernatural agents, generated by the unconscious mechanisms, these seem to be explained within CSRs research in ways that have a negative relevance to the rationale of keeping them as true. On the other hand, if we distinguish between beliefs that arise spontaneously and are not tested, and faith that is subjected to reflection and philosophical scrutiny, it is more problematic. The origin and characteristic feature of the former might be explained by the CSR, but it is not likely that CSR can provide an explanation of faith in the latter sense. CSR interest is based largely on unconscious cognitive processes. Explanations of such processes can be given. However, these explanations are not sufficient for an assessment whether the beliefs are rational to hold. In that respect, CSRs explanatory power is too limited.

Part 3

Main Models in the Dialogue between Science and Religion

INTRODUCTION TO PART 3

A recent Swedish book on the "come-back" of religion concludes with a conversation between three academics (an islamologist, a philosopher and a minster in the Church of Sweden) on the concept of religion, the need for religion, and religion and science with a professor of history as a moderator.[1] The discussion is (inadvertently?) an echo of David Hume's *Dialogues concerning Natural Religion* (1779) between Philo, Demea and Cleanthes. They were in a fundamental disagreement with each other, and so are the contemporary discussants. The part on science and religion is a good example. The *philosopher* and atheist argues that there is a fundamental and irreconcilable conflict between science and religion. One example is the biblical story that God created world in six days and that the first humans were Adam and Eve. Another example is belief in immaterial souls. The *minister* disagrees. There is no conflict, because religion is not about truth-claims. A religion with truth-claims is fundamentalism. The biblical story of creation is an expression of reverence for life. The Bible is one of many languages. It is about phantasy, intuition, the feeling of things that cannot be touched. The *islamologist* argues that there are different conceptions of the relationship between Islam and science. One position is that there is a permanent conflict, the second that religion and science are independent and separate, and the third is that religion and science—for the most time, at least— have lived in harmony. God is present in nature and is known through nature.

1. Björkman and Jarrick, *Religionen tur och retur*, 171–88.

Most Muslims are opposed to the first model, the second and third has been historically dominating, but the third is presently gaining momentum.

The conceptions of the islamologist are closely related to the three conceptions discussed in the first two parts of the present book. They will be further explored from a philosophical perspective in this third and last part. The distinction can be derived from various sources, but many of them are inspired by Ian Barbour's taxonomy. My primary reference will be the development of Barbour's proposal by Mikael Stenmark. Here is a short recap of his distinctions.

Conceptions of the relation between science and religion exist in the tension between what can loosely be described as (1) *the model of independence/separation* and (2) *the model of contact*. The model of independence interprets science and religion as two totally separated spheres with no contact or overlap. The model of contact/interplay acknowledges that science and religion interact with each other, come into interplay, and affect the shape of each other. Mikael Stenmark underlines that a model of contact can be focused on (a) harmony, (b) consistency, (c) tension, or (d) conflict. Emphasizing c) tension or d) conflict between science and religion does not amount to the idea of a deep-seated conflict, impossible to resolve. Such a more radical idea is advocated by proponents of a different model, namely, (3) the *model of incompatibility*, which states that there is an irreconcilable conflict between science and religion.

For the purpose of my analysis, some further clarifications are required. The first has to do with the very concept of model. It is a slippery concept used in everyday language as well as science and philosophy. In the present context, it is most fruitful to seek the place of the concept in cognitive science. Here it is used at different levels of thought. First, it might be used about ordinary cognition. In this sense science may be thought of as one model and religion as another. Once again, it is fruitful to link up with Steven Horst and his idea of mental models and their significance in human cognitive structure. Steven Horst writes:

> One feature of models that I emphasize in this book is that models are idealized. One way in which they are idealized consists simply in the fact that each model is a model of some specific domain and brackets everything else about the world. But most models are also idealized in the deeper sense that they provide good-enough ways of thinking and reasoning about their domains for some particular set of purposes without these being exact, unidealized, or context-neutral ways of representing their targets.[2]

---

2. Horst, *Cognitive Pluralism*, loc. 195–200.

Later in his book, Horst explains how we use different models in our interaction with the world "not merely simultaneously but cooperatively and in an integrated manner."[3] One way this is done is through *triangulation*, where two models are related to each other. One special case is when contradictions emerge from the use of different models. Such contradictions give rise to abstract questions "about the class of cases in which a model is aptly applied."[4] The result of such "metacognition"—thinking about models and not simply through them—might result in another type of models about the relationship between lower-level models. Horst has much to say about this type of metacognition in the last chapter of his book. He links up with the thought of Stephen Hawking and Leonard Mlodinow about "model-dependent realism" to solve the problem of different scientific models (such as the models of gravitation and the model of quantum physics) which resist smooth integration.

I would add that reflection on the relation between the models of science and religion might be one other form of metacognition, which may result in "metamodels" of the domains of science and religion. "Domain" is here primarily understood epistemologically in the sense of beliefs, but it may also be understood ontologically in terms of the state of affairs, which these beliefs refer to, or pragmatically to practices or actions constituting—for example—science or religion.

Metamodels of science and religion are "good-enough ways of thinking and reasoning" about the domains of science and religion and their relationship. This implies that they are in some sense simplifications open to further clarifications. Such clarifications will be given in the following chapters with focus on the three principal models of the relation between science and religion. (I will continue to use the term "model." even if the term "metamodel" might have been more appropriate.)

Chapter 12 concerns the *model of incompatibility*, i.e., that there is an *irreconcilable* conflict between science and religion. This model may come in many shapes and forms pending on (1) which *issues* science and religion are incompatible, and (2) the *kind* of argument for the incompatibility. The *issues* may have to do with truth-claims about human nature, the emergence of humanity, the emergence of life or the emergence of the universe, but it may also concern incompatibilities between scientific and religious practices. The *kind* of argument for incompatibility can also be different. It can be an argument to the effect that science is *undercutting* religious beliefs or practices or to the effect that science *rebutting* religious beliefs and practices. This distinction will be crossed with another distinction

---

3. Horst, *Cognitive Pluralism*, loc. 2885–86.
4. Horst, *Cognitive Pluralism*, loc. 3118.

suggested by William Alston. It concerns three types of possible conflict areas between science and religion, i.e., between (1) particular scientific results, hypotheses, or theories versus particular religious doctrines, (2) basic presuppositions of science and religious doctrines, and (3) scientific methods and religious practices. This results in six different areas of possible incompatibilities. The chapter provides examples of arguments within these six different areas and how these arguments can be assessed.

Chapter 13 focuses on the *model of independence*. I will distinguish between in two different types of models. *The first* can be describes as a *radical* form. Religion may appear to contain truth-claims, but in fact it does not. Religious beliefs have another "logic" or are part of a particular "context of meaning." Here we find Stephen Jay Gould's theory of nonoverlapping magisteria (NOMA) as well as Anders Nygren and his theory of the "atheoretical" religion. *Secondly*, the independence model comes in a more *modest* form. There is indeed a separation between science and religion, i.e., between religious statements and scientific statements, but this separation does not run as "deep" as Nygren and others propose. Religion does contain truth-claims, but the justification of religious truth-claims is fundamentally different from the justification of scientific truth-claims. One clear proponent of this view is Torsten Bohlin.[5] Bohlin argues that religious beliefs are (epistemically) justified by religious experiences in a way that is analogous to the way ordinary statements are justified by sense experiences. The American philosopher William Alston argues in a similar way, but has a different conception of religious experiences as perceptions and a more elaborated theory of the practices that contribute to the justification of different truth-claims.

Chapter 14 deals with the *model of contact*. Stenmark divides this type into four different subgroups: a) harmony, b) coherence, c) tension, and d) conflict. My focus will be on b) and departs from Robert John Russell's principle of Creative Mutual Interaction (CMI). I will focus on the Christian doctrine of creation in the form of God's initial creation (*creatio originalis*), ongoing creation (*creatio continua*), and final recreation (*creation nova*). There are models of *strong* contact, arguing that the doctrine of God's initial creation can be justified by the same type of evidence that is accepted in science such as the Big bang-theory. Models of *weak* contact suggest that scientific results or theories may confirm (or falsify) the universe or the evolution of life as "ambigouous objects" and thereby open for (or prevent) an explanation by transcending fundamental patterns. The American philosopher Thomas Nagel has suggested a reconsideration of teleological explanations as an alternative not only to contemporary neo-darwinism, but also to theism. But

---

5. See above, chap. 7.

such teleological explanation need not be contrary, but may be complementary to theism, and provide a reinterpretation of the Christian doctrine of God's continuing creation (*creatio continua*). The chapter will conclude with a discussion of the relation between science and *creatio nova*.

Mikael Stenmark has suggested a fourth type of model about the relationship between scence and religion, namely, *models of substitution*. They come in two contrasting types. On the one hand, there are those who argue that contemporary science works under certain presuppositions that should be substituted for theistic or Islamic presuppositions. This is the position of Alvin Plantinga and Mashdi Golshani, respectively. On the other hand, there are those who suggest that religious beliefs should be substituted by scientific statements. Stenmark argues that this is the position of Edward O. Wilson.

Before proceeding, three further things about the interpretation religious beliefs affect their relationship to science and need to be clarified. *First*, and in analogy to moral beliefs, religious beliefs may have a meaning and function that is obscured by their grammatical form. Analysis of the specific of meaning and function of moral beliefs is performed in the special branch of moral philosophy called metaethics. Similar questions are raises concerning religious beliefs. There, for example, are both descriptive and constructive metatheories of religious beliefs. Descriptive interpretations claim to describe what religious statements *actually mean*, while constructive interpretations are about the way they *should* be understood. *Secondly*, religious beliefs may have different localizations. They may have reference to the real world, reference to a fictive world, or reference to the real world via its reference to a fictive world (such as a drama, a novel or a Biblical myth). Anders Jeffner terms such statements "indirect." *Thirdly*, such indirect statements may or may not be translatable to direct statements (sometimes referred to as "irreducible metaphors"). That indirect statements may be untranslatable seems astonishing at first sight. It means that certain statements—such as "humans are wolfs" or "God will judge the living and the dead"—cannot be understood in the ordinary sense. They may be used to evoke and express sudden experiences of insight, but these insights are impossible to convey to persons who have not had the experience. These indirect religious statements are in some way fundamentally inadequate compared to other statements in, for example, weather reports, schoolbooks and new paper articles.[6]

These metatheoretical questions will surface at different points in the following analysis.

---

6. This view comes close to George Lindbeck's conception of religious doctrines as "experiential-expressive"—in opposition to the "cognitive-propositional" view that religious beliefs should be understood as assertions about the nature of spiritual reality, i.e., as direct statements that are basically like ordinary statements about the real world. See Lindbeck, *The Nature of Doctrine*, 17. The only difference is that ordinary statements refer to observable reality, while religious statements refer to a spiritual reality.

12

Models of Incompatibility

I shall now proceed to an analysis of some models of incompatibility. In the strong sense of the word, it implies that science and religion cannot both be true. But "incompatibility" may also cover cases where science provides more or less serious arguments against religion. Historically such conflicts are well known. In the 15th and 16th century, Copernicus and Galileo raised objections against the geocentric worldview. The conflict caused by Darwin's publication *On the Origin of Species* in 1859, is well-known.

William Alston has provided a useful overview of different types of conflict that have been discussed both historically and in the present. He distinguishes between three areas of conceivable "serious conflict" between religion and science:

1. Particular scientific results, hypotheses, or theories versus particular religious doctrines.
2. Religious doctrines and basic assumptions of science, and
3. Scientific and religious methods or procedures.[1]

The difference between (1), on the one hand, and (2) and (3), on the other, is clear. But what about the distinction between (2) and (3)? A reasonable interpretation is that (2) is about what there is, i.e., ontological questions (what are the basic categories of reality?) while (3) concerns epistemological issues (how do we know?).

---

1. Alston, *Perceiving God*, 240

The difference between (1), (2), and (3) can be combined with a second distinction between *undercutting* and *rebutting* defeaters.² Undercutting defeaters tend to weaken the connection between evidence and belief, while rebutting defeaters prevent the evidence from justifying a belief in a more direct and definitive way.³ I will call claims suggesting that science is undercutting religion *weak incompatibilities* and claims that suggest that science is rebutting religion *strong incompatibilities*. A combination of Alston's previously mentioned distinction between the three areas of serious conflicts between science and religion. This results in the following table of six different types of conflict—even if the borderlines between these different types are less distinct than the lines suggest:

Incompatibility between . . .	Weak incompatibility	Strong incompatibility
. . . particular scientific results, hypotheses, or theories versus particular religious doctrines[448]	(1)	(2)
. . . religious doctrines and basic assumptions of science	(3)	(4)
. . . scientific and religious methods or procedures	(5)	(6)

Table 14: Six models of conflict between science and religion[4]

I will now proceed by presenting and discussing some examples of each of these six different types of incompatibilities.

---

2. See Kelly, "Evidence." Pollock, *Contemporary Theories of Knowledge*.

3. Kelly gives the following example: "Thus, evidence which suggests that you are a pathological liar constitutes an undercutting defeater for your testimony: although your testimony would ordinarily afford excellent reason for me to believe that *your name is Fritz*, evidence that you are a pathological liar tends to sever the evidential connection between your testimony and that to which you testify. In contrast, a rebutting defeater is evidence which prevents E from justifying belief in H by supporting not-H in a more direct way. Thus, credible testimony from another source that your name is not Fritz but rather Leopold constitutes a rebutting defeater for your original testimony." See Kelly, "Evidence."

4. I will not take notice of the distinction between religious *beliefs* and religious *doctrines*. The concept of doctrine is obviously narrower that the concept of belief, but that is of no great significance in the present context.

WEAK INCOMPATIBILITY BETWEEN PARTICULAR SCIENTIFIC RESULTS AND RELIGIOUS DOCTRINES (1)[5]

Adherents of a weak incompatibility would argue that such scientific theory provides an argument that diminishes—or undercuts—the justification for belief in God. Paul Draper has forwarded such a weak argument against theism. He presents his argument in the following way (E stands for the theory of evolution, theism for the claim that "an omnipotent, omniscient, and morally perfect entity created the natural world," and naturalism for "the hypothesis that the natural world is a closed system, which means that nothing that is not a part of the natural world affects it."):

> First, I will show that, on the assumption that theism is true, we have good reasons to be surprised by E, reasons that we do not have when we assume that naturalism is true. Second, I will show that, on the assumption that naturalism is true, there are good reasons to expect E, reasons that we do not have when we assume that theism is true.[6]

In short, given that the evolutionary theory is true, one would not expect theism to be true. In contrast, given the truth of evolutionary theory, naturalism comes as no surprise. Draper concludes that "naturalism is much more probable than theism, *all else held equal.*"

Draper's argument has been critically assessed by Alvin Plantinga.[7] Plantinga's case focuses on the *ceteris paribus*-clause, i.e., "*if all else is evidentially equal*, theism is improbable."[8] Plantinga retorts that in fact *not* everything is equal. The existence of life on this planet as well as the existence of conscious beings seeking meaning in life through relationship with God are two phenomena which are comes as no surprise if indeed theism is true.

This argument notwithstanding, it does not diminish the force of Draper's argument *per se*. Is it not true that naturalism *given E* is much more probable than theism, all else held equal? This requires a closer look at E and the consequences of E for theism. Draper describes E in the following way:

> E: For a variety of biological and ecological reasons, organisms compete for survival, with some having an advantage in the struggle for survival over others; as a result, many organisms,

---

5. The number in parenthesis refers to the number in Table 14.
6. See Draper, "Natural Selection and the Problem of Evil." Draper, "Evolution and the Problem of Evil."
7. In Plantinga, *Where the Conflict Really Lies*.
8. Plantinga, *Where the Conflict Really Lies*, 51; my italics.

> including many sentient beings, never flourish because they die before maturity, many others barely survive, but languish for most or all of their lives, and those that reach maturity and flourish for much of their lives usually languish in old age; in the case of human beings and some nonhuman animals as well, languishing often involves intense or prolonged suffering.[9]

Is *E* surprising if theism is true and not if naturalism is true? I would argue that it depends on a number of other assumptions about good and evil. One such assumption has been called the "only way" argument. Christopher Southgate summarizes this argument in the following way:

> I acknowledge the pain, suffering, death, and extinction that are intrinsic to a creation evolving according to Darwinian principles. Moreover, I hold to the (unprovable) assumption that an evolving creation was the only way in which God could give rise to the sort of beauty, diversity, sentience, and sophistication of creatures that the biosphere now contains. As shorthand I call this the "only way" argument.[10]

Is the "only way" argument a valid argument against Draper's weak model of incompatibility? In the following, I will try to contribute to an answer to this question (with a concluding response in chapter 14).

Southgate writes about the *unprovable* assumption "that an evolving creation was the only way in which God" etc. Being an unprovable assumption, it has nevertheless been the object of argumentative reasoning from Augustine to Karl Rahner. The Catholic theologian Nathan O'Halloran has clarified this in an article, where he raises a series of objection to "only way" argument.[11] First, he seeks support in this from Robert John Russell, who argues that the "only way" argument introduces a "semi-Manichean" view of creation, i.e., a creation that is "partially evil at its most fundamental, material level, physics." Secondly, the only way-argument severely questions the omnipotence of God. O'Halloran quotes Russell:

> Why then did God choose to create *these* laws and constants knowing that they would then make Darwinian evolution possible and with it the sweep of natural evil . . . instead could there be another kind of universe in which life evolved *without natural evil*?[12]

---

9. Draper, "Natural Selection and the Problem of Evil."
10. Southgate, *The Groaning of Creation*, 16.
11. O'Halloran, "Cosmic Alienation and the Origin of Evil."
12. Quoted by O'Halloran with reference to Russell, "The Groaning of Creation," 124.

Michael Lloyd also points to the danger of this position for eschatology:

> Any metaphysical construct in which the creation of beings capable of freedom and love was only possible via predation and pain would be likely so to limit the power of God as to reduce eschatology to wishful thinking.[13]

I do not find any of these arguments against the only way-argument convincing. The charge of semi-Manicheanism—i.e. that the only way-arguments makes the universe "partially evil at its most fundamental, material level, physics"—rests on the assumption that there is no distinction between (1) what makes evil possible—contingency and matter—and (2) evil itself, suffering and pain throughout the evolutionary process. But matter and contingency need not be evil *in themselves* (which would be the Manichean position), even if their existence makes evil possible.

Admittedly, the omnipotence of God must be qualified. That is the whole point of the "only way" argument. O'Halloran illuminates this with a quote from Daryl Domning:

> God could no more make an unbreakable physical universe than a square circle. So I think that the physical "evil" of animal suffering is inseparable in principle from even a good creation, just as the real evil of sin is inseparable from a creation in which intelligent creatures have free will. God doesn't need to make excuses for this. It's part of the price we pay for all that is desirable in creation; and taking the package as a whole, it's worth the price.[14]

As for Lloyd's point that the "only way" argument undermines Christian eschatology, it has force only if God's eschatological perfection of creation would be logically impossible in the same sense as God creating life—i.e. Darwinian life—without suffering and pain.

There is, however, a stronger argument against the only way-argument, and it is the claim that the "only way" argument rests on a superficial understanding of evil. O'Halloran refers to Pierre Teilhard de Chardin:

> But is that really all? Is there nothing else to see [beyond this description of evil]? In other words, is it really sure that, for an eye trained and sensitized by light other than that of pure science, the quantity and the malice of evil *hic et nunc*, spread through the world, does not betray a certain excess, inexplicable

---

13. Lloyd, "Are Animals Fallen?"

14. Domning, *Original Selfishness,* 163. Quoted by O'Halloran, "Cosmic Alienation and the Origin of Evil."

to our reason, if to the normal effect of evolution is not added the extraordinary effect of some catastrophe or primordial deviation.[15]

It seems hard to deny this phenomenological analysis of evil. Evil displays "a certain excess" requiring some further explanation beyond the only way-argument. According to O'Halloran, this further explanation is has to do with a primordial fall of a spiritual creature "at the dawn of the creation of the universe." O'Halloran refers to St. Paul and his words about "principalities and powers." Nothing but an diabolical act of free will to do evil is sufficient to explain primordial evil and at the same time exonerate God.

This line of thought has its strengths, but also obvious weaknesses. Moreover, it could be part of a *reductio in absurdum* argument against theism. If theism can be coherent only under the claim that there exists a primordial metaphysical evil, then many would conclude that naturalism has the upper hand.

The health of theism requires some further reflection along the line of the "only way" argument. What is needed is a "theory of evil," which stays clear of mythological language, but at the same time develops an understanding of the "excess" not covered by the only way-argument. In chapter 14 I shall make an effort to contribute to a solution to this problem.

STRONG INCOMPATIBILITY BETWEEN PARTICULAR SCIENTIFIC RESULTS AND RELIGIOUS DOCTRINES (2)

At the outset it must be observed that strong incompatibility between "particular scientific results, hypotheses, or theories" and "particular religious doctrines" more often than not rests on a metatheoretical conception of religious doctrines, namely, that there is no significant difference between religious statements and other statements about reality. To be sure, religious beliefs are often indirect (symbolic, metaphorical, or analogical), but so are ordinary statements about the weather and about human behavior.[16] In other words, religious beliefs involve no *untranslatable* metaphors. They are statements of the same nature as ordinary statements (albeit referring to a supernatural reality).

One example of this position is found in the writings of the earlier mentioned Swedish theologian Harald Eklund. Eklund opposes what he calls the "faith-theological claims of the century," namely that religious

---

15. Teilhard de Chardin, *The Phenomenon of Man*, 312.

16. Models of weak connection allows for indirect *irreducible metaphors*, but this is essentially denied by the adherents of models of strong contact.

belief does not amount to truth-claims. His main argument is empirical. Religious belief as we meet it in concrete life includes elements of theoretical nature. It does not hover in an abstract sphere separated from evidence and argument. One of the main points in Eklund's philosophy of religion was that there is such a theoretical element in religion. His main thesis is that the very concept of belief implies knowledge of exactly the same theoretical meaning as all other knowledge, and that it is the task of doctrinal analysis to assess these claims without dogmatic prejudice. Eklund takes many examples of the object of this assessment from the New Testament, but also from later Christian tradition. Eklund argues that it is a fact that religious thinkers throughout history have tried to give reasons for their beliefs. It can even be questioned if religious beliefs are ever considered as an alternative to knowledge in religious contexts. In the New Testament "one believes because one has seen and experienced new and great things and draws conclusion from these."[17]

But if "one believes because one has seen and experienced new and great things and draws conclusion from these," what conclusions are rationally justified? If there is no difference between religious beliefs and ordinary beliefs, then one should expect that the same kind of evidence can be given for religious beliefs as the ones that that are given for ordinary statements. One could even make a stronger claim, namely, that one *should* give the same kind of evidence for a religious belief as it is given in empirical science.[18] This amounts to the position of "empirical theism" and one of its main proponents, namely, William Paley. The main question in the present context is this: can such evidence be provided?

The answer of many of the so-called new atheists is a clear and univocal "no." And this is what the strong conflict amounts to. In an interview with *Time Magazine*, Richard Dawkins makes the following univocal statement:

> TIME: Professor Dawkins, if one truly understands science, is God then a delusion, as your book title suggests?
>
> DAWKINS: The claim of the existence of God is a *purely* scientific one. Either it is true or not. A universe with God would be completely different from one without.[19]

---

17. Eklund, *Tro, erfarenhet, verklighet*, 48.
18. See Jeffner, *Filosofisk religionsdebatt*, 25–26.
19. My italics. Interestingly, in the same interview Francis Collins responds in the opposite. "TIME: Dr. Collins, you believe that science is compatible with Christian faith. COLLINS: Yes. God's existence is either true or not. But calling it a scientific question implies that the tools of science can provide the answer. From my perspective, God cannot be completely contained within nature, and therefore God's existence is outside

There is no doubt that, for example, the theory of evolution is verified by reliable evidence and not yet falsified, despite our best efforts. But is this theory conclusive evidence that the statement "God created the universe (including life on planet earth)" is false? This is what Dawkins' strong model of incompatibility amounts to. Is it a reasonable model of the relation between science and religion?

In his work *The Blind Watchmaker*, Dawkins departs from an understanding of religious belief derived from William Paley.[20] Paley begins his book in the following way.

> In crossing a heath, suppose I pitched my foot against a *stone*, and were asked how the stone came to be there; I might possibly answer, that, for any thing I knew to the contrary, it had lain there forever: nor would it perhaps be very easy to show the absurdity of this answer. But suppose I had found a *watch* upon the ground, and it should be inquired how the watch happened to be in that place; I should hardly think of the answer which I had before given, that, for any thing I knew, the watch might have always been there. Yet why should not this answer serve for the watch as well as for the stone?[21]

Essential to Paley's answer is the structural difference between object such as the watch and objects such as the stone is that objects such as the watch have a (1) goal-directed structure (i.e. consisting of smaller parts in an intricate interaction for a specific end), which (2) requires a designer. Now, central in Paley's argument is the claim that there is an analogy between things with a goal-directed structure (such as clocks, cars and computers) and certain natural objects (such as the human body, the solar system and the living cell). Based on this analogy Paley concluded that natural objects with a goal-directed structure must have a designer. And he identified this designer with God.

David Hume raised a number of well-known objections against this line of thought.[22] I will not reiterate those arguments in the present context, but focus upon the argument, which Richard Dawkins added. Dawkins argues that evolutionary theory provides us with a better explanation of natural objects with a goal-directed structure than the one that these objects are created by God. And the alternative explanation is

---

of science's ability to really weigh in." Dawkins, "God vs. Science." Collins view would amount to a model of independence or separation.

20. Paley, *Natural Theology*.
21. Paley, *Natural Theology*.
22. Hume, *Dialogues concerning Natural Religion*.

that they emerged through natural selection acting on randomly mutating organisms. Paley's argument from Design is thereby definitely refuted by the theory of evolution. In this sense, " . . . Darwin made it possible to be an intellectually fulfilled atheist."[23]

There are some weaknesses in this line of thought. *First*, Dawkins' argument is an argument of the second order, namely, an argument against an argument—Paley's argument from design. It is possible that there are other arguments for belief in God or that believers are justified to believe in God even if the argument from design is a failure. One failed argument is not the end of the story. *Secondly*, it is a matter of serious doubt if evolutionary theory refutes the argument of design. The argument from design may have many flaws and David Hume has exposed some of the most important. But is it true that the theory of evolution has provided us with an additional argument, let alone a conclusive argument against belief in God? It is not evident. Take one simple example: the first living organism. It is a paradigm example of a natural object with a goal-directed structure. It consists of many different parts that are linked together in an intricate system with the end to replicate itself. It emerged out of non-organic material, so it cannot be explained by evolutionary theory, which explains the development of life, but not the emergence of life itself.

I will return to this argument for the explanatory limits of the theory of evolution in chapter 14. Irrespective of its tenability, there are further reasons to question whether the theory of evolution is a serious blow to belief in God. These doubts concern Paley's and Dawkins' common point of departure, namely, empirical theism and the idea that the existence of God is a scientific hypothesis. In a later publication, Dawkins makes thoroughly clear his position that belief in God is a scientific hypothesis.

> God's existence and non-existence is a scientific fact about the universe, discoverable in principle if not in practice. If he existed and chose to reveal it, God himself could clinch the argument, noisily and unequivocally, in his favor. And even if God's existence is never proved or disproved with certainty one way or the other, available evidence and reasoning may yield an estimate of probability far from 50 percent.[24]

William Paley and Richard Dawkins disagree on the probability of the existence of God. Paley relies on the analogy between goal-directed and natural objects and claims that this makes the existence of God highly probable. Dawkins position is that "I cannot know for certain, but I think God is

23. Dawkins, *The Blind Watchmaker*, 6.
24. Dawkins, *The God Illusion*, 50.

very improbable, and live my life on the assumption that he is not."[25] But this disagreement hides a deeper agreement. Paley and Dawkins share the same point of departure—the existence of God is a scientific hypothesis. Or to put it more precisely; those who affirm belief in God, should give reasons for their affirmation, which are of the same kind as those given in empirical science.[26] This could be described as the *central tenet of empirical theism*. But why should we believe that this central tenet of empirical theism should be accepted?

Admittedly, many arguments in favor of interpreting belief in God in the form of a scientific affirmation have been proposed. Empirical theism subjects religious belief to the stringent canons of science. It has the methodological advantage of providing us with tools to solve the issue and not content ourselves with a lukewarm agnosticism. And—more importantly— it has been claimed that empirical theism is not foreign to religious believers. The most famous example is the Old testament-story of Elijah and the prophets of Baal. Elijah had a hypothesis about God and to verify it he sets up an experiment:

> Elijah went before the people and said, "How long will you waver between two opinions? If the Lord is God, follow him; but if Baal is God, follow him." But the people said nothing. Then Elijah said to them, "I am the only one of the Lord's prophets left, but Baal has four hundred and fifty prophets. Get two bulls for us. Let them choose one for themselves, and let them cut it into pieces and put it on the wood but not set fire to it. I will prepare the other bull and put it on the wood but not set fire to it. Then you call on the name of your god, and I will call on the name of the Lord. The god who answers by fire—he is God. Then all the people said, "What you say is good."[27]

Elijah had a hypothesis about God and to verify it he sets up an experiment. We all know the result of the experiment; the hypothesis of Elijah was verified. The procedure is in accordance with the basic tenet of empirical theism. Possibly, William Paley and Richard Swinburne (to take an example from contemporary philosophy of religion) would not go for such an experiment, but basically they would argue from nature to God as St. Paul did against the heathens in his Letter to the Romans:

---

25. Dawkins, *The God Delusion*, 50–51.
26. See Jeffner, *Filosofisk religionsdebatt*, 25–26.
27. 1 Kgs 18:21–24.

> For the invisible things of him from the creation of the world are clearly seen, being understood by the things that are made, even his eternal power and Godhead; so that they are without excuse.[28]

It is questionable if St. Paul had anything in the form of scientific evidence in mind. There are many forms of epistemic justifications for statements about reality and scientific evidence is just form of evidence. On the other hand, we need not go as far as Peter van Inwagen, when he claims that "for the unprejudiced know that the heavens are quite silent about the glory of God, and that the firmament displays nothing of his handiwork."[29] van Inwagen affirms that God is hidden and that St. Paul relies on a divine revelation for his affirmation. But it need not be the case that God is *totally* hidden, and that human beings by their unaided reason and experience are *totally* incapable of discerning the presence of God in nature.

Secondly, religious faith does not (*pace* Elijah and St Paul) appear as a scientific hypothesis in the lives and reasoning of most religious believers. In religion, it is praiseworthy, to hold on to your belief in face of evidence to the contrary. This is an attitude, which is foreign to the spirit of science. And, thirdly, a more radical critique against conceiving belief in God in the form of empirical theism is that God appears as a thing in the world. This is at odds with basic experiences of transcendence common in all religious traditions. These experiences support religious believers in their faith.

My conclusion is that empirical theism in the form of Elijah and William Paley fails to do justice to major forms of religious belief. Nevertheless, religious belief could be construed as a *non-scientific hypothesis* about reality as a whole. This comes close to arguments presented by Anders Jeffner and John Hick. Religious beliefs are appropriately termed *non-scientific*, because they are hypotheses about a systematically ambiguous reality. It is like an ambiguous situation, which admits many different interpretations—and none of them can conclusively be deemed justified according to canons of science or common sense. Writes Hick:

> It seems, then, that the universe maintains its inscrutable ambiguity. In some respects, it invites whilst in others it repels a religious response. It permits both a religious and a naturalistic faith, but haunted in each case by a contrary possibility that can never be exorcised. And realistic analysis of religious belief and experience, and any realistic defense of the rationality of

---

28. Rom 1:20–21
29. van Inwagen, *The Problem of Evil*, 135.

religious conviction, must therefore start from this situation of systematic ambiguity.[30]

If this position is correct, then empirical theism fails as an appropriate understanding of religious belief. And if empirical theism fails, so do some of the efforts to show that scientific evidence is that which ultimately makes religious beliefs rational or irrational. I write "some of the efforts," because there are others to which I will return shortly.

WEAK INCOMPATIBILITY BETWEEN THE PRESUPPOSITIONS OF SCIENCE AND RELIGIOUS BELIEFS (3)

Another kind of incompatibility may occur between religious doctrines and the basic assumptions of science. The nature and number of such assumptions have been discussed in philosophy of science. One of the most important is *realism*. Realism can be described as a conjunction of an ontological and an epistemological thesis. "To engage in scientific theorizing means presupposing that there is a real world of objective physical reality and that one can, at least to some extent, obtain information about the world that exists independent of the mind."[31] This kind of realism could be extended to a form of scientism, i.e., that there science is the only or most reliable way "to obtain knowledge about the world that exists independent of the mind." Religious belief claims knowledge of that, which find no support in science. Moreover, it does so without any warrant. One representative of this position is Bertrand Russell:

> God and immortality, the central dogma of the Christian religion, find no support in science . . . No doubt people will continue to entertain these beliefs, because they are pleasant, just as it is pleasant to think ourselves virtuous and our enemies wicked. But for my part I cannot see any grounds for either. I do not pretend to be able to prove that there is no God. I equally cannot prove Satan is a fiction. The Christian God may exist, so might the Gods of Olympus, Ancient Egypt or Babylon; but no one of these hypotheses is more probable than any other. They lie outside the region of provable knowledge and there is no reason to consider any of them.[32]

---

30. Hick, *An Interpretation of Religion*, 124. See also Jeffner, *The Study of Religious Language*, chap. 6 (with further references).
31. Moritz, *Science and Religion*, 70.
32. Russell, *Why I Am Not a Christian*, 50–51.

In this quote, Russell does not go as far as to suggest that science falsifies religious beliefs. He merely suggests that religious beliefs go beyond what science affirms of objective physical reality, and "that there is no reason to consider any of them."

Russell's position can be clarified by the concept of methodological naturalism (MN). MN dates back to the medieval philosophy, but it is claimed that the term was coined by religious scholar Paul de Vries in a paper from 1983.

> The naturalistic focus of the natural sciences is simply a matter of disciplinary method. It is certainly not that some scientists have discovered that God did not make phenomena occur the way they do. The original causes or ultimate sources of the patterns of nature are not proper concerns within any of the natural sciences — though they remain a wholesome and legitimate concern of many natural scientists. The natural sciences are limited by method to naturalistic foci. By method they must seek answers to their questions within nature, within the non-personal and contingent created order, and not anywhere else. Thus, the natural sciences are limited by what I call methodological naturalism.[33]

According to de Vries MN "signifies an agreement, on the part of the scientific community, to confine its explanations of phenomena to those which invoke contingent, non-personal factors—and *nothing else*."[34] (Furthermore, he distinguishes MN from *metaphysical naturalism* (MetaN), which simply denies the existence of God not only as guiding principle in science, but as an ontological statement.[35])

The exact nature of MN and the extent to which MN is a presupposition of science has been a matter of debate, into which I shall not enter in the present context.[36] The focus in the present context is weak arguments from presuppositions to science against religious beliefs. Can MN provide us with such an argument?

Ingemar Hedenius has presented such an argument. He speaks of the "methodological atheism of reasonable science,"[37] but for all practical

33. de Vries, "'Naturalism in the Natural Sciences," 289.

34. See Torley, "Is Methodological Naturalism a Defining Feature of Science?"

35. One version of metaphysical naturalism will be considered in the next section (4) of the present chapter on strong models of conflict between religious belief and presuppositions of science.

36. For a clear analysis, see Martin, "Justifying Methodological Naturalism."

37. Hedenius, *Tro och vetande*, 129–30.

purposes his position amounts to MN. Like Russell he does not claim that MN excludes the existence of God in the form of a transcendent reality, but MN is in conflict with the idea of God as *a reality that leaves traces in our world*. Moreover, MN explains the progress of science. As science establishes itself in new areas, it decreases the probability that there is a God active in the universe.

Does the progress of science presupposing MN provide an undercutting argument against religious belief such as belief in God and in God's activity in the world? A positive answer to this question would at least require that MN is a vital factor for this progress. Let us suppose that this is the case. Would this undercut belief in God? I don't think so. MN is a form of reductionism, i.e., limiting the range of explanatory factors for certain specific purposes such as prediction and control. That such purposes are successfully realized by such a reduction is not an argument to deny the existence of other factors. Take the example of a surgeon disregarding the personal properties of her patient and focusing on the patient's heart for the purpose of effectively curing a disease. The success of the surgery is no argument to deny that the patient is a person. Furthermore, MN also implies that moral arguments have no place in science, but the progress of science presupposing MN is certainly not an argument against morality.

One might also question whether MN really is the vital factor in the progress of science. The progress of science is determined by a number of other factors, such the ability of science to successful technological applications. Moreover, MN can simply be interpreted as an epiphenomenal expression of the scientific method requiring careful attention to facts and logical rigor.

Belief in miracles requires special attention. According to Hedenius, MN is compatible with belief in God as creator, but it is incompatible beliefs in miracles. de Vries agrees. MN does not imply that scientists must deny the God as an *ultimate* explanation of natural phenomena (even if belief to God must be a *philosophical* explanation and not a scientific one). But belief in miracles is incompatible with MN. I think that this is correct, but a provision needs to be added. Belief in miracles is incompatible with MN *in so far as miracles involve events in nature, which can be explored or explained by science*. There might be events in nature which cannot be explored or explained by science, and some of these events may be miracles. This is not incompatible with MN.

In another part of *Faith and Reason*, Hedenius raises additional objections against belief miracles. His arguments are taken from David Hume and claims that belief in is compatible with belief in miracles, in the sense of affirmation of occurrences within the physical universe, which are against natural laws. The support for such laws will always be infinitely stronger that any claim about the occurrence of events, which are against them.

It would require a longer analysis to arrive at a verdict of this famous argument. This will not be done in the present context. Suffice to say, that many philosophers and theologians would not be in agreement with Hume on his argument against the impossibility of a rational affirmation of miracles. Some of them would argue that the idea of miracle is incoherent on other grounds. William Alston is not one of them. Moreover, Alston argues that affirmation of miracles would not upset the normal course of scientific activity. Writes Alston:

> . . . since, as it is usually supposed religiously, miracles are most uncommon, the odd miracle would not seem to violate anything of importance for science. It would be quite a remarkable coincidence if a miracle should be among the minute proportion of cases . . . that are examined for scientific purposes . . . Thus it would appear that the belief in miracles fails to contradict anything that is of crucial concern for science.[38]

Alston's argument would remove one argument against the affirmation of miracles, but not Hume's argument. Alvin Plantinga has, however, suggested a serious argument against Hume. The essence of that is that we might have convincing and critically assessed experiences of occurrences, which go against the normal course of nature. Or we might have reports about such experiences, i.e., reports, which resist direct falsification. Are we rationally justified to hold on to these experiences or report? Alasdair MacIntyre has suggested we are as long as no natural explanation is verified.[39] This argument could mitigate the objection to religious beliefs about some, but surely not all, purported miracles.

STRONG INCOMPATIBILITY BETWEEN THE PRESUPPOSITIONS OF SCIENCE AND RELIGIOUS DOCTRINES (4)

There is not consensus on the nature and number of presuppositions of science. William Alston mentions causal determinism, materialism and mechanism, but general principles such as the uniformity of nature could be added. All these principles are in some way or the other embedded in scientific practice. Some of these presuppositions are regarded as evident, while others have been more or less seriously questioned (such as causal

---

38. Alston, *Perceiving God*, 244.
39. MacIntyre, *Difficulties in Christian Belief*, 49–50.

determinism). Furthermore, there are certain presuppositions which different philosopher have argued that they should be added. Ontological naturalism is one example.

One of the most clearly stated cases for ontological naturalism is Ernest Nagel's presidential address to the American Philosophical Association 1954.[40] He summarizes his version of naturalism in two different theses:

> The first is the existential and causal primacy of organized matter in the executive order of nature. This is the assumption that the occurrence of events, qualities and processes, and the characteristic behaviors of various individuals, are contingent on the organization of spatio-temporally located bodies, whose internal structures and external relations determine and limit the appearance and disappearance of everything that happens. . . . Naturalism does not maintain that only what is material exists, since many things noted in experience—modes of action, relations of meaning, dreams, joys, plans, aspirations—are not as such material bodies or organizations of material bodies. What naturalism does assert as a truth about nature is that though forms of behavior or functions of material systems are indefeasibly parts of nature, forms and functions are not themselves agents in their own realization or in the realization of anything else.

In an essay on naturalism, Jaegwon Kim notices likenesses between Nagel's position and Wilfrid Sellars (in a previous essay from 1927.)[41] Both emphasize the space-time causal world and its casual closure, i.e., that nothing outside the space-time casual world—transcendent deities, Kantian noumena, the Absolute, abstract universals—can determine anything within it. Furthermore, Nagel distances himself from reductive materialism. There are other things than material objects such as modes of action, meanings and experiences. But they are secondary emergences out of material things and "not themselves agents in their own realization or in the realization of anything else." This amounts to a denial of "downward causation" and equals epiphenomenalism.

The second thesis contains an explicit affirmation of spatio-temporal world as the only world there is:

> The second major contention of naturalism is that the manifest plurality and variety of things, or their qualities and their functions, are an irreducible feature of the cosmos, not a deceptive appearance cloaking some more homogeneous "ultimate reality"

---

40. Nagel, "Naturalism reconsidered," 7.
41. Kim, "The American Origins," 89.

or transempirical substance, that the sequential orders in which events occur or the manifold relations of dependence in which things exist are contingent connections, not the embodiments of a fixed and unified pattern of logically necessary links.[42]

In this second thesis, Nagel denies the Hegelian absolute as an "ultimate reality" beyond the relative reality of the empirical world. In this he is in obvious agreement with Axel Hägerström. But he immediately parts company with Hägerström with the latter part of the thesis. In contrast to Hägerström, Nagel denies that spatio-temporal reality is a "unified pattern of logically necessary links." Nagel makes a sharp distinction between logic and reality, while the anchorage of logic in reality is one of the cornerstones of Hägerström 's philosophy.[43]

Kim summarizes the essence of Nagel's argument in the following definition of metaphysical naturalism (MetaN):

> The spacetime world is the whole world. The entities, properties, events, and facts in spacetime are all the entities, properties, etc. of the world.[44]

It should be noted that MetaN is compatible with the existence of immaterial spirits, minds and even deities—as long as these are considered immanent to the spacetime world. Interestingly, MetaN would allow for the kind of spiritualism that a substantial minority of Swedes have sympathy for.[45] This would be in contrast to Nagel, who argues for "the existential and causal primacy of organized matter." Nagel would not allow any spiritual entities to be part of reality. Kim considers this claim to be "less than satisfyingly motivated" and "ad hoc." Similarly, many naturalists would consider it questionable to exclude "qualia" and intentional states.[46]

It can be argued that MetaN has to be supplemented with other principles to distinguish itself from any non-naturalistic implications. One such is the principle of casual and explanatory closure (PC) and the other the principle of science. I will deal with the principle of science in the last section of this chapter, but say a few words about PC here. Kim describes PC in the following way:

42. Kim, "The American Origins," 7–8.
43. See chap. 6. It should be noted that Nagel in the beginning of his career held a position close to Hägerström's (and in opposition to Dewey). In a short article from 1930 entitled "Can Logic Be Divorced from Ontology?" Nagel answered with a clear "no" (Pincock, "Ernest Nagel's Naturalism," 8).
44. Kim, "The American Origins," 90.
45. See above chap. 1.
46. Kim, "The American Origins," 91.

> The causal/explanatory closure of the spacetime world: the cause of any entity or event in the spacetime world is itself within the spacetime world; there is no causal intervention from outside the spacetime world. And no explanation of an event in the spacetime world need invoke any phenomenon or agency outside the spacetime world.[47]

Theological systems that allow miracles in the sense of divine causal interventions in the spacetime world are the primary example of systems that are incompatible with PC. But Kim argues that there may be theological systems affirming divine transcendence without suggesting any kind of divine intervention. This would be affirming PC, but denying MetaN. David Armstrong has an argument against this move. Either it is impossible for conceptual reasons for transcendent entities to causally act on the spacetime world, or it is just a matter of empirical fact that they do not. Either way, Occam's razor must be applied to "expunge" these entities. Furthermore, "anything that is acknowledged to exist must be in a causal relation with something in the spacetime world."[48] And if it is, it is itself part of the spacetime world.

I do not find Armstrong's argument convincing. Multiverse theories of the cosmos suggest that there might exist universes causally distinct from our universe. Moreover, one could admit PC and direct divine intervention, while accepting that a divine and transcendent creator could *indirectly* influence our spacetime world through implanting hidden potentialities in spacetime reality, which under specific circumstances might direct the course of events.[49]

Nagel and Armstrong make frequent references to empirical evidence in favor of MetaN or PC. Nagel claims that his first thesis about the existential and causal primacy of matter is "the best tested conclusions of experience." And Armstrong argues that casual closure is a matter of empirical fact.

There are two things that must be distinguished when it comes to these arguments. The *first* concerns the claims for empirical support for naturalism and the *second* concerns the refusal of naturalists to accept contrary evidence. When it comes to the first types of claims naturalism is treated as an adequately verified scientific hypothesis. It is one thing, A, to claim that reality as investigated in science has invariably discovered more or less complex organization of matter. It is another thing, B, to claim that reality as a whole is material or that there is the spacetime world and nothing more. To say that A is conclusive evidence for B, can

---

47. Kim, "The American Origins," 91.
48. Kim, "The American Origins," 92.
49. See further below.

be interpreted as an inductive inference from specific cases to a general principle. But in that case it is —as John Vickers famously shown—a weak or even incorrect inference. From the premise that "that all the swans we seen are white." we cannot conclude that "all swans are white." All swans *we seen* may have been white, but our observations may have been limited. Similarly, from Nagel's stated premise that "we found repeatedly that when we look closely [under which conditions things come into being and pass away] . . . we discover that those conditions invariably consist of some more or less complex organization of matter," we cannot conclude that "organized matter [is the existential and causal primacy] in the executive order of nature." But our "looking closely" might have certain limitations. Nagel's induction is questionable in the same way as it is incorrect to conclude that there is no life in the universe from the premise that we have found no life within 10 000 light year from earth.

The second thing concerns adherents of naturalism and their treatment of contrary evidence. This issue moves us from the ontological to the epistemic constraints on naturalistic metaphysics. We are told that evidence against naturalism—such as religious experience—are rejected not out of naturalistic prejudice, but "because independent inquiry fails to confirm it."[50] But what does such an independent inquiry amount to? Kim suggests that scientific method has two different elements. Science is (1) intersubjective, and (2) nomological. When it comes to alleged experiences of encounters with supersensory powers it is clear that they do not fulfill these scientific criteria of being experiences of something real. They are not intersubjective, at least not in the same way as ordinary sense experiences. And the causal behaviors of "supernatural entities or properties" are not governed by laws—or if they are, they cannot be scientifically investigated. If naturalism not only involves an ontological thesis such as MetaN and/or PC, but also an explanatory thesis, which claims that *only* scientific explanations are acceptable, then supernatural explanations must be rejected.[51] Kim admits that the explanatory thesis not only excludes supernatural entities, but also moral and mental properties, and that this is a continuing problem in contemporary analytic philosophy.

In conclusion, naturalism in the form of Nagel's two theses (supplemented with MetaN and/or PC) should not be treated as a scientific hypothesis. In contrast to his just mentioned inductive inference, this is admitted by Nagel when he writes that naturalism is not proposed as

---

50. Nagel, "Naturalism Reconsidered," 14.

51. Kim, "The American Origins," 95—96. On the epistemic constraints of naturalistic metaphysics, see further the last section in the present chapter.

"competitors or underpinnings for any of the special theories which the positive sciences assert," but as the product of some special philosophical mode of knowing."[52] How is naturalism as "a special philosophical mode of knowing" to be characterized? My proposal is that it should be described as non-scientific hypothesis about reality as a whole. In his theory of fundamental patterns, Anders Jeffner has explained the nature of such non-scientific hypotheses. "The difference between a scientific and a non-scientific hypothesis does not lie in the kind of explanation provided by the hypotheses, rather it lies in the objects they explain." Non-scientific hypotheses explain *ambiguous objects*, i.e., objects of which different people may have exhaustive knowledge, but might still "disagree as to the correct description of these objects."[53] Human nature and the empirical universe as a whole might be examples of such objects.

Following this line of thought, Ernest Nagel and David Armstrong can be interpreted as defending naturalism in the form of a *non-transcending* hypothesis about the world. Reality is basically matter or there is the spacetime—and nothing more. Occam's razor plays a certain role in the argument for such non-transcending hypothesis, but as Anders Jeffner argues we may ultimately arrive at an existential choice between this kind of hypothesis and *transcending* hypothesis such the hypothesis that the spacetime world is the creation of God.

I shall conclude this section with a discussion of an argument found in a well-known book by a another well-known American philosopher, Daniel Dennett. He argues that another form of naturalism is or should be a presupposition of science. Dennett links up with David Hume's *Dialogues*, where Hume formulates an alternative idea to the theism, which could explain the apparent design in and of the universe. Philo proposes a variation of the old Epicurean hypothesis:

> Instead of supposing matter infinite, as Epicurus did, let us suppose it finite. A finite number of particles is only susceptible of finite transpositions: and it must happen, in an eternal duration, that every possible order or position must be tried an infinite number of times. This world, therefore, with all its events, even the most minute, has before been produced and destroyed, and will again be produced and destroyed, without any bounds and limitations. No one, who has a conception of

---

52. Nagel, "Naturalism reconsidered," 8.
53. Jeffner, *The Study of Religious Language*, 122.

> the powers of infinite, in comparison of finite, will ever scruple this determination.[54]

Dennett argues that several interpretations of this presupposition have won further support from by physicists and cosmologists in recent years. Dennett mentions John A. Wheeler, who 1974 proposed "that the universe oscillates back and forth for eternity." Dennett continues:

> . . . a Big Bang is followed by expansion, which is followed by contraction into a big crunch, which is followed by another big bang, and so forth forever, with random variations in the constants and other crucial parameters occurring in each oscillation. Each possible setting is tried an infinity of times, and so every variation on every theme, both those that "make sense" and those that are absurd, spins itself out, not once but an infinity of times.[55]

Dennett argues that this Epicurean universe of eternal occurrence is a world picture, which sits better with Darwin's dangerous idea, than with any other world-picture, let alone the Christian or for that matter any other religious one.[56]

Let me make three comments to Dennett's Epicurean argument. The first concerns the scientific evidence he cites, the second is about the relationship between the Epicurean and the religious hypotheses and the third about Dennett's arguments against the religious hypothesis.

1. Dennett is a proponent of ontological naturalism in the form of the Epicurean hypothesis and claims that there is scientific evidence for this hypothesis. "Consistency and simplicity are in its favor" writes Dennett.[57] But this cosmology has in fact been cast into serious doubt by the physicist (and Nobel laureate 2011) Saul Perlemutter and his colleagues. In 1998 through studies of super novae they discovered that rather than slowing down, the expansion of the universe is accelerating. This could be described as evidence against the Epicurean hypothesis that each and every

---

54. Hume, *Dialogues concerning Natural Religion*, 52.

55. Dennett, *Darwin's Dangerous Idea*, 179. Dennett refers to Wheeler, "Beyond the End of Time."

56. Moreover, Dennett argues that it is hard to see how the existence of God could explain anything. The ultimate metaphysical question—Why is there something rather than nothing?—may not even make an intelligible demand at all. And if it does, the answer "Because God exists" is probably as good an answer as any. "But look at its competition!" Dennett, *Darwin's Dangerous Idea*, 180–81.

57. Dennett, *Darwin's Dangerous Idea*, 179. The theory of multiverse might (admittedly) give an alternative support for Dennett's argument. See further below, chap. 14.

possible setting of matter occurs infinitely many times. The universe has a beginning, but also an end. In a press release, the Swedish Royal Academy of Sciences states:

> For almost a century, the Universe has been known to be expanding as a consequence of the big bang about 14 billion years ago. However, the discovery that this expansion is accelerating is astounding. If the expansion will continue to speed up the Universe will end in ice.

Needless to say, this gloomy outlook challenges traditional Christian eschatology.[58] But it is also at variance with the Epicurean hypothesis as well as with Wheeler's and Dennett's idea of eternal recurrence.

2. What is the relationship between the Epicurean and Christian belief in the world as God's creation? Dennett presupposes that the Epicurean hypothesis implies that the universe is eternal, which the Creation hypothesis denies. But this is not necessarily so. Thomas Aquinas discussed the possibility that the world was eternal and argued that, if so, it was not at variance with the claim that God exists and created the world. The essential point in the belief that the universe is the creation of God, is not that the universe had a beginning or it had its origin in the Big Bang, but rather that it is—eternally, at every moment—kept in existence by God.[59] In fact, Aquinas believed that the universe did have a beginning, but this belief was not based on reason. Aquinas believed that God in the Bible revealed this. But Biblical scholars, for example Gerhard von Rad, have challenged this interpretation.[60]

3. Dennett has another argument against the Creation hypothesis. He suggests that the Creation hypothesis is unintelligible. It is possible that he means that the Creation hypothesis is unintelligible *as a scientific hypothesis*, i.e., that it is not clear what actual or possible facts could count as evidence for or against belief in God or for the world as a divine creation. Here, I would argue, Dennett is correct. Again, I would argue the possibility that the Creation hypothesis is a non-scientific hypothesis. I will elaborate this possibility further in chapter 14.

---

58. See further below, chap. 14, where some further arguments for a universe without beginning and end will be considered.

59. Thomas Aquinas, *De Aeternitate Mundi*, 178–80.

60. von Rad, *Genesis*, 51.

WEAK INCOMPATIBILITY BETWEEN SCIENTIFIC METHOD AND RELIGIOUS PRACTICES (5)

The origin of science and scientific method has been a matter of controversy among historians and philosophers of science.[61] Enlightenment philosophers have traditionally been emphasized, but modern research have brought medieval philosophy and theology to the foreground. Roger Bacon (c. 1220–1292) was a pioneer for observation of the experimental method as was also Islamist thinkers such as Ibn Al-Haytham (Alhazen, c. 965–c. 1040). Later important contributions came from Francis Bacon, John Locke and Isaac Newton.

Historical research into the origin of science has also entered the philosophical discussion on science and religion. One controversial issue concern the significance of Christian belief for the historical emergence of the scientific method. Alfred North Whitehead argued for importance of medieval theology and other Christian theologians have underlined the concept of natural law and its background in the Christian understanding of God. Theologian William Lane Craig underlines the apologetic importance of such historical studies.

> Furthermore, the whole scientific enterprise is based on certain assumptions which cannot be proved scientifically, but which *are* guaranteed by the Christian world view; for example: the laws of logic, the orderly nature of the external world, the reliability of our cognitive faculties in knowing the world, and the objectivity of the moral values used in science. I want to emphasize that science could not even exist without these assumptions, and yet these assumptions cannot be proved scientifically. They are philosophical assumptions which, interestingly, are part and parcel of a Christian world view. Thus, religion is relevant to science in that it can furnish a conceptual framework in which science can exist. More than that, the Christian religion historically *did* furnish the conceptual framework in which modern science was born and nurtured.[62]

Other Christian theologians have suggested similar arguments, for example Alvin Plantinga. He argues that "it is important to see that our notion

---

61. In the following, I am relying on the Wikipedia article on "Scientific Method and Religion." Some critical questions have been raised to this article, but I find it sufficiently reliable for my present purposes.
62. Craig, "What Is the Relation between Science and Religion?"

of the laws of nature, crucial for contemporary science, has this origin in Christian theism."[63]

These statements by Craig and Plantinga are clearly intended to lend support to Christian theism. They suggest that even if science and religion have occasionally come into conflict, there is a deeper harmony between these domains of belief, more specifically, between the scientific method and the Christian worldview. This harmony lends a certain support to the Christian worldview, even if it is not to be regarded as conclusive evidence. I shall call it "the argument from the origin of science."

The argument from the origin of science has been questioned in different ways. Some of these critical questions would amount to a whole-sale rebuttal of the argument, while others would be limited to an undercutting of the harmony between science and religion. I will depart from a weak interpretation of the critical arguments. They may take many different forms. One is to question the historical accuracy of viewing the Christian worldview as the metaphysical "cradle" of the scientific method. Another avenue is taken by those who admit the historical premise of the argument from the origin of science, but question the conclusion. I will present and review these argument in turn.

Bertrand Russell and his book *The Scientific Outook* (1931) can serve as an illustration of the first line of critique. It was through the Arabs that the tradition of civilization was carried on, and it was largely from them that Christians such as Roger Bacon acquired "whatever scientific knowledge the later Middle Ages possessed." Russell suggests that "scientific method, as we understand it, came into the world full-fledged with Galileo (1564–1642), and, to a somewhat lesser degree, in his contemporary, Kepler (1571–1630)."[64] Furthermore, "Galileo questioned both Aristotle and the Scriptures, and thereby destroyed the whole edifice of mediaeval knowledge."[65] This summary of mediaeval history is not evidently true. Later historical research describes a different picture than Russell, and suggests that medieval universities and theology laid the foundation for modern science.[66]

Another route is taken by those who acknowledge the contribution of the Christian worldview to the emergence of the scientific outlook, but question the conclusion of the argument from the origin of science, i.e., that this lend support for the religious beliefs from which science originated. The

---

63. Plantinga, *Where the Conflict Really Lies*, 276.
64. Russell, *The Scientific Outlook*, 22.
65. Russell, *The Scientific Outlook*, 33.
66. See Grant, *The Foundations of Modern Science*, especially chap. 8. Hick comes close to Russell's position in Hick, *An Interpretation of Religion*, 327–29.

origin of ideas must be distinguished from their justification, and the scientific method need not be justified by the context from which it emerged. The scientific outlook could be removed from its origins and transplanted into another context, for example into ontological naturalism. Or it might simply discard a metaphysical transplantation altogether and be substantiated pragmatically by its very success.[67]

I think that this reasoning is largely correct, but that it must be pointed out that if a pragmatic substantiation can be shown to be insufficient, then a metaphysical backing for the scientific method could be required. As discussed in the earlier section, ontological naturalism presents itself as a prime candidate for such a backing. But what about ontological naturalism itself? The problem is that reference to empirical support "may smack of circularity."[68] It has been claimed that Christian theism can justify the scientific method—and the assumption about the uniformity of nature—in a non-circular way.[69] If this is the case, then the case for the argument from the origin of science might be strengthened.

It needs to be added that even if it is the case that the argument from the origin of science has some force, it does not make the historical conflicts between science and religion disappear. But it may transform these conflicts from a matter of conflict between two independent domains into a "domestic" conflict within a religiously determined culture. This places the usual examples of Galileo Galilei and Charles Darwin in a different light.

STRONG INCOMPATIBILITY BETWEEN SCIENTIFIC METHOD AND RELIGIOUS PRACTICES (6)

It is frequently argued that there is a more basic conflict between science and religion. Besides—or behind—the conflict between well-established scientific results, hypotheses, or theories, there is a *strong* conflict between scientific methods and religious practices. This conflict has been mentioned earlier in this chapter, when we discussed Jaegwon Kim and his analysis of the epistemic constrains of naturalistic metaphysics. This idea comes in a related form in the writings of the American scientist Jerry A. Coyne. Coyne argues that these constraints are not part of a naturalistic metaphysics, but of science itself.

---

67. As argued by Sidney Hook and Kai Nielsen. For references, see "Scientific Method and Religion."
68. Coyne, *Faith vs. Fact*, 209.
69. Hanes, "The Presupposition of Science-Based Atheism."

> The different methods that science and religion use to ascertain their "truths" couldn't be clearer. Science comprises an exquisitely refined set of tools designed to find out what is real and to prevent confirmation bias. Science prizes doubt and iconoclasm, rejects absolute authority, and relies on testing one's ideas with experiments and observations of nature. Its sine qua non is evidence—evidence that can be inspected and adjudicated by any trained and rational observer. And it depends largely on falsification. Nearly every scientific truth comes with an implicit rider: "Evidence X would show this to be wrong." Religion begins with beliefs based not on observation, but on revelation, authority (often that of scripture), and dogma. Most people acquire their faith when young via indoctrination by parents, teachers, or peers, so that religious "truths" depend heavily on who spawned you and where you grew up. Beliefs instilled in this way are then undergirded with defenses that make them resistant to falsification. While some religious people do struggle with their beliefs, doubt is not an inherent part of belief, not is it especially prized. No honors accrue to the Southern Baptist who points out that while there is plenty of evidence for evolution, there is none for the creation story of Genesis.[70]

William Alston's has a sanguine response to this kind of reasoning. "[W]e must be careful to distinguish conflict from difference." To be sure, Christian belief cannot be supported in the same way as hypotheses in science, but this does not mean there is a serious conflict or incompatibility. The Christian practice is simply different from scientific practice. One can imagine Coyne responding that the methods of science are "the only way of finding out anything." Continuing this imagined dialogue, Alston retorts that this is just a bit of "epistemological imperialism." "It is not the sort of thing that could itself be established, or even supported, by the procedures of science; it does not fall within the purview of science."[71]

But if Coyne's "epistemological imperialism" does not fall within the purview of science, it might nevertheless fall into the purview of philosophy. This brings us back to the famous faith and reason-debate between Swedish theology and Ingemar Hedenius 1949 and the years that followed. Central to Hedenius was the *principle of intellectual morality*: you should not believe anything, for which there is no rational reason to suppose it to be true. Rational reasons are for Hedenius the same as scientific reasons.[72]

---

70. Coyne, *Faith vs. Fact*, 65–66.
71. Quotes taken from Alston, *Perceiving God*, 242.
72. Hedenius, *Tro och vetande*, 83.

Another term for this position could be "evidentialism" (or more precisely "scientific evidentialism"). One important wording of this principle was formulated by the American philosopher William Clifford: "It is wrong always, everywhere, and for anyone to believe anything on insufficient evidence."[73] The literature on Clifford's principle and evidentialism is immense and this is not the place to dive deep into this issue.[74] A question central in this context is the following: is scientific evidence necessary for rational religious beliefs? Coyne gives a clear and unambiguous answer to this question. It is necessary—and lacking. And not only is necessary scientific evidence lacking. Moreover, we find a big hole: "the absence of evidence when the evidence *should be there*." Coyne takes this point from the well-known astronomer Carl Sagan. Ancient scripture should have given us scientific evidence for God, for instance presented information not known to humans when the sacred texts were written such as unequivocal information about DNA, evolution, and quantum mechanics. Or bright light in the heavens with solemn music and blaring trumpets heard everywhere. Coyne gives many other examples of what would give him scientific evidence for the existence of God. The absence thereof is sufficient to reject the existence of any supernatural beings of powers.

Coyne's argument rests on human ability to discern what it is proper for God (in the Christian sense of the word) to do, i.e., if God exists and is the creator of our world. This ability could and has been thought to be an important part of human rationality. An objection could be raised against Coyne's argument at this point. Belief in this form of human rationality is not part of Christian faith. The Cross of Christ is foolishness.[75] What is proper for God is beyond human comprehension.

Coyne's (and Hedenius') claim that it is reasonable to demand *scientific* evidence for religious beliefs—for example, belief in the existence of God—could, however, be supported in another way. It seems reasonable to argue that science is constituted by methods, which are the most reliable methods for arriving at the truth about the world. We may simply point to the track-record of science, or to its baffling success to reshape our world and improve human life. It has been argued that from this we can conclude that science is in the possession of the most reliable methods there is. And because this is so, the "epistemological imperialism" of science is justified.[76] This leads

73. Clifford, *The Ethics of Belief*, 5.
74. For a good overview, see Chignell, "The Ethics of Belief."
75. 1 Cor 1:18.
76. "Epistemic imperialism" could be described as a value-laden expression of a model of the relation between religion and science which Mikael Stenmark termed the model of substitution, where science occupies the territory of religion.

into a larger philosophical issue, which was at the center of William Alston, namely, the question of epistemic reliability in general.[77]

First, it is necessary to make clear what we mean by saying that a method is reliable. When speaking about science and religion we are not speaking of any kind of practice, but of *doxastic* practices, i.e., "a way of forming beliefs and epistemically evaluating them." Needless to say, religious practices are much more than doxastic practices. Religious practices are mainly *soteriological* practices, i.e., practices aimed at finding and strengthening community with God or the gods. But this does not exclude them being doxastic practices as well. Moreover, realizing peace with God requires at least some kind of reliable knowledge of God. A religious doxastic practice (RP) aims at providing such knowledge, i.e., *"it would yield mostly true beliefs in a sufficiently and varied run of employments in situations of the sort we typically encounter."*[78] Empirical science could be described as a specific sort of sense perceptual doxastic practice (SP)—or the most reliable SP, i.e., the SP that yield more true beliefs than any other SP.[79]

Secondly, we must make clear that the reliability of science pertains to one particular domain, namely, the natural world accessible by our sense perceptions. It is in *this* domain that science is the most reliable SP. To be sure, there is an overlap between this domain and the transcendent world, which concerns RP. This is a source of potential conflicts (for example, miracles), which may be solved by giving priority to SP in all or some of the cases where SP and RP come into conflict. The two extremes are, on the one hand, a weaker kind of epistemic imperialism and an epistemic "two state-solution" on the other.

But there is also a stronger form of epistemic imperialism—i.e. substituting the SP of science for *all* forms of religious RP, including the Christian mystical doxastic practice (CMP). This is the position of Hedenius and Coyne. Christian belief in God is either logically contradictory or empirically unwarranted (dependent on which concept of God is presupposed). Other religious beliefs such as the belief in life after death may not be contradictory, but they are simply scientifically indefensible. This is the case with belief in the resurrection of the body.[80] Moreover, scientific physiology falsifies immortality

---

77. Alston, *Perceiving God*.
78. Alston, *Perceiving God*, 104–5.
79. Alston's basic thesis is that there is no non-circular way to show that SP is a reliable.
80. I will return to eschatological issues towards the end of chap. 14.

of the soul[81] and the resurrection of Jesus is falsified by the well-established natural law that dead persons do not return to life.[82]

In conclusion, it is a matter of continuing discussion how the overlap between the scientific SP and CMP should be managed. But epistemological imperialism in the stronger form—to claim that the scientific SP is methodologically justified to pass a final verdict on CMP or similar religious doxastic practices—is unwarranted.

Two final points are necessary. The first concerns an argument against evidentialism voiced already by William James. His basic argument was that there are situations in which it is reasonable to believe certain things about the world without evidence, let alone scientific evidence. If that is the case and if some of these situations are religious situations, then it is reasonable to entertain certain religious beliefs *without evidence* in these situations. If this is the case, we have a different model between science and religion, i.e., a model of separation.

A similar model of separation emerges if one—as William Alston and Torsten Bohlin—departs from a wider concept of evidence. Scientific evidence is not the only kind of evidence. There is also religious, spiritual or moral experience. This will be an important issue in the next chapter.

CONCLUDING REMARKS

The study of models of incompatibility is a difficult task. Such models are often advanced with a heavy dose of polemics. This was also the case with Ingemar Hedenius in his critique of religion and Christianity. Closer studies of Hedenius reveal that he was at the same time well acquainted with the writings of his opponents and with contemporary theology in general. He was knowledgeable in classical Greek and the New Testament. Such theological knowledge is not always present in contemporary American critique of religion. But there are some exceptions. One is Michael Ruse. Summarizing an article on atheism and science, Ruse concludes:

> One could go on for a very long time tossing up arguments supposedly showing that science refutes religion, and I am sure one could go on an equally long time hitting them out the ballpark.[83]

Ruse states that one can criticize the Christian and even if the argument from evil is difficult to refute "one cannot refute the believer absolutely on ground of science." My conclusion is about the same. Needless to say, my

---

81. Hedenius, *Tro och vetande*, 233–67.
82. Hedenius, *Tro och vetande*, 108–18.
83. Ruse, "Atheism and Science," 86.

assembly of arguments for incompatibility is not exhaustive and much remains to be said about the arguments discussed. But the analysis indicates the presumption that the model of incompatibility in general should be taken with a grain of salt.

13

Models of Independence

One way to avoid conflicts between science and religion is to view the two enterprises as totally independent and autonomous. Each has its own distinctive domain and its characteristic methods that can be justified on its own terms. Proponents of this view say that there are two jurisdictions and each party must keep off the other's turf.[1]

This classical characterization of the model of independence is sharp and distinct. That serves an important purpose, but as an application is made to different thoughts and theories several clarifications are called for. For example, one general distinction could be made between a *radical* separation between science and religion and a more *modest* one. A radical separation is achieved by denying that religion contains any truth-claims at all. This idea has reoccurred in different forms during the last century. One example is Stephen Jay Gould's idea about science and religion as "nonoverlapping magisteria." This will be more closely considered in this chapter. Furthermore, I will revisit Anders Nygren's theory of the atheoretical religion. It could be described as a *logical model* of independence in so far as it excludes even the possibility of a contact or conflict.[2] After the rejoinder

---

1. Barbour, *Religion and Science*, 84.

2. Another way establish such a logical model of independence would be a thoroughgoing application of *methodological* naturalism. But such a principle would result in a one-sided model of independence. Models of independence are usually models of *mutual* independence. It is not only the case that science should be independent of religion but also that religion should be independent of science.

with Anders Nygren, I will reconnect with another Swedish theologian—Torsten Bohlin. He advances a model, which is very different from Nygren's atheoretical idea, but nevertheless of a similar logical character drawing inspiration not from Kant, but from Kierkgaard and his distinction between subjective and objective truth. While affirming Christian faith as containing truth-claims based on religious experience, Bohlin advocates a sharp separation between religious and scientific truth-claims.

Logical models differ from *empirical* models of independence, which allow for the possibility of both contact and conflict, but claim that no such contact or conflict does in fact obtain. The American philosopher William Alston writes: "I will not go so far as to maintain that there *can* be no contradiction between scientific results and the central tenets of Christianity."[3] But Alston finds no such conflict. In contrast to Nygren, Alston's theory of independence implies that a religion such as Christianity involves truth-claims and that such truth-claims are central in Christian belief. But these truth-claims are not undergirded by, nor incompatible with, scientific truth-claims. Alston's model of independence is entwined with his more general theory of doxastic practices, which will be presented and discussed. The chapter will be concluded with a comparison between Alston's and Bohlin's theories of independence and separation.

NOMA—NONOVERLAPPING MAGISTERIA

Models of independence usually depart from a particular interpretation of religious beliefs. These interpretations are advanced on a high level of abstraction and tacitly or explicitly implied in different models of independence. One often cited example is Stephen Jay Gould's theory of Nonoverlapping magisteria (NOMA).

> The net of science covers the empirical universe: what it is made of (fact) and why does it work in this way (theory). The net of religion extends over questions of moral meaning and value. These two magisteria do not overlap, nor do they encompass all inquiry (consider, for starters, the magisterium of art and the meaning of beauty). To cite the arch clichés, we get the age of rocks, and religion retains the rock of ages; we study how the heavens go, and they determine how to go to heaven.[4]

Gould's theory of NOMA is open to different interpretations, most reasonably not interpreted descriptively, but constructively as a theory about how

---

3. Alston, *Perceiving God*, 240.
4. Gould, "Nonoverlapping Magisteria," 16–22.

sentences and how scientific sentences *ought to* be used. Scientific sentences can be used as statements but religious sentences should not be used in that way. They ought rather to be used as moral statements. If they are, we avoid a lot of problems, the most important of which is that we avoid stating things about the world, which are obviously false. Furthermore, scientific sentences should not be used to make moral statements. Gould writes:

> NOMA also cuts both ways. If religion can no longer dictate the nature of factual conclusions properly under the magisterium of science, then scientists cannot claim higher insight into moral truth from any superior knowledge of the world's empirical constitution.[5]

Gould makes an interesting application of this theory in his discussion of the Catholic doctrine of the soul. Gould writes that he does not personally accept this view. But this factual religious belief could be reinterpreted as a metaphorical sentence "expressing what we most value about human potentiality." If sentences about the human soul are reinterpreted as moral sentences, then Gould finds a legitimate and important use for such sentences expressing "all the ethical and intellectual struggles that the evolution of consciousness imposed upon us."

One might ask why Gould thinks that—for example—the Catholic doctrine of the soul is false (used as a factual statement of the world). He gives a clear answer to this. He writes that the Catholic doctrine of the soul is merely "a device for maintaining a belief in human superiority within an evolutionary world offering no privileged position to any creature." One interpretation of this is that the Catholic doctrine of the soul is false because it is incompatible with the theory of evolution. The context shows that Gould has the idea of an infusion of the soul in mind, but is that idea incompatible with evolutionary theory? A version of this theory has been argued by—for example—Sir John Eccles. Eccles has received scant support for his dualistic theory of the mind, but it is not obvious that it is incompatible with evolutionary theory (even if it may be wrong for other reasons).

If the NOMA-theory should be implemented according to Gould's suggestion, many problems would be solved. Religious beliefs would find an unproblematic use as moral statements and science delimited to factual statements. The relation between religion and science would be pacified. But for religion there is an obvious cost—the NOMA-theory implies a reduction of religion to morality and doctrines reinterpreted to moral assertions. A

---

5. Gould, "Nonoverlapping Magisteria," 16–22. The phrase "moral truths" suggests some kind of moral objectivism. But this could be an overinterpretation. Gould might merely suggest that some moral prescriptions are basic and ultimately important.

more sophisticated version of this kind was proposed by the British philosopher R. B. Braithwaite. Religion should mainly be understood as having a moral function supported by certain myths and parables.[6] One critique against Braithwaite's theory is also relevant against Gould's. Belief about God not only strengthen practical behavior with a psychological reinforcement. Beliefs in God also make a certain way of life attractive and rational.[7] If beliefs in God are reinterpreted as moral statements, this background motivation for Christian morality is eliminated.

I have argued that Gould's NOMA-theory is most reasonably interpreted constructively and not descriptively. But there is another version of the model of independence, which is descriptive and not normative. It is time to revisit the theory of atheoretical religion of Anders Nygren.

NYGREN'S THEORY OF ATHEORETICAL RELIGION—A REJOINDER

As noted earlier, there are at least three important differences between Gould and Nygren. First, Nygren does not use the concept of magisteria. Instead he speaks about categories or contexts of experience or meaning. Anders Nygren's basic philosophical outlook is shaped by Immanuel Kant, even if he goes beyond Kant in providing religion with a special a priori. Secondly, Nygren writes that there are (at least) *four* different contexts of experience: the theoretical, the ethical, the aesthetical, and the religious. Each of these is determined by specific presuppositions. The theoretical field (primarily science) is linked to the question of truth, the ethical to the question of goodness, the aesthetical to the question of beauty and the religious to the question of the eternal.

A third difference concerns another aspect of Gould's NOMA-theory. Gould understands his theory as compatible with the interaction between the claims of the magisteria. ". . . in fact, the two magisteria bump right up against each other, interdigitating in wondrously complex ways along their joint border." Gould gives one example of a question that involves both evolutionary facts and moral arguments:

> Since evolution made us the only earthly creatures with advanced consciousness, what responsibilities are so entailed for

---

6. Braithwaite, *An Empiricist's View*. For a discussion and critique, see Hick, *Philosophy of Religion*, 80–84.

7. See further Hick, *Philosophy of Religion*, 83–84.

> our relations with other species? What do our genealogical ties with other organisms imply about the meaning of human life?[8]

This quote is important, because it questions whether it is appropriate to classify the NOMA-theory as a model of independence. It would seem that it is closer to the model of contact, which—in the words of Mikael Stenmark—affirms that while science and religion are fundamentally two different activities, they may nevertheless come into contact with each other and affect each other's content. Or in the words of Ian Barbour:

> . . . science and religion are considered as relatively independent sources of ideas, but some areas overlap in their concerns. In particular, the doctrines of creation, providence and human nature are affected by the findings of science.[9]

Possibly, Gould would not go as far as Barbour in suggesting that "some areas overlap in their concerns." Gould goes beyond the independence model in suggesting an interaction between the two magisteria, but he would not go as far as the integration model (i.e. the fourth model of Barbour). Gould's concern would rather come close to Barbour's third model of dialogue. Questions might occur on the boundary between science and religion. One example cited by Barbour is David Tracy, who mentions ethical uses in science.[10] This comes rather close to the kinds of questions Gould has in mind.

Returning to Nygren, his theory of independence has traditionally received a radical interpretation. The different contexts of meaning have been understood as sharply distinguished from each other. One example is Paul R. Clifford's review of Nygren's *Meaning and Method*: "Nygren invokes philosophy as the handmaid of isolationism, not of dialogue."[11] This verdict is not without justification. In chapter 10 of *Meaning and Method* Nygren makes no compromise in his distinction between the different contexts of meaning. But another voice is speaking through his early writings. Here Nygren is very clear about an overlap between theoretical knowledge and religion:

> . . . the differentiation is not absolute, for there are significant overlaps between religion and the other spheres. We can even say that there is no religion that is not, at the same time, knowledge. This is an important point, for there is a strong tendency today to contest this view; the opponents wish to claim that "religion is not theory, but life." Of course, on its positive side, this is a true

---

8. Gould, "Nonoverlapping Magisteria."
9. Barbour, *Religion and Science*, 100–101.
10. Barbour, *Religion and Science*, 92.
11. Nygren, Review of *Meaning and Method*, 496.

statement. But it is also true that the religious life penetrates into theoretical considerations. For example, religious life expresses itself in conceptions of faith. It would be incorrect to separate these spheres absolutely.[12]

Similarly, there is a significant overlap between religion and morality. "All religion gives expression to an ethical ideal, which ideal is expressed in a distinctiveness form of life." A few lines later Nygren hastens to add that the relationships between the meanings of contexts are secondary to their autonomy. In his later writings, the autonomy of religion completely overshadows its relationship to science and morality.[13]

Concentration on the overlap between science and religion can lead to an *intellectualization of religion*. Religion in the form of Christianity is not an ideology, is not *primarily* doctrine or a set of dogmas, but a total life-relationship, something that concerns "our whole human existence."[14] This total relationship cannot be reduced to feelings or emotions, but Nygren does not clearly acknowledge that reason and understanding also belongs to a total life-relationship.

THE INDEPENDENCE-MODEL OF TORSTEN BOHLIN

Nygren's insistence on the independence and autonomy of religion is shared by other Swedish theologians, but not construed in the same way as Nygren. The model of independence does not come in one piece—there are many varieties.

On the one hand, there is the theory of Hjalmar Lindroth. In his writings during the 1930s, Lindroth leans towards an understanding of Christianity in the form of *credo quia absurdum*. This can be described as a radicalization of Nygren, but without Nygren's idea of religious beliefs as atheoretical. Lindroth affirms that God's love in Jesus Christ is an incomprehensible and impenetrable mystery. God is the Irrational Unknown and the theologian has to reckon with this against all logic and all the comments from reason.[15]

---

12. Nygren, *Anders Nygren's Religious apriori*, 48–49.
13. Nygren, "Tro och vetande."
14. Nygren, "Tro och vetande," 168.
15. As noted earlier in chapter 7, Lindroth later came to take another point of departure.

On the other hand, there is Torsten Bohlin and John Cullberg. Bohlin is in contrast both to Anders Nygren, John Cullberg, and Hjalmar Lindroth.[16] Contra Nygren and Cullberg, Bohlin affirms the significance of objective truth—and contra Nygren the significance of personal religious experience.

Bohlin was heavily influenced by Kierkegaard, but not the Kierkegaard of Karl Barth. Bohlin emphasized Kierkegaard's distinction between objective and subjective truth. But interestingly he understood both kinds of truths as two branches on the same stem of the correspondence theory of truth. The following quotation from Bohlin is representative of his view.

> Christianity rests on the conviction that the innermost of reality is a creating and good will, that the Father of Jesus Christ is love. Is this conviction truth, i.e., does this certitude correspond to a supernatural entity, or is it just a reflection of the wishes and dreams of human heart under duress?[17]

In his dissertation from 1996, Mikael Lindfelt has clarified Bohlin's response to this question.[18] Bohlin's primarily affirms that the fundamental point of departure for Christian theology is an experience of a personal relationship with God. This experience comes to expression in statements about God, the world and humanity. These statements should be understood as truth-claims, even if they come in the form of symbols and paradoxes. Important for Bohlin is that these truth-claims—as far as they are Christian truth-claims—are made in interplay with the Bible interpreted as a divine revelation. Bohlin's theological method is hermeneutical long before the term became fashionable in philosophy. The relation between text and reader is understood as a dialogue, which is determined by the horizons of the text and the reader, which gradually may merge with one another.

From this point of view, it is not surprising that historical science—and not physics or biology—is of primary importance for Bohlin. But it is historical science in conjunction with an affirmation of a transcendent divine reality, which gradually reveals itself in the history of Israel and ultimately in Jesus Christ. *Central to Bohlin is this interplay between the historically transmitted revelation in history and the personal religious experience.* Objective truth is may correct subjective truth and vice versa. This basic theological

---

16. This is clarified in an illuminating article by Jeffner, "Torsten Bohlin som teolog," 21–27.

17. Bohlin, *Den ofrånkomlige*, 60. Quoted by Lindfelt in *Teologi och kristen humanism*, 133.

18. Lindfelt, *Teologi och kristen humanism*, 130–57.

idea qualifies his model of independence. This qualification raises three fundamental problems.[19]

The first concerns the content and function of the subjective religious experience. Lindfelt finds that Bohlin is remarkably silent on the content of this religious experience. Is it empirical or emotional, an experience of value or an experience of mystical unity? Bohlin does not seek to answer these questions. The most that can be said that he seems to have some kind of experience of value in community with God in mind.[20] This experience has the basic function of identifying that which is genuine Christianity in history. The genuinely Christian in history, acts as correcting counter instance to the subjective experiences of the individual Christian. This is an important point for Bohlin.[21] It defines his position against Nygren, but also against the experiential theology of the so-called Erlangen school of theology.[22] Bohlin wants to correct what he takes as the subjective onesidedness of this school and make room for Christian tradition. This brings us to the second problem.

The second problem concerns the nature of the Christian tradition which stand and must stand in an interplay with the personal experiences of the individual Christian. The Christian tradition has its center in the salvation history revealed in the Bible. This revelation is not connected with certain propositions, but rather with certain events in the history of Israel and ultimately in the history of Jesus Christ. There are well-known problems with this theological approach, but I will limit myself to one remark. Historical events understood as revelatory events come with a Christian reading attached. In order for such an interpretation to be credible, a critical reading of Christian interpretations of the Hebrew Bible as well as the New Testament is necessary. Such a critical reading is acknowledged by Bohlin, but his emphasis is upon a subjective interpretation of the historical facts assisted by the personal religious experience. This seems problematic for two reasons. First, it has the flavor of circularity. God's revelation in history is discerned by a personal religious experience, which in turn stands corrected by God's revelation in history. Secondly, it could be argued that the enterprise of an adequate theological explanation the major events in salvation history "require that the facts in question do not come with a Christian

---

19. Lindfelt, *Teologi och kristen humanism*, 148–52.

20. According to Lindfelt, there is a clear influence from Nathan Söderblom on this point. See Lindfelt, *Teologi och kristen humanism*, 106n58.

21. Lindfelt, *Teologi och kristen humanism*, 141.

22. Lindfelt, *Teologi och kristen humanism*, 150.

reading attached."[23] Moreover, such an enterprise could break the circularity between revelation and experience.

A third problem has to do with the *interplay between the historically transmitted revelation in history and the personal religious experience*. Which instance is given priority if there is a contradiction between revelation and personal experience? Lindfelt makes clear that Bohlin opts for revelation. This seems reasonable, but raises the problem of the criterion itself. Revelation is not a static, but a dynamic concept. The understanding of revelation has undergone significant changes over the centuries, for example during the reformation in the 16th century. Revelation may function as an "over-rider-system" in "normal religion," but the criterion does not account for the process of transformative changes.

In sum, religious experience and revelation are the ultimate warrants for Christian faith. These warrants are of another kind than ordinary sensual perceptions, which are the ultimate warrants for truth-claims in science. Bohlin's theology rests on a model of independence, but it is qualified by the reference the God's revelation in history.

ARE RELIGIOUS EXPERIENCES ABOUT REALITY?

At this point, it is profitable to take a step back and consider a fundamental question for Bohlin's model of independence: can specific experiences of transcendence—such as revelatory or mystical experiences—be evidences for religious truth-claims in a parallel way to which observations and perceptions are evidences in science?

There are arguments against the claim that transcendence experiences could provide such evidences. They are uncontrollable and can be used to justify completely conflicting teachings. At this point, it might be illuminating to revisit a more general argument against the evidential value of transcendence experiences formulated by Anders Jeffner:

1. There are people who have had strong experiences of transcendence, but these experiences have not clearly been experiences of something real. In fact, they have shown to be illusions.
2. Those who claim to have knowledge of external reality as a result of transcendence experiences are therefore required to present a method by which one can distinguish between knowledge and illusion.
3. The presented method does not work.

---

23. Alston, *Perceiving God*, 298.

4. Experiences of transcendence are not reliable reasons (evidence) for claims of external reality.[24]

The first two premises—(1) and (2)—are evident. There are many kinds of religious experiences, which are illusory. We can only think of eccentric experiences of spirits, angels, demons, saints and gods, which we do not have any reason to believe being experiences of something real. If anyone has knowledge claims based on such experiences, it is reasonable to call for some sort of method that separates knowledge from illusions. By (3), Jeffner claims that there are no such methods. Strictly speaking, this general verdict cannot be denied. Experiences of transcendence cannot be tested in the sensual perceptions are tested. Still, the American philosopher William Alston has argued that there is at least an analogy between the practices whereby ordinary perceptions are tested and the practices whereby experiences of transcendence are assessed. His elaboration of this analogy involves a specific model of independence between science and religion and encounters similar problems as the ones we met in Torsten Bohlin's theology. To get a grip on this specific model, we need to take a closer look at Alston's general theory.

ALSTON'S THEORY ON RELIGIOUS EXPERIENCE AND HIS ALTERNATIVE MODEL OF INDEPENDENCE

The epistemological background theory of Alston's work is reliabilism. According to reliabilism it is reasonable to have certain beliefs, if they are formed by a reliable process, i.e., a process that maximizes the true beliefs about reality and minimizes false. Sensory Perception Doxastic Practice (SP) is the paradigmatic example of a belief forming habit or mechanism. Admittedly, we cannot avoid a certain circularity. We are forced to rely on at least some of the experiences whose reliability must be presupposed. But this circularity is inherent in all belief-forming processes. This is also the case with Mystical Perceptual Doxastic Practices (MP) and Christian Mystical Perceptual Doxastic Practice (CMP).

MP, for example, is a special form of life, which evokes mystical experiences justifying certain religious beliefs. MP opens a reality beyond ordinary reality and provides evidence that that they are experiences of something real.[25] Moreover, it could be claimed claimed that MP provides a method which could help "to distinguish between knowledge and illusion."[26]

---

24. Jeffner, *The Study of Religious Language*, 109.

25. Alston, *Perceiving God*.

26. Broad, *Religion*, 192. It should be noted that Broad does not think that mystical experiences in any way warrants the belief that the reality experienced in mystical

If I want to assess the epistemic value of transcendence experiences, I must get involved and realize the special way of life taught by mysticism. This includes insulation (to achieve so-called sensory deprivation), repetitive prayer (such as the Jesus Prayer in the Eastern Orthodox tradition), and meditation. The goal of MP can be summarized to that which in Sufism is called the perfected human, i.e., a human being who has fully realized his essential unity with the divine.

This experience of undifferentiated unity is something other than the "normal" Christian experience of God, which Alston characterizes as a "non-sensory appearance of a purely spiritual deity . . . as loving, powerful, . . . good."[27] Let us call them *numinous experiences*.[28] The conditions for numinous experiences are different than the conditions for mystical experiences. They are not based on isolation and one-sided concentration on the soul's communion with God, but rather a way of life in personal communion with other people and with God in worship and prayer. We experience God in the midst of, and not separate from, ordinary reality. In contrast to mystical experiences in a strict sense, these experiences are not ineffable, even if they are communicated through analogies and metaphors.[29]

A main point of Alston is that numinous experiences play a part similar role in the CMP as sensory experiences play in the SP. Numinous experiences may be dim, meager and obscure, but deserves nevertheless to be characterized as perception, i.e., they have a specific content. The main question is: are they sufficient warrant for certain types of truth-claims about God?

Alston gives an affirmative answer to this question and justifies this positive response in two steps. First, he argues that CMP is an adequately designed belief forming practice. There is a similarity between CMP and SP in the *structural* sense. And secondly, he addresses series of objections that CMP is not sufficiently reliable, i.e., that there is a significant analogy between SP and CMP in the *functional* sense.[30]

Alston claims that CMP is truly a belief forming practice. His argument for this is based an analogy between CMP and SP in what Zachariason calls the *structural* sense. To be sure, there are differences between the two doxastic practices, but that is what one might expect, given that in the

---

experiences is personal.
27. Alston, *Perceiving God*, 20.
28. Zachariasson, "A Problem with Alston's Indirect Analogy-Argument," 331.
29. Alston, *Perceiving God*, 29–32.
30. The distinction between structural and functional senses is taken from Zachariasson, "A Problem with Alston's Indirect Analogy-Argument," 335.

one case is about the physical reality, and in the second about God. For example, it is not possible without modification to transfer our criteria of reliable sensory experiences to the criteria of reliable perceptions of God. Despite this, Alston argues that there are sufficient parallels between the SP and the CMP for CMP to be a "real" belief forming practice. Like the SP, it is the socially established, albeit weakened by secularization, but still nonetheless an available and effective practice for the formation of belief. CMP is different from SP in that CMP requires a more active involvement in prayer, worship, confession, and praise. As with SP, a person is often engaged in this practice early in life, but in some cases, it may also occur later in life in a critical way. In both SP and CMP, we accept some perceptions as authentic and genuine. Moreover, both in both SP and CMP there exists "overrider systems" consisting of general knowledge of the larger world and by which one correct different experience. In SP the basic overrider system is science—in CMP it is theology. There are some differences of opinion between the various Christian traditions of how such a theological overrider system should be designed; Protestant theology emphasizes the New Testament scriptures with testimony of Jesus Christ at the center, while the Catholic theology emphasizes the doctrinal tradition of the Church. Both these systems are based on transcendence experiences. These experiences are the starting point for those overrider systems as well as subjected to test by the same overrider systems. In this respect, CMP has a certain circularity. But a similar circularity characterizes SP.

Secondly, there is according to Alston a *functional* similarity between SP and CMP. Both these doxastic practices assist us in revising our beliefs about some object of perception. In short, "one encounters what one has been brought up to expect to encounter."[31] The content of the numinous experience is determined by one's antecedent beliefs. Alston argues that there is no difference here between numinous experiences and ordinary sense-eexpereinces.

> When I look around me I typically make use of my familiar and much used concepts of houses trees, grass, and so on, in recognizing what I see and forming beliefs about them. And so it is with CMP.[32]

On a second charge that a religious perceiver does not learn anything new in her numinous experiences, Alston admits that CMP does not typically alter the major outlines of a person's faith." Nevertheless, numinous

---

31. Alston, *Perceiving God*, 205.
32. Alston, *Perceiving God*, 206.

experiences may provide additional justification for old beliefs deepen our religious understanding, and give us additional "insights into facets of the scheme." Zachariasson is not satisfied with this argument. SP functions critically in a way that CMP does not. "CMP works in a way which systematically excludes the possibility of new information that challenges or corrects received opinion."[33] Nor does he find the explanation from the differences between the nature of SP's and CMP's object of experiences convincing.[34]

Zachariasson's objections undermine the reliability of CMP. How serious are they? Presupposing that the objections concern empirical questions about the lack of flexibility of CMP in comparison to SP, I have three considerations. The first concerns the Swedish psychologist of religion, Hjalmar Sundén, and his theory of role-taking. Zachariasson makes a reference to Sundén. Sundén's theory emphasized the significance of proceeding expectations on subsequent religious experiences. Interestingly, he derived this theory from a more general theory of perception, which underlined the importance of an acquired "frame of reference" for the occurrence of perceptions.[35] Sundén's theory strengthens the analogy between ordinary and religious experiences. New knowledge modifies previous experiences, but radical changes in our acquired frames of reference are rare. Scientific observations are another matter. Through special training in scientific method, expectations of deviating observations are raised. Secondly, some of the empirical findings in the first part of the book is relevant. When asked about the statement "when religious faith and science contradicts each other, religion is always right," 84 percent of a representative sample of the Swedish adult population and 63 percent of the Swedish clergy disagree. It is easy to read too much into these figures, but they suggest what I would call a readiness for religious change in the "overrider-system" of Christian belief. To be sure, this is not a change due to certain numinous experiences. Some other figures are more directly relevant on this issue. In 1986, 399 Swedes were asked about if they had the experience of being close to a powerful, spiritual life-force. 16 percent answered in the positive. In a follow-up question they were asked if this experience had changed their view of life. About half of them answered in the positive. It might be argued that this is irrelevant to the question of the degree of flexibility of the CMP. The figures merely indicate an experience-induced change from a secular to a religious worldview and not within a religious worldview. This may be correct, but I would nevertheless argue that

---

33. Zachariasson, "A Problem with Alston's Indirect Analogy-Argument," 336.

34. In the present context, I will bypass the question about ethical criteria and "fruit tests" as a possible solution to the problems mentioned.

35. Sundén, *Religionen och rollerna*, 47–51.

the figures suggest that Zachariasson's objections from the inflexibility of CMP may not be entirely conclusive.

Summing up Alston's theory, what kind of model of independence emerges? In contrast to Nygren, it could be interpreted as a "soft" model of independence. In the sense of doxastic practices, science and religion are autonomous, Alston writes:

> . . . we are just not in a position to get beyond, or behind, our familiars practices and criticize them from a deeper and more objective position. Our human cognitive situation does not permit it.[36]

But this model of independence is qualified in various ways. Alston underlines that there are "various grounds for criticism of existing practices, as a result of which they can be rationally modified or, in extreme cases, abandoned." As a matter of fact, there are no serious conflicts between science and religion, but Alston does not exclude the possibility. Should conflicts occur, they need not amount to the abandonment of, say, CMP. There is always the possibility of an modification in CMP:s overrider-system.

A COMPARISON BETWEEN BOHLIN'S AND ALSTON'S MODELS OF INDEPENDENCE

At this point, we have sufficient material for a reasonable comparison between Alston's model and Bohlin's. The result can be summarized in the following five points.

1. Science and religion are independent and autonomous. Bohlin writes that faith is "different in kind from exact knowledge and belief in probabilities."[37] Furthermore, Bohlin claims that science can never explain the questions about the nature of reality or about the beginning of life, its meaning and purpose: "they fall beyond its working area."[38] Alston also separates science and religion as being on two different turfs. But for him it is more a matter of fact than of principle. Nonetheless, it is not just an accidental feature of the present situation. There are deeper reasons. "Science is essentially concerned with the structure of the physical universe, including embodied human beings and their social organization." Religion—in contrast—is concerned "to

36. Alston, *Perceiving God*, 150.

37. Bohlin, *Kristen livssyn*, 19. But a few lines later he remarks that religious belief is on the same plane as beliefs in probabilities in so far as certainty about its truth can never be finally established.

38. Bohlin, *Kristen livssyn*, 19–20.

determine what external source, if any, that universe depends on, or what that source is like and how it is related to its creation."[39]

2. The starting point and epistemic basis for Christian belief is awareness of God. But this awareness is described differently. Bohlin refers to Nathan Söderblom's thesis that in the "consciousness of infinity and the urge to the ideal, the wellsprings of religion flows forth upon the earth out of the innermost depths of the Divine being." To this Söderblom adds—and Bohlin would concur—"God reveals himself to us."[40] The difference between Söderblom and Bohlin on the one hand, and Alston on the other, is that Alston emphasizes the perceptual (albeit non-sensory) character of the awareness of God. There is also another difference when it comes to the epistemological framework. Bohlin works from an extended form of evidentialism, which takes a certain form of personal experiences of transcendence as evidence for Christian beliefs. Alston takes his epistemological point of departure in reliabilism. The belief forming practices of Christianity are in the foreground.

3. Both Bohlin and Alston acknowledge the important role of analogy, symbol and metaphor in religious language—even if paradoxes and "tensions" seem to be more in the foreground in Bohlin's thought. Alston has some elaborated thoughts on this and there is an interesting difference between Alston and Bohlin. Bohlin suggests that the metaphors used by Christian believers are irreducible and that they cannot be translated into plain literal language.[41] But Alston explicitly questions this position.[42]

4. An interesting similarity and difference between Bohlin and Alston becomes evident when one looks at the question about the ultimate warrant for religious statements. Alston writes that "[t]he most obvious candidates for a noncircular support for the reliability of MP come from natural theology and revelation."[43] In the section where Alston comments on this statement, Alston is skeptical of natural theology in the form of the traditional philosophical arguments. He argues that natural theology cannot provide any noncircular support for the reli-

---

39. Alston, *Perceiving God*, 241.
40. Bohlin, *Kristen livssyn*, 21–22; Söderblom, *The Nature of Revelation*, 123.
41. Jeffner, "Torsten Bohlin som teolog," 24.
42. This is at the core of an earlier book by Alston, i.e., Alston, *Divine Nature and Human Language*, especially chap. 1 on irreducible metaphors.
43. Alston, *Perceiving God*, 144.

ability of any mystical doxastic practice.[44] In the final section of his book, Alston is more positive to natural theology arguing that "theism is explanatory superior."[45] Bohlin is categorically dismissive of natural theology and draws a sharp line between religion and metaphysics. The relationship between humans and God is a personal relationship and not a relationship between things.[46]

5. When it comes to revelation, there are both similarities and differences between Alston and Bohlin. Initially, Alston is critical of revelation as an ultimate noncircular warrant for a mystical doxastic practice. He writes that "revelation, as a source of information is just one kind of theistic perception."[47] However, in the last section of his book, he is more positive to both natural theology and revelation as providing important contributions to a "cumulative case for religious belief" (together with religious experiences). In this, there is a certain similarity between Alston and Bohlin. Alston makes a distinction between three different forms of divine revelation: (1) messages, (2) divine inspiration, and (3) divine action in history. Bohlin—influenced by Einar Billing and Nathan Söderblom[48]—emphasizes (3). What does it mean to affirm a revelation in the form of divine actions in history? Alston suggests that it means, "accepting certain events in the world as brought about by God for certain purposes." Moreover, this implies that "those events are best explained in this way."[49] Messages from God—i.e. (1)—could support such explanations, but, obviously, it is a tall order to discern such messages and distinguish them from human constructions.[50] Bohlin does not succeed in meeting this order. But it should be noted that Bohlin advances a more informal understanding of revelation. It is discerned in a dialectical hermeneutical process,

---

44. But Alston's reasons may be different than Bohlin: "[E]ven if we can establish the existence and basic nature of God without reliance on MP, how do we get from that conclusion to the informational efficacy of MP? Natural theology operates at a too high level of abstraction to enable us to do this job." Alston, *Perceiving God*, 144.

45. Alston, *Perceiving God*, 298.

46. For references, see Lindfelt, *Teologi och kristen humanism*, 98.

47. Alston, *Perceiving God*, 144–45.

48. Lindfelt, *Teologi och kristen humanism*, 100–112.

49. Alston, *Perceiving God*, 291–92.

50. "However revelation is to be understood, the medium through which it comes to us is so much a part of our ordinary world that we have to asses it, to judge it, to evaluate it—and at once all the problems of starting from out finite knowledge of this finite world which it was intended to bypass come flooding back inescapably." Wiles, *The Remaking of Christian Doctrine*, 24.

where the horizon of the contemporary interpreter gradually merges with the horizon of the Biblical message.[51] Such an elaborated hermeneutical theory is absent in the thought of William Alston.

CONCLUDING REMARKS

The independence or separation model has a long and respectable history, and has a resonance among Swedish clergy and the general public in Sweden. One example is the strong voice of Swedish minister referred to in the introduction to present part of this book. She can be regarded as a proponent of Gould NOMA-model, which has certain similarities to other models eliminating or at least devaluing the significance of truth-claims in religion. I think that such models of separation fail on various grounds. As descriptive models, they are in glaring opposition to obvious facts of how religion works and is understood in history and in the present. As normative models, they are open to other objections. The ethical significance of religious beliefs consists "in the way in which they render a certain way of life attractive and rational."[52] If belief in God is not understood as a truth-claim (possibly, not a scientific truth-claim, but a truth-claim nonetheless), it would be impossible to extract an ethic from, for example, Christian belief.

Alston's and Bohlin's version of the independence-model is another matter. The main distinction is not between religious belief and secular/scientific truth-claims, but between religious truth-claims and scientific truth-claims. Alston cannot find any conflict nor connection between them and Bohlin is openly critical to the whole idea of a connection. That would amount to metaphysics. But other philosophers and theologians are more positive and advance a model of contact in the sense of a creative interaction. Varieties of this model will be further discussed in the next and final chapter.

---

51. Jeffner, "Torsten Bohlin som teolog," 23.
52. Hick, *Philosophy of Religion*, 82–83.

14

Models of Contact

In the last part of this book, I have analyzed two models of science and religion, i.e., models of conflict and models of independence. Different varieties of these models have emerged. When it comes to the model of independence there is the a priori-version of Anders Nygren as well as the empirical model of William Alston. Alston relies heavily on the special characters of the awareness of God, but he and—to some extent also Torsten Bohlin—recognizes the significance of taking a broader spectrum of data into consideration. Both Bohlin and Alston mentions Biblical revelation and Alston broadens the field even further to natural theology as a part of a cumulative argument for Christian belief (even if there is a disagreement on what components are involved in such a cumulative argument).

I concur with this program of forming a cumulative argument. The need for such a program can also be derived from another observation. Certain forms of religious beliefs seem to have a closer relationship to general statements about the world rather than to personal religious experiences about specific events. Examples of such general statements are beliefs that the world is the creation of God and that it ultimately will be recreated by God. These beliefs belong to the Christian doctrine of creation

The doctrine of creation is central in the Christian tradition, but theologians are not in agreement concerning its interpretation. I will begin with a short analysis of one major fault-line in this discussion, and then move on to a presentation of the program of Creative Mutual Interaction (CMI) between science and religion. The main part of the chapter concerns the specific relation between science and the Christian doctrine of creation. It is sometimes interpreted in a *strong* sense, i.e., science being taken as evidence for beliefs about God's creation. But it may also be taken in a *weak* sense, where scientific results and theories are understood as "clues" or (under a specific interpretation) as indicators of such beliefs.

The Christian doctrine of creation may be divided into three parts, i.e., *creatio originalis, creatio continua,* and *creatio nova*.[1] *Creatio originalis* concerns the very beginning of the universe, usually in the form of *creatio ex nihilo*, i.e., creation out of nothing. *Creatio continua* refers to God's constant conservation of the universe as well as God's ongoing creation of new possibilities for the world. *Creatio nova* has to do with the future fulfilment of creation, God's realization of a new heaven and a new earth. These different parts of the Christian doctrine of creation are interpreted in different ways and at times reinterpreted in a way that simply dissolving the distinction altogether. But I will begin with some reflections (1) on one major clash of interpretations concerning the Christian doctrine of creation, and (2) on the general idea of Creative Mutual Interaction between science and theology.

HEILSGESCHICHTE VS. EXISTENTIAL INTERPRETATION

How are cosmological Christian beliefs to be interpreted? The well-known Swedish theologian Gustaf Wingren makes a comparison between Rudolf Bultmann and another German theologian, Oscar Cullmann.[2] Wingren speaks about Cullmann's "long" line of *Heilsgeschichte* (history of salvation) between creation and judgment in contrast to Bultmann's existential interpretation and his concentrated attention to individual existence. Wingren's aligns himself with Cullman and writes:

> The Creation of the world and the Judgment of the world have reference to more than my personal existence. Cullmann's line with its events, which are external to us, cannot be "demythologized" and then incorporated by the individual into himself.[3]

Wingren writes that Cullmann's "long line" cannot be "demythologized. A more appropriate translation from the Swedish original is "should not." But why not?

One strong argument in favor of Bultmann's demythologizing program is that they make a reinterpretation of problematic Christian beliefs about the beginning and the end of the universe. Several other problematic beliefs remain such as sentences about God's action in the world and it is a

---

1. See further, Moltmann, *God in Creation*, 207–11.
2. Wingren, *Creation and Law*, chap. 3.
3. Wingren, *Creation and Law*, 83–84.

matter of discussion if Bultmann stopped short of realizing his program of demythologization.[4] But he clearly insisted on relocalization of Christian cosmological beliefs about the emergence of the universe and its ultimate destiny to the here and now of our individual existence.

Bultmann's program of demythologization is most reasonably interpreted as a constructive proposal. The assets of this proposal are clear, but there is nevertheless one serious shortcoming. It is a proposal too far removed from the ordinary understanding of Jewish, Christian, and Muslim belief. Wingren has related concern.

> It cannot be denied that these events on Cullmann's long time line are distinct events of importance in their own right. If we deny this, faith loses the possibility of being faith.... To deny that God has dealings with the world (i.e. with more than me personally) is to deny the very ground of faith, and ultimately to render impossible the "pro me" to which the Creed points.[5]

I think that there is also a third argument against Bultmann's proposal. And it is that it is intolerably anthropocentric. Past and future, universe and nature fall outside the scope of religious belief. *Recognizing this and taking the move outside the narrow circles of my individual existence, will inevitably bring us in contact with scientific theories and results.*[6]

THE METHODOLOGY OF CREATIVE MUTUAL INTERACTION (CMI)

Many philosophers and theologians have underlined the significance of developing a closer interplay between science and religion. One of the most influential is Robert John Russell and his "methodology of creative mutual interaction" (CMI). One of the basic characteristics of CMI is going beyond "a one-way relation in which theology's sole role was to interpret science."[7] Is it possible that theology could offer "creative suggestions in the form of questions, topics and conceptions of nature which scientists might find helpful in their research?"[8] Russell has gradually developed his methodology and presented the state of the art of his methodology. Russell suggests that theology can indirectly influence science as a whole, but also inspire

---

4. See Bultmann, "Zum Problem der Entmythologisierung," 196.
5. Wingren, *Creation and Law*, 87–88.
6. This is also an argument against Gould's theory of Nonoverlapping magisteria (NOMA).
7. Russell, *Cosmology: From Alpha to Omega*, loc. 460.
8. Russell, *Cosmology: From Alpha to Omega*, loc. 467.

specific research programs in science. In detail, he has applied his interactive methodology to the challenge of Christian eschatology by scientific predictions for the cosmic far future. Russell understands "research program" in the specific sense of Imre Lakatos. He writes that "[a]ll scientific research programmes may be characterized by their 'hard core.'"[9] Encountering contrary evidence, a "protective belt of auxiliary hypothesis" should be articulated and refined. These auxiliary hypotheses may be adjusted or readjusted, or completely replaced. A research program is successful if all this leads to a progressive problem-shift and unsuccessful if it leads to a degenerative problem-shift.

Russell has gathered various articles on his research program in his *Cosmology: From Alfa to Omega* (2008). From the introduction of this volume it is clear that he relies heavily om Lakatos and his philosophy of science. He cites various theologians advocating the theological appropriation of lakatosian methodology (primarily Clayton, Murphy, Hefner, and Peters) and regards CMI as the "culminating approach."

This is not the place to make a detailed exploration and evaluation of Lakatos' philosophy of science, nor of the theological application of his method. But it is significant to draw attention to a recently published article on the theological appropriation of Lakatos by Victoria Lorrimar. It is especially devoted to Philip Hefner's theology, but her main conclusions are relevant also for Russell.

> While a comprehensive exploration of epistemology is beyond the current scope, the unresolved questions concerning the validity of Lakatos' model represent a potential vulnerability in Hefner's work. Various insights drawn from a Lakatosian understanding of theory acceptance, including the notion of explanatory power, are potentially useful for theological enquiry, however the rational component of Lakatos' thought must be tempered with the acknowledgement of the epistemic constraints under which scientific investigation actually occurs. Yet, a rejection of any form of realism along with foundationalism is also troubling. As Hefner himself does not address these questions, and other theologians in favour of applying a Lakatosian framework to theological statements have not supplied a compelling case for doing so, we must read his theological anthropology, dependent as it is on his scientific methodology, with caution.[10]

---

9. Lakatos, *The Methodology*, 48.

10. Lorrimar, "Are Scientific Research Programmes Applicable to Theology?" 198–99.

I would add another reason for caution in this area. And the reason is that theological methodologies inspired by Lakatos invites what could be described as a *customization of science* to fit specific worldviews.[11] Customization of science may take different forms and driven by economic and/or ideological interests. Mikael Stenmark has termed the latter form "worldview-customized science." A *strong form of worldview-customized science* would allow for worldviews and ideologies influencing scientific activity not only in the problem-stating, development, and application phases, but also in the justification phase. Science would be transformed from an autonomous activity into a set of practices driven by corporate, political, ideological and religious considerations. Worldviews would be invited to shape scientific theories about . . .

> . . . IQ and race, sex and gender, free will and consciousness, free markets an economic growth, environmental emissions and planetary weather, genetically modified organisms and global climate change, cloning and stemcells, health and insanity, nuclear power and radioactive waste disposal, evolution and intelligent design, the Big Bang and the multiverse, and about morality and religion . . .[12]

Such a transformation of science would be problematic. And Lakatos' philosophy of science is not sufficiently guarded against pleas for such a transformation.

A *milder form of worldview-customized science* (mWCS) would restrict customization to certain phases of scientific activities. Stenmark has an argument for this. Under certain conditions, religious and ideological considerations would be allowed to shape the problem-stating and the development phases of science. One such condition is that a plurality of ideological and religious points of views are allowed to flourish within the scientific community. Furthermore, the application phase of science should be open to religious influences pending on the cultural context.[13]

When it comes to the justification phase of science, the situation is different. Stenmark is univocally clear. "Science should be religiously neutral in the sense that it ought not to presuppose the truth of any particular worldview, religion, or ideology such as Christianity, marxism, feminism, or

---

11. See further Fuller, Stenmark, and Zackariasson, eds., *The Customization of Science*.

12. Stenmark, "The Customization of Science," 1–2.

13. Stenmark, *Rationality in Science*, 229.

naturalism *in the justification phase.*"[14] Scientific theories should be evaluated by norms internal to science, and only by such norms.

Later in this chapter I will present Thomas Nagel's critique of neo-darwinism and his plea for an alternative neo-naturalistic worldview. This plea could be supplemented with an elaborated research program about teleological explanation in the biological sciences. Such a research program would differ from established research, but it would for this very reason be of considerable interest. A comparison with Russell's program is illuminating and illustrates an important aspect of the present problem. If resources are to be allocated to Nagel's or Russell's research, then research council and universities need to consider these proposals in competition with other research proposals. The criteria of excellence in traditional "normal" science are well established, but what are the criteria of excellence when comparing different research programs working within explicitly new paradigms? This is an underestimated problem by both Nagel and Russell.

Allowing religion, worldviews, and political ideologies more space in the problem-stating, development, and application phases of science, may not only be a problem in theory, but also in practice. Powerful institutions may distort scientific education and narrow the attention to knowledge that questions certain basic religious or ideological belief. This requires institutional reforms. If worldview influenced research programs such as the one Russell and Nagel propose are to be accepted and supported by universities and research councils, scientific education need to extend scientific education about the influence of worldviews and "develop a better and less naïve understanding of how their own and other peoples ideological and religious commitments interact with scientific practice at different levels."[15]

This brings me to a constructive suggestion about the relevance of mWCS in evaluating religious beliefs such as the Christian doctrine of creation. Central to this suggestion is Jeffner's theory of fundamental patterns. In short, *mWCS would be open to a broader rational assessment if certain well-established scientific facts or theories allow for certain transcending fundamental patterns.* In this sense, there is room for an interplay between science and religion, where non-transcending and transcending fundamental patterns would be judged according to their ability to incorporate well-established scientific results or theories. This would amount to a reinterpretation of CMI.

Russell wants to assemble for "vital clues to the new creation." i.e., scientific evidence from special relativity, quantum mechanics, chaos theory

---

14. Stenmark, *Rationality in Science*, 249; my italics.
15. Stenmark, *Rationality in Science*.

and thermodynamics. He argues that science could assist theology in identifying the "elements of continuity" in our universe to the new creation. This could involve a deeper analysis of temporality and relationality, and aspects of nature that physics so far has overlooked. Furthermore, "elements of discontinuity" between creation and new creation could be explored. This could involve deeper insights into "natural evil," which Christian eschatology expects to be no part of new creation.[16]

What would Russell's research program amount to in the perspective of the theory of fundamental patterns? "Vital clues to the new creation" and "elements of continuity" can be interpreted as scientific results and theories which would not be evidence in the scientific sense, but would rather enhance the ambiguity of the universe. Such "clues" would add to the possibility of the transcending fundamental pattern of Christian eschatology. On the other hand, it should not be excluded that scientific research could reduce the ambiguity and make such a transcending fundamental pattern less credible—and a non-transcending fundamental pattern more convincing.

One implication of interpreting Russell's eschatology as a transcending fundamental pattern could be that alternatives to this fundamental pattern would be an important part of an adequate research program. Russell explores and dismisses these alternatives such as the dysteleologies of Bertrand Russell and Steven Weinberg, the "physical eschatology" of Freeman Dyson and others, but also different whole-sale or partial models of separation between science and religion. I do not find a research program dismissing alternatives to its main thesis attractive. The comparative perspective is vital to the health of a program on such central existential issues as Christian eschatology.

Certain other arguments against the CMI-methodology and its relationship to the model of connection have to be considered.[17] The main import of the model of contact (in the form of coherence) is not the epistemic contribution of theology to natural science, but the contribution of science to theology. Advocating the former type of contribution of theology to science, risks going against the Jeffner's earlier mentioned principle of the *primacy of empirically based knowledge*, which states that no other data than sense data are relevant for the truth or falsity of propositions and theories about empirical reality.[18] I agree with Stenmark that religion or other worldviews should not be relevant in the justification phase of science. In this sense, scientific research should be shielded from the influence

---

16. Russell, "Eschatology in Science and Theology," 550.
17. Ward, "Cosmology," 261–63.
18. See above.

of religion and ideologies. But there are legitimate ethical concerns about the selection of scientific problems and acceptable scientific practices. In this sense, religion as well as other worldviews is of significance in (1) the problem-stating phase, (2) the development phase and/or (3) the application phase of science.[19]

Furthermore, Jackelén's plea for a role of theological hermeneutics is significant. Science is increasingly important as a deliverer of interpreting patterns for human life and the world as a whole. This has been exemplified in the first part of this book, especially in Chapter 1 and the influence of biologism in the Swedish public arena. This influence need not be taken as something negative in principle. It is a legitimate human endeavor to try to grasp the ultimate context of our existence. This is achieved by borrowing phenomena from ordinary life and scientific knowledge to assist us in our efforts to understand what kind of world we live in. One problem is that these patterns are not always helpful and do not "fit" ordinary experience and scientific knowledge. Another problem is that this "migration of concepts" bring with the false claims of scientific warrant. It is an important task of theology to reveal and express these ideological shortcomings. It is also a theological task to be of assistance of forming more convincing fundamental patterns, be they religious or secular.

I shall return to Robert Russell's thought on eschatology later in this chapter. First, I will discuss the relationship between science and two other aspects of the Christian doctrine of creation, namely, the doctrine of *creatio originalis* and *creatio continua*.

CREATIO ORIGINALIS AND CONTEMPORARY CSOSMOLOGY

The traditional interpretation of *creatio originalis* asserts that the existence of universe is neither necessary nor by chance. Furthermore, the universe has a single beginning; it is not the case that different parts ultimately came about in different ways. *Creatio originalis* means that the world emerged through a singular event by an act of God. In this section, three questions will be discussed: how is this theological idea related to modern cosmology and the Big Bang-theory? Secondly, what is the theological significance of the so-called fine-tuning of the universe, i.e., that the life-sustaining properties of our universe are dependent upon a surprising combination of initial conditions at the very beginning of our universe? Thirdly, are there any alternatives to the stronger (and not obviously valid) arguments from cosmology, which could be relevant to the doctrine of *creatio originalis*? If

19. See Stenmark, *How to Relate*, 216–17.

so, how could such models of weak contact between cosmology and God's initial creation be construed?

The Big Bang-theory

The origin of the Big Bang-theory is frequently associated with the American astronomer Edwin Hubble, but the honor should be shared with the Belgian priest and physicist Georges Lemaître. In turn, Lemaître relied on data from another American, Vesto Slipher, and ultimately on Albert Einstein and his theory of general relativity.[20] In 1931, Lemaître voiced his proposal that the *universe* expanded from an initial point of infinitesimal density, a "primeval *atom*" or "the Cosmic Egg, exploding at the moment of the creation."[21] But it was over 25 years later—in 1949—that the term Big Bang was coined, originally in a sarcastic remark by Fred Hoyle, who was an adherent of the competing theory of steady state (later abandoned by Hoyle). One of the reasons for Hoyle's resistance to the Big Bang-theory was that it suggested an updated version of the Biblical myth of creation. (Lemaître was himself careful to distinguish between faith and science.)

Robert Jastrow was a well-known American astronomer, who was intrigued by the parallels between the Biblical story of Creation and the Big Bang-theory. He declared himself as an agnostic, but explored the metaphysical implications of the Big Bang in—for example—*God and the Astronomers* (1992). A quote from this expresses his bewilderment about the Big Bang:

> At this moment it appears that science will never be able to raise the curtain on the mystery of creation. For the scientist who has lived by his faith in the power of reason, the story ends like a bad dream. He has scaled the mountains of ignorance; he is about to conquer the highest peak; as he pulls himself over the final rock, he is greeted by a band of theologians who have been sitting there for centuries.[22]

As will be explained shortly, most contemporary cosmologists do not share this bewilderment, nor Jastrow's belief that the Big Bang hides an inaccessible mystery. But let me first give a short insight into a more formal reasoning of the theological relevance of the Big Bang-theory. The purpose is not to give an exhaustive analysis, but to serve as an entry to an exploration into the difficulties of arguing from science to religion.

---

20. Gustafsson, *Svarta hål*, 134–37.
21. Mastin, "Important Scientists."
22. Jastrow, *God and the Astronomers*, 116.

William Lane Craig argues that there is strong justification for belief in God. The Big Bang-theory offers such a justification. "[T]he evidence of contemporary cosmology actually renders God's existence considerably more probable than it would have been without it."[23] More precisely, he elaborates this claim in the form of the following *Kalām* cosmological argument:[24]

1. If the universe began to exist, then there is a transcendent cause, which brought the universe into existence.

2. The universe began to exist.

3. Therefore, there is a transcendent cause which brought the universe into existence.

Craig justifies premises (1) and (2) in the following way:

> I take it that (1) is obviously true. Rather the truly controversial premise is (2). Traditional supporters presented philosophical arguments in support of (2), which, for me, constitute its primary warrant. But they're not the subject of tonight's debate. Rather what's emerged during the twentieth century is remarkable empirical confirmation of the second premise from the evidence of astrophysical cosmogony. Two independent but closely interrelated lines of physical evidence support premise (2): *evidence from the expansion of the universe* and *evidence from the second law of thermodynamics.*[25]

In another debate with the physicist Sean Carroll, Craig defended this argument and the argument from the fine-tuning of the universe. I find many of Carroll's objections to Craig worth serious consideration.[26]

Let's start with premise (1). The only possibility that the universe once began to exist is a transcendent cause, i.e., a cause that is beyond time and space. If the universe once started, then it must have been God who created the universe "out of nothing." Is there no alternative explanation? One such thing might be that the universe arose by chance, i.e., without any cause. According to some interpretations of quantum physics, there are events

---

23. Craig, "God and Cosmology."

24. Emerging from the *Ilm al-Kalām* —Arabic for "the study of words" —in medieval Islamic scholasticism.

25. Craig, "God and Cosmology." A more thorough exploration of the *Kalām* argument is found in Craig, *The Kalām*.

26. Craig and Carroll, "God and Cosmology." Carroll returned to some of these arguments in Carroll, "Does the Universe Need God?" For further critique of the *Kalām* argument, see Pitts, "Why the Big Bang Singularity."

without cause in the subatomic world and the rise of the universe could be another example. In any case, it cannot be excluded that there are such non-causal events. It is an empirical question and, as a matter of fact, it could be that the origin of the universe was such a non-causal occurrence.

Another objection to the premise (1) is that there is a reason for the emergence of the universe, but that is not because God created the world "out of nothing." Rather, it is so that the universe's origin "out of nothing" is a completely natural process. This thought has been developed by the American physicist Lawrence Krauss. It is the main idea of his book *A Universe from Nothing* (2012): "... we exist now and now because of quantum fluctuations in virtually nothing."[27] This is difficult to comprehend and we who are unskilled in physics need help with simplified models and descriptions. First, we need to distinguish between the so-called Standard Model of the Big Bang and more recent Quantum Models. According to the traditional Standard Model everything began with the Big Bang, the so-called "Singularity" where the density and temperature were almost infinite. From this state, time and space developed. But during the last decades this Standard Model has been revised. Ulf Danielsson writes:

> Now we know better. There seems to be something even before the heat hit. . . . we see a new darkness that stretches backwards, a darkness that is a breathtakingly rapid, expansion—inflation— from which our universe was born.[28]

Lawrence Krauss writes that this inflation "before the heat hit" is driven by large, constant vacuum energy. This energy is associated with subatomic events, and the energy of the "false" vacuum is released in the form of heat upon expansion. And it's a completely natural process.

In an aftermath to Krauss' book, Richard Dawkins writes that if Darwin's *Origins of Species* was the hardest hit of biology against supernatural ideas, Krauss' book might be perceived as cosmology's equivalent. The title—*A Universe from Nothing*—should be interpreted exactly as it stands. And what's in there is "devastating." Why "devastating?" Because modern cosmology gives us a natural explanation for the rise of the universe, just as Darwin gave a natural explanation of life.

The American astronomer Alex Vilenkin, like Krauss, believes there is reason to believe that the world had a beginning, without assuming a transcendent cause. Together with two other leading cosmologists—Arvind Borde and Alan Guth—he formulated the so-called BGV Theorem (Borde,

---

27. Krauss, *A Universe from Nothing*, 97.
28. Danielsson, *Mörkret*, 22.

Guth and Vilenkin), which supports this thought. In contrast to Craig, Vilenkin believes that it does not say anything about God's existence. In correspondence with Craig after the discussion in Sydney, he writes:

> Whatever it's worth, my view is that the BGV theorem does not say anything about the existence of God one way or the other. In particular, the beginning of the universe could be a natural event, described by quantum cosmology.[29]

Many physicists and philosophers have objected that the natural explanation of the universe's origin—quantum fluctuations in vacuum—is not the same as nothing in the literal sense of the word "nothing." A distinction must be made between nothing 1, which equals vacuum, i. e. a space without matter in any form, and nothing 2, which is space without both matter and fields of quantum energy. Creation from nothing 1, is the creation of something, or as Krauss himself writes, "a boiling bubbling brew of virtual particles and wildly fluctuating fields." It is something other than creation from nothing 2. It is likely that our universe emerged from through a natural process out of nothing 1. But in that case, the main question is just pushed a step back. Where did these events in nothing 1 come from? The quantum events that caused our universe are, presumably, be governed by laws that also apply in our universe. In a popular science book, Alex Vilenkin writes the following:

> It follows that the laws should be 'there' even prior to the universe itself. Does this mean that the laws are not mere descriptions of reality and can have an independent existence of their own? In the absence of space, time and matter, what tablets could they be written on? The laws are expressed in the forms of mathematical equations. If the medium of mathematic is the mind, does that mean that mind should predate the universe?

Vilenkin immediately withdraws from his question and finishes his book with the following reflection:

> This takes us far into the unknown, all the way in the abyss of great mystery. It is hard to imagine how we can ever get post this point. But as before, this may just reflect the limits of our imagination.[30]

The Big Bang theory usually assumes that time and space occurred at $t = 0$. But the quantum interpretation of the Big Bang makes it natural to imagine

---

29. Craig, "'Honesty, Transparency, Full Disclosure' and the Borde-Guth-Vilenkin Theorem."
30. Vilenkin, *Many Worlds in One*, 205.

that the universe *emerged* in time. The quantum fluctuations that gave rise to our universe arose in another universe and in another space, which preceded ours. There are many theories how such a so-called multiverse could look like. Theories of the multiverse is a main theme in described in the thought of Max Tegmark.[31] He suggests multiverse consisting of four different levels. "Our" universe is on Level I. At Level II is another universe where natural laws can look completely different. The strange world of quantum physics belongs to Level III. Tegmark also calculates a Level IV. It is mathematics's world. Mathematical structures are fundamental in existence (but there are different structures for different universes). Level IV reminds of Platon's world of ideas, but much more extensive. In a radio conversation with Max Tegmark, Ulf Danielsson has proposed another level in the multiverse, namely, Level V, which consists of worlds beyond mathematics, i.e., worlds which are completely unstructured and thus conceivably unthinkable. Physics transforms into metaphysics.

According to another theory, our universe is really a "kick off" of another universe, which in turn blown up from another, etc. The Russian-American cosmologist Andrej Linde has developed a theory of such multiverse, which resembles a cosmic seaweed. Linde describes a world of "self-producing chaotic inflation."[32] Such a multiuniverse theory could also provide a natural explanation for the so-called fine-tuning of the universe. I will return to this idea in below.

How is the premise (2) in Craig's reasoning? He himself is very determined at this point. He writes that there is "a remarkable empirical confirmation of the second premise from the evidence of astrophysical cosmogony." But that's not true. There are many possible scientific alternatives to the perception that the universe began to exist. The first option—a universe without beginning or end—can be illustrated by a well-known anecdote, narrated by Stephen Hawking at the beginning of his well-known book *A Brief History of Time* (1989).

> A well-known scientist (some say it was Bertrand Russell) once gave a public lecture on astronomy. He described how the earth orbits around the sun and how the sun, in turn, orbits around the center of a vast collection of stars called our galaxy. At the end of the lecture, a little old lady at the back of the room got up and said: "What you have told us is rubbish. The world is really a flat plate supported on the back of a giant tortoise." The scientist gave a superior smile before replying, "What is the tortoise

---

31. Tegmark, *Our Mathematical Universe*, chaps. 6, 8, and 12.
32. Linde, "Inititial Conditions."

standing on?" "You're very clever, young man, very clever," said the old lady. "But it's turtles all the way down!"[33]

The old lady still had one point. If you think of the turtles, the essence of her objection can be interpreted as the universe without beginning and end. The universe is endless both in time and space. The thought is bewildering, but it's not logically impossible—or is it? Mathematical infinity is nothing strange, but what about real infinites?

There is a mathematical paradox called Hilbert's hotel (according to his author, German mathematician David Hilbert). It's about a strange hotel with an infinite amount of rooms. A beautiful princess arrives exhausted after a long journey and the receptionist would love to give her a room. But the hotel is unfortunately fully booked. The sign "No vacancies" is switched on. Unfortunately! But after some reflection the receptionist hits upon a brilliant idea. Since the hotel has an infinite number of rooms, just move the guest in room no. 1 to No. 2 and the guest in room 2 to room 3 etc. Then the beautiful princess also gets the place!

Craig uses the paradox to show the impossibility of a universe without beginning and end. If the universe never began to exist, then before the present there existed an actual infinite number of previous events. But Hilbert's hotel shows how absurd this thought is. Actual infinity is an impossibility, whether it concerns a hotel or the universe!

Another way to get away from the idea that the universe has a beginning is that the very concept of "the beginning of the universe" is theoretically without meaning. One must distinguish between Big Bang as (1) a time of creation (*creatio in tempore*) or (2) the time when time began (*creatio cum tempore*)? A common perception within cosmology is (2), i.e., that not only space but also time emerged from the Big Bang. This means that it has no meaning to say that there is moment "before" Big Bang in the sense of $t = 0$. Stephen Hawking refers to St. Augustine (370–430), who had an opinion close to (2).

> What did God do before God created heaven and earth? I'm not responding smoothly to this question. He prepared hell for those who ask such questions. For at no time God had not done anything, because time itself was created by God.[34]

Thus, there was no time before creation. Not only space, but time emerged from the singularity at $t = 0$. This can also be explained as follows. An event begins if there is an event that preceded the first. My life starts on March 28,

33. Hawking, *A Brief History of Time*, 1.
34. Citation in Hawking, "Quantum Cosmology," 651.

1945. That means my life was not there at March 27th. If the universe starts at time t, this means that there was a time t minus, let's say, 2 minutes when the universe was not there. However, according to the Big Bang theory, the universe does not have a beginning in this sense. Time—and space—emerge in connection with the Big Bang. Therefore, it is theoretically meaningless to claim that the universe has a beginning.

There is another and more radical alternative to the thought of the beginning and end of the universe. This so called "no-boundary" proposal has been developed by James Hartle and Stephen Hawking. In an encyclopedia article, the Big Bang is explained as follows.

> . . . the beginning of the universe can be characterized by a simple geometric state, a multi-dimensional spherical space where time has, in some sense, has spatial charcteristics and where a special starting point cannot be identified.[35]

Hawking's "no-boundary" proposal amounts to the idea that the universe is finite but has no boundary (in imaginary time). Space is a more fundamental feature of existence than time. Time is simply a secondary property that occurs because of certain features of space. A simple explanation may be derived from on our usual timeline extended from the beginning of the universe until its end. Now, let the two ends join each other you get a circle, or a closed circular timeline, which is at the same time without limits. It has the strange consequence that a point in time may lie before and after another. All this is hard to understand for those who are not familiar with modern quantum physics and mathematics, but we can recieve some help from St. Augustine. He also thought about the time and the different tempus forms, imperfect, present and future. Augustine found that they exist only in the consciousness and not in reality:

> Perhaps it might be said rightly that there are three times: a time present of things past; a time present of things present; and a time present of things future. For these three do coexist somehow in the soul, for otherwise I could not see them. The time present of things past is memory; the time present of things present is direct experience; the time present of things future is expectation.[36]

If St. Augustine lived today, he might have said that time is a "construction." Hawking comes to a similar conclusion. He compares the universe with a

---

35. See the *Swedish National Encyclopedia*, s.v. "stora smällen".
36. Augustine, *Confessions* 11.20.26. Quoted by Hawking in "Quantum Cosmology," 651.

globe, where the north pole is the Big Bang, which is then expanded to a maximum size at the equator, and shrinks towards the South Pole, the Big Crunch (the Great Crusher). The latitudes are as small as the time "in reality" and neither 90 degrees north nor 90 degrees south. Looking for these degrees is as stupid as looking for the axis of the Earth. Degrees and axis are just our way of constructing the globe.

According to Hawking, the no-boundary proposal has theological consequences. As long as we assumed that the universe had a beginning, we could believe that it had a creator. But if the universe is completely self-sufficient and without borders or edges, then it cannot have a beginning or an end: it would only "be." "What's place, then, for a creator?" Hawking's no-boundary proposal can be interpreted as a refutation of premise (2) in Craig's argument. Universe would "only be." Out of habit, we apply our everyday timeline beginning and end, before and after on the whole universe. But if we are to understand the universe, we need to acquire another way of thinking. It has no meaning to talk about beginning or end. And then the need for a creator lapses and an innovator at the end. Just as in the Hindu creation story, it is all about renewing.

The no-boundary proposal is related to complex philosophical and scientific issues. Particularly difficult is Augustine and Hawking's thought that time is not a fundamental property of physical existence at all. Seemingly, it contradicts to Einstein's general theory of relativity according to which the universe consists of the three dimensions of the space and time as the fourth dimension. On the other hand, physics has not yet been able to find an integration of relativity theory and quantum physics. The no-boundary proposal lacks a coherent theoretical framework.

In sum, an important thought in Jewish, Christian, and Muslim faith is that God created the world "at the beginning" (out of nothing). William Lane Craig assumes that this can be proven as the starting point in two premises: (1) If the universe began to exist, there is a transcendent cause that created the universe, and (2) the universe began to exist. However, both these premises can be questioned in different ways. The universe may have a beginning, but it does not have to be transcendent. It may also be that the universe does not have a beginning. There may be no beginning, or it is meaningless to talk about both beginning and end. The discussion shows, among other things, that it is difficult—and often outright misleading—to apply our everyday concepts to the issue of the universe's origin. The problem is well-known from theology. When we use our everyday concept of God, it is easy to end up in contradictions and obscurities. Symbols and metaphors are inevitable both in theology and physics. But everything in physics and theology cannot be symbols (at least not in the same sense).

"God is the Maker of heaven and earth" can be a symbolic statement. If Hawking is right, this symbolic statement cannot be interpreted as the fact that God once created heaven and earth because it has no meaning to speak of a beginning (and an end). In that case, the creatio originalis must be reinterpreted. The German-American theologian Paul Tillich said that there is only one non-symbolic statement about God: God is the ground of being. Hawking believes that the universe "only is." But Hawking does not comment as a physicist. The question of the universe "exists only" or if the universe is because God is beyond physics. But if God is not found in the world of physics, what kind of world is there room for God? The belief in God as a foundation is a central point in the teaching of *creatio continua*, the ongoing creation. If Hawking is right—and it is not at all certain—then perhaps the doctrines of the creation at the beginning and of God's ultimate renewal of the world must be reduced to the doctrine of God as the ground of being. This is not a scientific hypothesis, but a transcending fundamental pattern. It is way of expressing our experience and knowledge in a way that points beyond these experiences and knowledges. God is not part of a scientific theory, but part of a more comprehensive interpretation of the world. The atheist and theist could have knowledge of the same scientific laws and theories, but still disagree with the design. The question is this: is it possible that quantum cosmology and multiuniverse theory along with the description evolutionary theory of the origin of life into a pattern that points beyond this world? That question cannot be answered in general. There is a large room for philosophical and theological responses, but ultimately it is a matter of personal and subjective decisions.[37]

The fine-tuning argument

The so-called fine-tuning argument is the claim that many of the initial parameters of our universe appear to be carefully adjusted to a narrow range, which makes life possible. God initiated the universe in such a way that life, consciousness and human beings should arise. Craig attaches significant importance to this argument. God is the "fine-tuner" of the universe. He suggests that there is a science-based justification for the conclusion that the universe is designed and presents the argument as follows.

1. The fine-tuning of the universe is due to either physical necessity, chance, or design.
2. It is not due to physical necessity or chance.

---

37. Jeffner, *The Study*, 128–31.

3. Therefore, it is due to design.[38]

I shall briefly consider these three statements in turn, but first it should be observed that there are number of questions about the phenomena of fine-tuning in the first place. The physicist, astronomer, and philosopher Victor J. Stenger argues that "(m)any of the examples of fine-tuning found in theological literature suffer from simple misunderstandings of physics."[39] The examples Stenger provides may support his thesis, but there may be other instances of find-tuning which do not amount to such an misunderstanding. The American philosopher Kelly James Clark writes that "[s]cientists have collected over two-dozen cases of fine-tuning."[40] He claims would be surprising if all of these are "simple misunderstandings" especially in view of the fact that many trained and renowned cosmologists affirm that fine-tuning is real.

A more serious argument against (1) is that Craig limits the number of explanations to "physical necessity, chance, or design." There are at least two other options. First, fine-tuning might indeed be a fact, but we do not know what different alternative explanations are possible, nor which one of these is true. We do not know at present and/or we may not come to know in the foreseeable future, if ever. Second, fine-tuning might have come out of nothing. Theologians and philosophers since antiquity have claimed that *ex nihil, nihil fit*, i.e., out of nothing comes nothing. But this is denied by some cosmologists and most notably by Lawrence Krauss. The central claim of his book is that that the universe originates from "density fluctuations" or "quantum fluctuations" in this "quantum nothingness."[41] Clark cites an interview where Krauss summarizes his claim: "Not only can something arise from nothing, but most often the laws of physics require that to occur."[42] Nevertheless, there is no consensus on this among cosmologists, let alone among philosophers.[43]

What about the design option? What does it amount to? Craig is surprisingly brief on this central point. From what he writes, the design option simply amounts to an empirical hypothesis that the world has a creator or designer. There are well-known problems with this. First one might ask if belief in a creator or designer is an *empirical* hypothesis in the first place. And, if so, why exclude a design hypothesis in the sense of the world being

   38. Craig, "God and Cosmology."
   39. Stenger, *God—the Failed Hypothesis*, 145.
   40. Clark, *Religion and the Sciences of Origins*, 192.
   41. Albert, Review of *A Universe from Nothing*.
   42. Clark, *Religion and the Sciences of Origins*, 194.
   43. Albert, Review of *A Universe from Nothing*.

created by a whole ensemble of gods—or by a single god, but not the one affirmed by theism?

Let us grant that Craig might have justification for limiting his options to necessity, chance or theistic design. But why is fine-tuning from necessity or chance unlikely alternatives? This brings us to (2). In short, Craig dismisses the necessity option on the ground that constants and quantities are independent of the laws of nature, and the chance hypothesis because it would require a plurality of universes, a multiverse, which is highly improbable. There are many cosmologists that would not concur with this analysis, but let us suppose that Craig has the arguments on his side.

This brings us to the real problem, namely, the step from (1) and (2) to (3) and the conclusion that the fine-tuning is due to design. As already explained, (3) amounts to a version of empirical theism. I have already expressed a number of arguments against such an interpretation of religious belief. Kelly Clark has more to say:

> Belief in God is not . . . a scientific theory held tentatively or not at all until the available evidence piles up to confirm God's existence. Theism is not a scientific theory in competition with other scientific theories such as evolutionary theory. Even if evolutionary theory were well supported by the evidence, rational belief in God would not be precluded by it. Of course, various religious believers such as young earth creationists and Intelligent Design theorists do conceive of God as a scientific hypothesis in competition with evolutionary theory; such believers do, indeed, have a problem.[44]

Arguing from a fine-tuning universe to God in the way Craig does, certainly looks like treating belief in God as a scientific hypothesis. There are serious arguments for the theologian to tread carefully in these areas. Science changes and what kind of concept of God is implied in the conclusions?

Interestingly, Clark takes the argument from fine-tuning very serious. Nevertheless, he avoids treating belief in God as a scientific hypothesis. In a footnote, Clark, clarifies that "none of this should suggest that theism should be treated as a scientific theory that makes predictions about our universe or multiverse. Theism is not a scientific theory. But it does lead us to expect a life-sustaining universe."[45] But how is it possible to affirm that belief in God as the creator of the universe is a non-scientific hypothesis and at the same time seek for confirmation of the expectations this hypothesis raises? The answer to this question is that a non-scientific hypothesis raises

---

44. Clark, *Religion and the Sciences of Origins*, 6.
45. Clark, *Religion and the Sciences of Origins*, 254.

another kind of expectations than a scientific theory. We need to revisit the theory of fundamental patterns.

The theory of fundamental pattern and models of weak connection

Basil Mitchell—a well-known British philosopher of religion—once in dialogue with Antony Flew and Richard Hare conceived the parable of the stranger and the partisan. The main purpose was to illustrate how trust in God can be sustained even when evidence or experience points to the contrary (such as suffering and evil in human history):

> During the time of a war a Partisan meets a stranger claiming to be the leader of the resistance. The stranger urges the Partisan to have faith in him, even if he is seen to be acting against Partisan interests. The Partisan is committed to a belief in the stranger's integrity, but his friends think he is a fool to do so. The original encounter with the stranger gives the Partisan sufficient confidence to hold onto his faith in him even when there is evidence to the contrary.[46]

The Stranger's acting against the cause of the resistance movement frustrates the expectations of the partisan. This notwithstanding, the acts of the Stranger occasionally meet the expectations of the Partisan. "[S]ometimes the Stranger is seen helping members of the resistance, and the partisan is grateful and says to his friends, 'He is on our side.'"[47]

The connection between religious belief, expectations and evidence is a main point in Clark's interpretation of the Big Bang-theory and the fine-tuning argument. When we weigh the evidence in favor of theism or naturalism, we use a common and plausible method called the *expectation principle*. He gives the following example:

> Suppose you are the parent of a young child, one prone to reckless behavior with sporting equipment. While sitting inside your home you hear a loud noise. You know your child is playing outside, near your garage, with a tennis racquet and ball, and the sound you heard was shattering glass. Your child walks in, and you ask what happened. He sheepishly looks down and replies, "Nothing." Unconvinced, you think to yourself, "Dang it, he did

---

46. See "The Falsification Principle."
47. Mitchell, "Theology and Falsification. C.," 103–5.

it again. Evan just broke the garage window!" When you formed that belief, you were using the *expectation principle*.[48]

It is important to note that applying the expectation principle and assessing the plausibility of different hypotheses, we do this against our general background knowledge and our basic beliefs about what exists and how things work in the universe. "So while a small alien spacecraft would fully explain the broken window, it fails the likelihood test because it doesn't match up with our understanding of reality."[49]

Once again, we need to consider Jeffner's theory of fundamental patterns. Many of our beliefs about reality are informed by well-established scientific knowledge, but there are what Jeffner calls "ambiguous objects" such as the universe as a whole, human nature or the evolutionary process. Science informs us about unambiguous phenomena and "the difference between a scientific and a non-scientific hypothesis does not lie in the kind of explanation provided by the hypotheses, rather it lies in the objects they explain." Non-scientific hypotheses—such as the creation hypothesis—explain ambiguous objects. Different persons may have common and sufficient knowledge about the details of these ambiguous objects, but might still "disagree as to the correct description of these objects."[50]

Scientific and non-scientific hypotheses differ in the object they explain, but their justification is not totally dissimilar. In fact, there is a gradual difference. Basil Mitchell gives the following explanation.

> (There is) no sharp dichotomy between scientific explanation and other kinds of explanation . . . instead, there is a continuum of rational disciplines from physics and chemistry through the biological and social sciences to the humanities and metaphysics . . . At each stage of the continuum, not excluding the rest, there is discernible a broader type of rationality in which rival explanations are canvassed and defensible choices made between them. The degree of analogy between each stage and the next, and the evident conduit between them, make it highly implausible to suggest that at some point in the sequence we encounter a decisive break.[51]

---

48. Clark, *Religion and the Sciences of Origins*, 187–88.

49. Clark, *Religion and the Sciences of Origins*, 188.

50. Jeffner, *The Study of Religious Language*, 122. See above, chap. 10 for a more detailed explanation.

51. Mitchell, *Faith and Criticism*, 77. Quoted from Brown, "Basil Mitchell 1917–2011," 313.

The existence of an analogy between scientific explanations and "other kinds of explanation" implies both *differences* and *similarities*.

In what sense are there differences? Comparing scientific explanation and transcending fundamental patterns such as theism reveals one important difference. Transcending fundamental patterns require certain ultimate existential choices for which there are no guarantees. Jeffner suggest three such choices. The *first* is a choice about the "truth-reaching resources" of transcending fundamental patterns. Are there truths about human beings and the world, which go far beyond the limits of science? The answer of the metaphysical optimist is yes and the answer of the pessimist is no. The *second* choice concerns whether it is worthwhile to seek the answer to questions, which in principle are beyond science or not. There might be many things about the world that never will be known through scientific research. But should we seek those answers or not? The *third* choice concerns Occam's famous razor, i.e., to be economical when it comes to truth claims in our explanations of events and processes. On the other hand, even if Occam's razor has been a useful principle in science, is there anything that speaks in its favor outside the scientific realm?[52]

On the other hand, there are also similarities. One similarity between scientific and non-scientific hypothesis is that they raise certain expectations. The scientific hypothesis of general relativity raised the expectation that the light from stars passing the sun should be affected by the gravity of the sun. This was verified during observation at an eclipse of the sun in 1919.[53] Non-scientific hypothesis about ambiguous objects also raise certain expectations. While ordinary hypotheses raise expectations, which strengthens or weakens the hypotheses, fundamental patterns raise expectations that *certain experiences or scientific theories will confirm ambiguity of, for example, the universe as a whole and allow for an explanation by a transcending (or non-transcending) fundamental pattern.*

The difference between the expectations raised by scientific hypotheses and those raised by fundamental patterns, can be linked to Ernan McMullin's plea for "consonance" instead of "coherence" between science and religion. He takes the relationship between the Big Bang-theory and the Christian doctrine of creation as an example.

> What one *could* readily say, however, is that if the universe began in time through the act of a Creator, from the vantage point is would look something like the Big Bang that cosmologists are now talking about. What one cannot say is, first, that the

---

52. Jeffner, *The Study of Religious Language*, 128–31.
53. Gustafsson, *Svarta hål*, 117–23.

> Christian doctrine of creation "supports" the Big Bang model, or second, the Big Bang model "supports" the Christian doctrine of creation.[54]

Big Bang model and the Christian doctrine of creation are not of such a character that they are or are not *coherent*, but they may or may not be *consonant*. The Big Bang model—or the fine-tuning of the universe— is consonant *with God's creation originalis if these scientific theories confirm the ambiguity of the universe that allows for an explanation in terms of a transcending fundamental pattern*. The main question is: do they?

As explained earlier, there are theories about the Big Bang and the fine-tuning of the universe, which claim that the universe is not ambiguous in the sense that it allows for a transcending fundamental pattern of God's *creatio originalis*. A creation out of nothing through quantum fluctuations may explain the emergence of our universe, so there is no mystery here. And the fine-tuning of the universe may seem astonishing and require a special explanation until we realize that our universe is just one of a huge number of universe of which some are life-permitting and some are not. It is just a matter of chance that we live in one which is life-permitting.

Is this theory of the unambiguity character of the universe convincing? Does the theory of our universe arising out of nothing in the form of a field of quantum potentialities eliminate the ambiguity of our universe? Is the multiverse-theory eliminating the enigmatic fine-tuning of the universe? This is not evident. Our universe having a natural quantum-physical explanation beyond the Big Bang and the fact (if it is a fact) that we live in one of a huge number of universe, only pushes the original question one step back. Richard Swinburne explains:

> It may well be that our universe belongs to a multiverse, in which there are many universes governed by different lower-level laws. But our only grounds for supposing that there is such a multiverse are that the postulation of such a multiverse governed by a highest-level law and having some general physical features (such as an eternally expanding vacuum space) explains the coming into being of our universe. Yet in order for this to happen, the multiverse must have a certain sort of general law and general features. For there are innumerable logically possible multiverses governed by different laws with different general features which would never produce a universe in which there is a planet where human beings could evolve. So if there is a multiverse, what science will never explain is why the multiverse is

---

54. McMullin, "How Should Cosmology Relate to Theology?," 39.

of such a kind as to produce a universe "fine-tuned" to produce humans.[55]

A similar case could be made for the Big Bang-theory supplied with a quantum-physical explanation. Discounting the multiuniverse-theory, we don't know if this quantum-physical explanation is true, and but even if it is, it requires some general laws and features that function universe-generating. Lawrence Krauss makes this clear:

> The laws themselves are all that exist. These laws themselves require our universe to come into existence, to develop and evolve, and we are an irrevocable by-product of these laws. The laws may be eternal, or they too may have come into existence, again by some yet unknown but possibly purely physical process.[56]

My point is not some kind of God of the gaps-argument, but rather that the ambiguity of the universe is not easily eliminated. If there is a quantum-physical explanation of our universe, science will never explain is why there are such general laws that account for the quantum-physical processes that gave rise to our universe.

In sum, there are no strong arguments from cosmology to the effect that God as the creator of world exists. But there are weaker arguments, which in the basis of the theory of fundamental patters show that (1) the Big Bang and the fine-tuning of the universe confirm the ambiguity of our universe, and (2) that this ambiguity can be explained by a transcending fundamental pattern according to which God is the creator of the universe. To be sure, a non-transcending fundamental pattern is also possible, as is no fundamental pattern at all. My claim is simply that the universe is ambiguous and that cosmological science has not shown otherwise.

In the next section, I will proceed to the second idea within the Christian doctrine of creation, namely, the idea of continuing creation (*creatio continua*). This idea has many components, one of which is that God continuously sustains the universe. Another component is that God continuously renews creation and provides for the well-being of God's creation. I will concentrate on the latter idea and explore the question if evolutionary science may contribute to such an interpretation. This is strongly denied by many philosophers, but I will suggest otherwise.

---

55. Swinburne, "The Limits of Science." Further arguments for the compatibility of theism and the multiverse-theory can be found in Page, "Does God Love the Multiverse?"

56. Krauss, *A Universe from Nothing*, 142.

CREATIO CONTINUA AND THOMAS NAGEL'S NEO-NATURALISTIC WORLDVIEW

A theistic worldview interpreted as a non-scientific understanding of the universe as a whole is not only related to the past, but also to the future. Future may here be understood as the events beyond the present, but also as the last things in the eschatological sense. In this context, I will be concerned with events beyond the present, but will conclude with some reflections on the doctrine of Christian eschatology.

My point of departure will be Thomas Nagel and his thought on evolution and its driving forces. It should be underlined that Nagel is not a theist. Nevertheless, he has suggested some ideas that are related to the Christian idea of *creatio continua*. Nagel has suggested that the universe is moving towards an extended expansion of organization and consciousness. He claims that besides pure chance, creationism and directionless physical law, there is a fourth alternative, namely, a "conception of some increase in value through the expanded possibilities provided by the higher forms of organization toward which nature tends."[57] Let me call this a *neo-naturalistic worldview*.[58]

Nagel writes about the yearning for "cosmic reconciliation . . . has been part of the philosophical impulse from the beginning."[59] The paradigmatic example of this religious temperament is Plato, who sought a view of the world, which connected human beings to the whole of reality. "How can one bring into one's individual life a recognition of one's relation to the universe as a whole—whatever that relation is?"[60]

Not believing a religious answer to this question being available, Nagel turns to secular responses. He distinguishes between those who simply dismiss the yearning for cosmic reconciliation and those who accept it, but finds it impossible to fulfill it. The dismissive response is founded on the conviction that the universe revealed by chemistry and physics is incapable of meaning. But for Nagel the cosmic question does not go away when science replaces the religious worldview. He discerns three less dismissive alternatives to the default position of affectless atheism. The *first* is the response of humanism in a Kantian or existentialist version. This means working "from the inside out" with our own feeble resources. "It is we who give sense to the universe, so there is no need for a higher principle to give

---

57. Nagel, *Mind and Cosmos*, 91–92.
58. Nagel, *Secular Philosophy and the Religious Temperament*. Nagel describes his view as a form of "evolutionary Platonism."
59. Nagel, *Secular Philosophy and the Religious Temperament*, 3.
60. Nagel, *Secular Philosophy and the Religious Temperament*, 3–5.

sense to us."[61] But whether we like it or not, we are embedded not only in a human order, but also a natural order. The *second* less dismissive alternative is evolutionary naturalism is working from the outside-in of this natural order. Nagel's primary example is Nietzsche. His philosophy is a secular response to the cosmic question, which is more than merely a biological form of understanding. It tells us how to live our lives as an expression of the fundamental forces of nature. But Nagel finds it hard to see ourselves as expressions of the will to power and suggests—very tentatively—a *third* alternative, "evolutionary Platonism."

> Each of us . . . is part of a lengthy process of a universe gradually waking up. It was originally a biological evolutionary process, and in our species, it has become a collective cultural process as well. It will continue, and seen from a larger perspective, one's own life is a small piece of this very extended expansion of organization and consciousness.[62]

Nagel describes an alternative to theism. Nevertheless, his theory does not contradict theism. Theism could be described as an extension of Nagel's worldview. This is also acknowledged by Nagel himself in the conclusion of his summary in *The New York Times*.

> I would add that even some theists might find this acceptable; since they could maintain that God is ultimately responsible for such an expanded natural order, as they believe he is for the laws of physics.[63]

I am such a theist and would like to highlight Nagel's suggestion of a theistic expansion of his neo-naturalistic worldview. To get a firmer grip on Nagel's basic proposal, we need to distinguish between three different elements.

Three elements of Nagel's neo-naturalistic worldview

The *first* element of Nagel's neo-naturalistic worldview is linked to the result of his efforts to solve the mind-body problem, which he has explored in earlier books and articles.[64] And it is his conviction that reductionist physicalism is false. Mental phenomena cannot be reduced to physical phenomena because of "the subjective character of experience." Because of this "it is useless to base the defense of materialism on any analysis of mental

---

61. Nagel, *Secular Philosophy and the Religious Temperament*, 11.
62. Nagel, *Secular Philosophy and the Religious Temperament*, 17.
63. Nagel, "The Core of 'Mind and Cosmos'"; See also Nagel, *Mind and Cosmos*, 95.
64. In, for example, Nagel, "What Is It Like to Be a Bat?"

phenomena that fails to deal explicitly with their subjective character." And the reason why it is impossible to make such a reduction is "that every subjective phenomenon is essentially connected with a single point of view, and it seems inevitable that an objective, physical theory will abandon that point of view."[65] Nagel's famous example is the experiences of bats. "The essence of the belief that bats have experience is that there is something that it is like to be a bat."[66]

Nagel's argument spurred a flurry of comments, which has continued into the present. The discussion has for the most part centered around the concept of qualia, which has been explained by the philosopher Amy Kind in the following way:

> Qualia are the subjective or qualitative properties of experiences. What it feels like, experientially, to see a red rose is different from what it feels like to see a yellow rose. Likewise, for hearing a musical note played by a piano and hearing the same musical note played by a tuba. The qualia of these experiences are what give each of them its characteristic "feel" and also what distinguish them from one another.[67]

In 2013 Nagel summarized his argument in *The New York Times* that the essence of the mental is subjective feeling, being an irreducible aspect of reality and requiring a new form of scientific explanation.[68] This is denied by those who (a) identified the mental with some aspect of the physical be it patterns of behavior or patterns of neural activity as well as (b) those who reject it (being a kind of illusion) as a part of the reality at all. It is also denied by those who (c) regard it as an unexplained extra property of certain physical organisms or (d) that it had a non-scientific explanation based on the idea of divine intervention. Nagel's own position is that science may be expanded to include the mental in the form of a non-materialist theory of the immanent order of nature. He concludes:

> Mind, I suspect, is not an inexplicable accident or a divine and anomalous gift but a basic aspect of nature that we will not understand until we transcend the built-in limits of contemporary scientific orthodoxy.[69]

---

65. Nagel, "What Is It Like to Be a Bat?," 437.
66. Nagel, "What Is It Like to Be a Bat?," 438.
67. Kind, "Qualia."
68. Nagel, "The Core of 'Mind and Cosmos.'"
69. Nagel, "The Core of 'Mind and Cosmos.'"

The *second* element of Nagel's neo-naturalism is concerned with that which Nagel calls the *constitutive account* of "how certain complex physical systems are also mental" (in contrast to the third element, i.e., the *historical account* "how such systems arose in the universe from its beginnings"). The essence of Nagel's constitutive account is that the human mind is a clue to the basic nature of reality. As clarified above, Nagel is foreign to a reductive account of the mental. Such an account cannot do justice to the subjective character of experience. But Nagel is equally foreign to an emergent account. Emergence is a trivial phenomenon and is often exemplified by liquidity, which is an emergent property of water basically composed of water-molecules. But subjective experiences in the form of mental states cannot emerge from physical phenomena in this way, because consciousness is something completely new. Rather, he is leaning to a special form of *neutral monism* or *panpsychism*. Nagel writes:

> Consciousness is . . . not, as in the emergent account, an *effect* of the brain processes that are its physical conditions; rather, those brain processes are *in themselves* more than physical, and the incompleteness of the physical description of the world is exemplified by the incompleteness of their purely physical description.[70]

This may be understood as a form of reductive account of the mental, but different from the one found in ordinary physicalism, which understands matter as that which is described in physics *and nothing more*. But Nagel's analysis proceeds on the assumption that "everything, living or not, is constituted from elements having a nature both physical and non-physical."[71] This amounts a form of *panpsychism*: all the element of the physical world are also mental. In an earlier work Nagel defines panpsychism in the following way:

> By panpsychism I mean the view that the basic constituents of the universe have mental properties, whether or not they are parts of living organisms.[72]

In *Mind and Cosmos* Nagel refers to various philosophers, among others Galen Strawson, but also process philosophers such as Charles Hartshorne and Alfred Whitehead.

Nagel also uses another concept to describe his position, namely, neutral monism. To be sure, Nagel's kind of neutral monism is of a different

70. Nagel, *Mind and Cosmos*, 57.
71. Nagel, *Mind and Cosmos*, 57.
72. Nagel, *Mortal Questions*, chap. 13.

character than the one classically advocated by Bertrand Russell.[73] Russell found the neutral "stuff" in the content of our experiences, while Nagel argues that reality (including human consciousness) is built up out of "protomental" elements that are somehow unified simultaneously into an organism and a self."[74]

The *third* element consists of Nagel's *historical account* of how complex physical-cum-mental systems arose in the universe from its beginnings. He discerns three different possibilities: (1) a casual historical account, (2) a teleological account, and (3) an intentional (theological) account. Nagel opts for the teleological account. He is convinced that such an account—in the form of a theory of teleological laws—is coherent. The existence of such laws implies that things happen "because they are on the path that leads to certain outcomes—notably, the existence of living, and ultimately of conscious, organisms."[75] Admittedly, the idea of teleological explanations presents some serious problems, but so do the other alternatives. Nagel suggest a form of natural teleology (i.e. a teleology without an intending agent), where a certain value occurs in the *explanans*. This implies (1) that ordinary causal laws are not fully deterministic and (2) that certain states in nature have "a significantly higher probability than is entailed by the laws of physics alone."[76] Essentially, they would be laws of the self-organization of matter. Nagel concludes:

> A naturalistic teleology would mean that organizational and developmental principles of this kind are an irreducible part of the natural order, and not the result of intentional or purposive influence by anyone.[77]

At another point in his argument he summarizes his worldview together with an important qualification.

> The story includes huge quantities of pain as well as pleasure, so it does not lend itself to an optimistic teleological interpretation. Nevertheless, the development of value and moral understanding, like the development of knowledge and reason and the development of consciousness that underlies both of those higher-order functions, forms part of what a general conception

---

73. Russell, *The Analysis of Mind*.
74. Nagel, *Mind and Cosmos*, 83.
75. Nagel, *Mind and Cosmos*, 67.
76. Nagel, *Mind and Cosmos*, 93.
77. Nagel, *Mind and Cosmos*, 93.

of the cosmos must explain. As I have said, the process seems to be one of the universe gradually waking up.⁷⁸

Natural teleology and creatio continua

What are the fault-lines between his neo-naturalistic position and Christian theism? An answer to this question could be formulated against the background of Jeffner's theory of fundamental patterns. Nagel's neo-naturalism could be interpreted as an non-transcending fundamental pattern. Nagel concludes from a certain formation of available knowledge the existence of something, which *later* might could be verified and found to be a fact, namely, teleological principles. If that were to happen, it should rather be classified as a scientific theory than a fundamental pattern. If it is not, it may still be rational to affirm it—but not on the basis of science. It would rather be a question of metaphysics.⁷⁹ In that case, the vital question would be the following: *are the scientific evidence for the evolutionary process allowing for a description of this process as an ambiguous object?* This is what a naturalistic interpretation of the evolutionary process denies. Nagel can be interpreted as making the case for a positive answer to this question.

For the theist, an interesting possibility could be considered, namely, the transformation of Nagel's non-transcending fundamental pattern into a transcending fundamental pattern. On the assumption that the teleological principles assumed by Nagel are real, it is possible supplement these principles with a theistic fundamental pattern. They may conceive as "seeds" implanted by God in creation. This comes close to an idea in the thought of St. Augustine. Referring to the first chapter of Genesis, St. Augustine gives the following example in his work on the Trinity:

> But, in truth, some hidden seeds of all things that are born corporeally and visibly are concealed in the corporeal elements of this world. For those seeds that are visible now to our eyes from fruits and living things, are quite distinct from the hidden seeds of those former seeds; from which, at the bidding of the Creator, the water produced the first swimming creatures and fowl, and the earth the first buds after their kind, and the first living creatures after their kind.⁸⁰

---

78. Nagel, *Mind and Cosmos*, 117.

79. For a classical treatment of teleology as metaphysics, see Taylor, *Elements of Metaphysics*, book 3, chap. 5.

80. Augustine, *On the Trinity* 3.8.13.

Theists who assume Nagel's teleological principles believe "that God is ultimately responsible for such an expanded natural order, as they believe he is for the laws of physics." Furthermore, such a form of theism would be future-including suggested by some version of the doctrine of *creatio continua*.

Nagels's theory of teleological laws involves "some conception of an increase in value through the expanded possibilities provided by higher forms of organization toward which nature tends . . . "[81] It is not in the rituals and proclamation of the Church that these expanded possibilities are realized. It is *nature* that is—in short—"biased toward the marvelous." Ecclesiologically speaking, the Church is reflecting and celebrating these possibilities of the future.

Evil and hope

At this point there is place for a short suggestion how Nagel's neo-naturalism could be supplemented with a theory of evil. Nagel underlines that the emergence of value is the emergence of both good and evil. There is not only "a purely benign teleological explanation: a tendency to the good."[82] This seems to imply that there are malicious teleological explanations, tendencies toward evil. In chapter 12, I discussed the need of a "theory of evil," which stays clear of mythological language, but at the same time develops an understanding of the "excess" not covered by the so-called "only way-argument." Could this be what we are looking for? Nagel suggests naturalistic teleology, i.e., organizational and developmental principles as irreducible part of the natural order, but not the result of intentional or purposive influence by anyone. These teleological principles could possibly be divided into tendencies towards good and tendencies towards evil. Extending the "only way argument," these tendencies could be linked together in such a way that (1) tendencies toward the good are ontologically prior to tendencies to evil, and (2) tendencies toward the good makes tendencies toward evil inevitable. Augustine might have had something like this in mind when he wrote:

> Unless . . . something is good, it cannot be corrupted, because corruption is nothing more than the deprivation of the good. Evils, therefore, have their source in the good, and unless they are parasitic on something good, they are not anything at all. There is no other source whence an evil thing can come to be.[83]

---

81. Nagel, *Mind and Cosmos*, 91–92.
82. Nagel, *Mind and Cosmos*, 122.
83. Augustine, *Enchiridion* 4.14.

Notwithstanding the speculative nature of this amendment to Nagel's neo-naturalism, it amounts to a theory of natural evil which could be described as a demythologization of the Biblical language, but at the same time providing us with a description of the excess not covered by the "only way argument" as it is usually construed. This theory of good and evil also provides an inroad to a Christian view of history. By many theologians, the work of Christ has been understood as a struggle between good and evil. This is central to the view of history in early Christianity and a significant theme in the work of the Swedish theologian Gustaf Aulén. He argues that the entire work of God in God's creation must be seen as a dramatic struggle against evil.[84]

This dramatic understanding of history is accompanied by a basic mood of hope. The contemporary threat of climate change makes such a hope existentially relevant. Two thirds of the Swedish adult population are in agreement with the claim that humanity is facing a self-inflicted environmental disaster. The universal and apocalyptic proportions of scope of this disaster has been described in a well-known article by David Wallace-Wells.[85] In the next century—if not sooner—climate change might bring heat death, famine, global plagues, unbreathable air, perpetual war, poisoned oceans, and permanent economic collapse. In face of such global afflictions, it is tempting to shut your ears and close your eyes. The ancient wisdom has it that when Pandora opened her famous box, a nameless number of evils flew out. But she was quick enough to close the box and preserve Elpis, the Greek deity of hope.

What does such a hope amount to in the presence of contemporary climate change? The English theologian John Macquarrie has provided some important reflections on hope that I have found helpful in a response to this question. Hope is found on many levels from the hope implicit in everyday life, to the hope of world leaders and prophets. One of the most striking features is that hope on the whole predominates over fear. Ultimately, hope is total hope for nothing less than the human race, a hope for the world itself. Is such a hope just a grandiose delusion? It is not, argues Macquarrie.

---

84. Aulén, *Christus Victor*, 1931. See further Jeffner, *Theology and Integration*, 62–63.

85. Wallace-Wells, "The Uninhabitable Earth." The article has caught wide attention and acclaim, but also been described as a hyperbole and a gross exaggeration, fostering gloom and hopelessness. Expert have criticized Wallace-Wells for factual errors. Professor David Archer (University of Chicago) concludes that factual errors notwithstanding "I feel that the overall thrust of the article is not wrong, wildly misleading, or out of bounds of the discussion we should be having about climate change" (see Climate Feedback, "Scientists Explain").

> ... it is only by the introduction of the religious dimension that one could rescue the total hope of an ideal human society from the dreamland of utopia ... The symbol 'kingdom of God' does not stand for a perfected human society but for a transformed reality, that is indeed a work of God, though a work that God will do only with human consent and cooperation.[86]

This puts hope into the context of *creatio continua*. Maybe that is the proper place of Christian hope today. Wallace-Wells concludes his article with some interviews of senior climatologists. He summarizes their worldview in the following way.

> [W]hen we do truly see the world we've made, they say, we will also find a way to make it livable. For them, the alternative is simply unimaginable.[87]

This is consonant with the perspective of total hope. It is linked not only to the doctrine of *creatio continua* but also to the ultimate hope of Christian eschatology. According to Willem Drees, eschatology has at least three functions: 1. Judgement on the present: it is not as it ought to be; 2. Appeal to action as response to that judgement; and 3. Consolation in contexts of injustice, failure and suffering.[88] All these points apply to the quest for hope in the face of the looming climate-change. But is there also a fourth function, having to do with not only present transcendence, but also with future transcendence, i.e., the renewal of creation in a more radical sense? This is an important question in the next section.

CREATIO NOVA, SCIENCE AND RELIGION

I have argued that Thomas Nagel's metaphysical thought is of relevance for the contemporary interpretation of the doctrine of God's continuing creation. But what is the relevance of Nagel's thought for the doctrine of *creatio nova*, i.e., Christian eschatology? As I see it, there are at least two possible answers to this question. The *first* answer would be that the doctrine of eschatology is most properly interpreted as an expression of divine providence. What is the new heaven and the new earth, but an ultimate act of God that creates eternal well-being for God's creatures and God's creation? In that case, Nagel's idea of "an increase in value through the expanded possibilities provided by higher forms of organization toward which nature tends" points beyond itself towards an ultimate fulfillment of value

---

86. Macquarrie, *In Search of Humanity*, 247.
87. Wallace-Wells, "The Uninhabitable Earth."
88. Drees, *Beyond the Big Bang*, 93 (with reference to Carl Braaten).

and community. It has been called "physical eschatology," i.e. an immanent perfection of creation.[89]

A *second* alternative would be that providence and eschatology should be kept apart, and that God's ultimate act of judgment and fulfillment is not just an actualization of possibilities arising out of the created order. The ultimate eschatological visions of the Christian faith are transcendent in a radical sense and not the result of human achievements. They are—as John Wisdom dryly remarked—"logically unique expectations."[90]

Antje Jackelén is critical to the physical eschatologies of Tippler and Dyson. They are too speculative and cites John Polkinghorne's harsh verdict when he describes this kind of physical eschatology as "a cosmic Tower of Babel" and "the most extreme *reductio in absurdum* of an exclusively evolutionary optimism."[91] This is not the place for a longer discussion. I must simply admit that I find the central claims of physical eschatology such that "Christianity is not a mere religion but an explanatory testable science" and that "the laws of physics require life to survive"[92] beyond comprehension. Moreover, they suffer from the weakness that evil being an integral part of our world; physical eschatologies preclude a state of perfection in the sense of classical eschatologies. I must leave the matter at that and explore Jackelén's alternative, which focuses on eschatology as a recreation of existence in continuity and discontinuity with the past. She writes:

> The dynamic nature of the "already" and the "not-yet" also holds together ethics and expectation, struggle and hope. This dynamic relation is the quintessence of Christian eschatology. It can be reconciled with findings from the field of natural science, but it cannot be derived from them.[93]

This quote from Antje Jackelén's dissertation highlights a basic puzzle for the Christian theologian, who wants to proceed with the idea of *creatio nova*. The "already" of Christian eschatology could, possibly, be clarified with reference to *creatio continua* (possibily updated with assistance of Nagel's metaphysical thought). But how do we understand the "not yet?" That phrase has a reference to the future in the radical sense of the ultimate

89. The most prominent proponents are the American phycisists Freeman Dyson and Frank Tippler. See further Halvorson and Kragh, "Cosmology and Theology"; and Drees, *Beyond the Big Bang*, chapter 5.

90. Wisdom, "Gods," 186.

91. Jackelén, *Time and Eternity*, 208 (quoting Polkinghorne, *The Faith of a Phycisist*, 165).

92. Tipler, *The Physics*, 3 and 62.

93. Jackelén, *Time and Eternity*, 221.

future. When it comes to understanding the ultimate future, there is an even greater problem. Robert John Russell explains why.

> When we expand the domain of eschatology from the anthropological and even eco-terrestrial context of 'world history' to the cosmos itself, we encounter science's grim prediction that all life must inevitably be extinguished as stars go supernova and following this that the far future of the universe is either endless cold or unimaginable heat.[94]

After reviewing some less helpful alternatives, Russell argues "we may yet discover some vital clues from scientific cosmology for our constructive attempt to relate it to Christian eschatology."[95] Russell has elaborated his idea of *creatio nova* in a more extensive work.[96] In short, his proposal involves two assumptions, one philosophical and one theological:

> The first assumption is *philosophical*: namely that the events science predicts must come to pass. Instead we can suppose that the laws of nature are descriptive, not prescriptive. In this case predicted events do not necessarily occur. The next step is *theological*: namely that the efficacy described by the laws of nature does not reside in the processes of nature which science describes, but instead it is the result of God's ongoing and faithful action as Creator. Finally, we can believe theologically that God is free to act in radically new ways, not only in human history but also in the ongoing history of the universe. Because of this, we can claim that the scientific predictions are right but inapplicable since God did act in a radically new way at Easter and will continue to act to bring about the new creation. In doing so, we are not in a conflict with science but with a philosophical interpretation brought to science. In short, the future of the universe would have been what science predicts (i.e., 'freeze' or 'fry') had God not acted at Easter and did God not continue to act in the future.[97]

Russell's proposal involves several philosophical and theological issues and I will highlight some of these issues in turn.

---

94. Russell, "Eschatology in Science and Theology," 544.
95. Russell, "Eschatology in Science and Theology," 547.
96. Russell, *Time in Eternity*.
97. Russell, *Time in Eternity*, 549; my italics. It should be noted that Russell explicitly advocates that his research progam should presume methodological naturalism for past and present events. Russell, *Comology*, chap. 10, guideline 2. Divine and natural causalities should be regarded as distinct. But what about future events?

1.. Philosophical issues. Basic to Russell's research program is the claim that the universe is "transformable by God's action."[98] Such a transformability is suggested by the belief in the resurrection of Jesus Christ. Russell argues that the transformability of the universe by God's action would require a revising not of the scientific method itself, but the philosophy of space, time, matter, and causality underlying contemporary physics.[99] He boldly presupposes that such a revision is possible and that "[b]cause of Easter and God's promise for its eschatological completion . . . the freeze or fry predictions will not come to pass."[100]

In addition to divine acts to which the acronym NIODA refers (i.e. non-interventionist objective special divine action in the subatomic, indeterminstic world), Russell defends miracles "in the Humean sense of miracle as a violation of the laws of nature."[101] The resurrection of Jesus and *creatio nova* are such miracles. Russell argues that these miracles are not in conflict with science; they are only in conflict with the philosophical assumption that the events predicted by science are *bound to happen*. Refuting this assumption seems to be one of the most challenging part of his project. But Russell suggests a simple solution: "[W]e can suppose that the laws of nature are descriptive and not prescriptive . . . predicted events do not necessarily occur."[102]

The idea that laws of nature are descriptive suffers from one major weakness. Interpreting natural laws as descriptive implies that it is impossible to distinguish them from accidental generalizations.[103] Russell may hope for a solution of this basic philosophical problem, but there are others. What are the grounds for supposing that *creatio nova* will occur in the first place? Scientific prediction about the end of the universe are neither necessary (the *creatio nova* might occur long before the predicted end) nor sufficient (the "freeze" or "fry" is followed by no divine recreation). So what is the basis of Christian hope in the sense of *creatio nova*?

---

98. Russell, "Eschatology in Science and Theology," 550; Russell, *Cosmology*, chap. 10.

99. Russell, *Cosmology*, chap. 10.

100. Russell, *Cosmology*, loc. 5324.

101. Russell, *Cosmology*, loc. 2313.

102. Russell, "Echatology in Science," 549.

103. "Suppose that everyone here is seated . . . Then, trivially, that everyone here is seated is true. Though true, this generalization does not seem to be a law. It is just too accidental. Einstein's principle that no signals travel faster than light is also a true generalization but, in contrast, it is thought to be a law; it is not nearly so accidental. What makes the difference?" (Carroll, "Laws of Nature").

Hjalmar Lindroth—a relentless defender of the eschatological motif—has an answer. God wants life for this world and every human being and has therefore set a goal, which ultimately consists of God's recreation of the universe. This gives an otherwise meaningless life a meaning.[104] John Macquarrie has a similar response. Human life is ingrained with a "total hope." And such a total hope can be fulfilled only a recreation of the world in which all humans participates.[105]

I will return to these ideas, but first consider Russell's response. His ground has to do with the theological part of his proposal, centering on the resurrection of Jesus Christ.

2. *Theological issues.* According to Russell, Christian eschatology is based on the knowledge what God did on Easter. Russell distinguished between two different interpretations of resurrection of Jesus: the *subjective* and the *objective*. According to the subjective interpretation, the Biblical language describing the appearances of Jesus and the empty tomb should be interpreted as a pictorial expression of the strengthened faith of the disciples and their conviction that the activity of Jesus goes on. Jesus died and his body decomposed as any other dead human body. The objective interpretation means that "God raised Jesus from the dead." He describes this event as the "bodily resurrection of Jesus" interpretation of the New testament texts on the resurrection.[106] "[S]omething happened to Jesus of Nazareth which cannot be reduced entirely to the experiences of the disciples."

Russell dismisses the subjective interpretation and opts for the objective. Furthermore, he regards the bodily resurrection of Jesus Christ as the central event in the history of salvation culminating in the divine recreation of a "new heaven and new earth." This understanding of Christian faith has been supported by many, for example, the BioLogos foundation.[107] Criticism has been also been voiced by Klaus Nürnberger, a South-African Biblical Scholar. Nürnberger is echoing the central claims of Rudolf Bultmann summarized in the beginning of the present chapter. In an article in *Zygon* 2012, Nürnberger suggests that classical Christian eschatology arose from a specific historical situation, namely, apocalyptic visions occasioned by heavy suffering in postexilic times. Moreover, these visions have proved to be unreliable.

> The expectation (rather than the vision) of a new creation without evil, suffering, and death is not constitutive for the

---

104. Lindroth, *Filosofiska*, 228–29.
105. Macquarrie, *In Search of Humanity*, 246.
106. Russell, *Cosmology,* chap. 10.
107. See http://biologos.org.

substantive content of the biblical message as such. Biblical future expectations must be reconceptualized in terms of best contemporary insight and in line with a dynamic reading of the biblical witness as God's vision of comprehensive optimal well-being that operates like a shifting horizon and opens up ever new vistas, challenges, and opportunities.[108]

When it comes to the interpretation of the resurrection of Jesus, Occam's famous razor must be applied. The easiest explanation is that the different stories of the resurrection emerged on "the level of structured individual and collective consciousness, rather than at the physical and biological levels."[109] Arguably the story of the empty tomb belongs on this level. At least, following Occam's razor, it is by far the easiest explanation. Nürnberger comes close to the subjective interpretation. The resurrection of Jesus must be separated from the notion of the general resurrection, ultimate judgment and the imminent apocalyptic transformation of the world. The ultimate validity lies in the affirmation of Jesus as the redeeming love of God opening up fellowship with God to universal participation. Biological life ends in death. Eternal life means an authentic human life participating in God's creative and redemptive project.

Nürnberger suggests an understanding of Christian faith based upon a thorough application of an historical critical method and a more sustained effort to formulate a hermeneutical interpretation of the Christian tradition, i.e., to reinterpret the Christian message against the background of our contemporary worldview. In a short response, Russell agrees that large parts of the biblical worldview must be rejected. But in contrast to Nürnberger, Russell wants to preserve the idea of an objective divine action in nature including the bodily resurrection of Jesus as a prolepsis to a radical recreation of an entirely new universe.[110] This includes "a full and transformed psychosomatic destiny for humanity and for all life in the universe."

Needless to say, Russell and Nürnberger are not the only alternatives when it comes to the understanding of the resurrection of Jesus and the *creatio nova*. There are other "objective interpretations" besides the "bodily resurrection of Jesus" interpretation. Such an alternative interpretation has

---

108. Nürnberger, "Eschatology," 970.

109. Nürnberger, "Eschatology," 986.

110. Russell, "Eschatology and Scientific Cosmology," 1010. Russell refers to NIODA, even if the miraculous resurrection of Jesus belongs to another category of divine acts.

been articulated by the Swedish exegetical scholar Erik Aurelius. The following is a summary of a shorter article.[111]

Aurelius departs from the oldest text in the New testament, i.e., 1 Cor 15. The foundation for the continued existence of Jesus Christ is that he appeared to his disciples.[112] There is no mentioning of an empty tomb, not in any of the Pauline letters. On the contrary, towards the end of the cited chapter St. Paul emphasizes that "flesh and blood cannot inherit the kingdom of God, nor does the perishable inherit the imperishable."[113] The death of Jesus Christ is an example of ours. It is not the physical body of Jesus, which is brought back to life, but that his spiritual body appeared and lives in eternity. And so will we.

In later texts, St. Paul does not uphold the distinction between physical and spiritual bodies. Moreover, the Gospels tell us that the physical body of Jesus did not remain in the tomb. Thereby the identity between the crucified Jesus and the resurrected Jesus is emphasized. But this identity can be acknowledged without support stories about the empty tomb and without abandoning the basic principle that flesh and blood cannot inherit the kingdom of God. There seems to be many different (objective) interpretations in the New testament—and different such interpretations could be equally possible in the Christian church.

I have referred to Aurelius' argument for one particular purpose. If the resurrection of Jesus is understood as Aurelius understands it, the argument that "[b]ecause of Easter and God's promise for its eschatological completion ... the freeze or fry predictions will not come to pass" is weakened. The linkage between the resurrection of Christ and the eschatological transformation at the end of the universe is dissolved.[114]

It is not difficult to imagine the response of Russell to this alternative. He might suggest two different objections. First, such a take on the Christian hope would include only a shadow of "our full humanity, our psychosomatic unity of persons in the world."[115] And, secondly, without a *bodily* resurrection, the continuity between the person in this world and the person in the word to come would be lost.

---

111. Aurelius, "Olikheter ryms i Nya testamentet."

112. 1 Cor 15:3–5, 8.

113. 1 Cor 15:50.

114. This does not mean that Christian eschatology in terms of God's re-creation of the universe has to be abandoned. It may be claimed that it has an alternative anchorage in the parables of Jesus and his proclamation of the kingdom of God. See Meier, *A Marginal Jew*, 2:312n8.

115. Russell, "Eschatology and Scientific Cosmology," 1011.

These arguments can claim an impressive theological provenience, but are nevertheless not convincing. The English theologian Maurice Wiles reminds us that "God is bodiless and inconceivable; but this is not (in Christian eyes at least) ground for denying his existence or regretting its form." Secondly, "it is eternal life, life in relation to that God, of which the Christian is trying to speak in all his talk about life after death." And it is not clear "that to speak of life after death in terms of a body makes such speech either more intelligible or more attractive."[116] A relation to God seems sufficient for fullness of existence after death. In what way is a bodily existence required?

Furthermore, Wiles has serious misgivings about Russell's second objection that a bodily resurrection is necessary for continuity. For however that continuity is conveyed, it is precisely *not* conveyed through the body.[117] For similar reasons, Wiles has an argument against the view that the resurrection of the body is necessary to express a positive attitude to the physical aspects of our present life. This is hardly the case if the resurrection body is neither part of nor derived from our present material existence.

Maurice Wiles has many other significant things to say, but he is silent on the relationship between eternal life and *creatio nova*. This may be due to his theology being explicitly guided by two general objectives: coherence and economy. Coherence has to do with the consistence between different doctrines, and economy "the pruning of some luxuriant growths that have come to be highly valued."[118] The doctrine of *creatio nova* might be considered to be such a luxuriant growth.

Nevertheless, respect for economy may not dismiss one question concerning Christian doctrine and the ultimate future. What can the theist expect from scientific cosmology? Many theologians have responded that the answer depressing. The "freeze or fry" scenarios would falsify the existence of God. Russell refers to Ted Peters:

> Should the final future as forecasted by the combination of big bang cosmology and the second law of thermodynamics come to pass . . . we would have proof that our faith has been in vain. It would turn out to be that there is no God, at least not the God in whom followers of Jesus have put their faith.[119]

This answer is uncompromisingly honest, but is nevertheless somewhat off the mark. One is reminded of Sir Arthur Eddington's rhetorical

116. Wiles, *The Remaking*, 140.
117. Wiles, *The Remaking*, 141.
118. Wiles, *The Remaking*, 18.
119. Peters, *God as Trinity*, 175–76, quoted in Russell, "Eschatology in Science," 546.

question: "Since when has the teaching that 'heaven and earth shall pass away' become ecclesiastically un-orthodox?"[120] To be sure, the end of the all-enveloping death of the universe is clearly consistent with ontological naturalism as is individual human death. In contrast, Christian theology has always suggested that the end of the universe as well as the end of human life, is not the end of everything.

When St. Paul and Aurelius speak about the appearances of resurrected Christ as an encounter with a spiritual body, they face an obvious challenge: what kind of experience are they talking about? To be sure, they suggest an experience of something real. But what is the nature of this reality? Some theologians would argued that it is the same kind of experience of the presence of Christ that Christian believers have experienced through the ages. This suggests that that we are speaking of a certain kind experience of God. And this in turn could be linked to the kind of process metaphysics advanced by Marjorie Suchocki. This is not the place to enter into the details of her proposal. I will limit myself to one short and speculative suggestion. One of Suchocki's central ideas in *The End of Evil* (1988) is her affirmation of subjective immortality. Not only are the facts of our lives preserved in the memory of God.[121] Our subjective awareness is preserved in God. Death is the transition from our earthly, finite forms, "outside" of God to an existence "inside" of God. Experiencing the presence of Christ is at the same time experiencing the presence of God, who—according to classical Christian theology—can be identified as the second person of the Trinity.

CONCLUDING REFLECTIONS

The Ariadne thread of the present chapter has been the idea that the evolutionary process and the universe are "ambiguous objects." In this perspective, the question is not whether modern cosmology supports or contradicts God's *creatio nova*, but rather *if it confirms the ambiguity of the ultimate future such that it allows for a transcending fundamental pattern of God's recreation of the universe*. Present cosmological opinions include affirmations of a universe with a beginning as well as an end, and suggest that in this sense the ambiguity remains. Other cosmological theories suggest no beginning and no end. On Stephen Hawking's hypothesis of the closed universe, it would have neither beginning nor end: it would simply "be." "What place, then, for a creator?" Hawking's hypothesis notwithstanding, the mystery remains. Why is there a universe in the first place?

120. Eddington, *New Pathways*, 59. Quoted in Halvorson and Kragh, "Cosmology."
121. Suchocki, *The End of Evil*, chap. 5.

One theological consequence of Hawking's "no-boundary" proposal would be a reduction of not only of (1) *creatio originalis*, but also of (2) *creatio nova*, to the doctrine of *creatio continua*. The implications such reductions open up for issues well beyond the present context. I will restrict myself to one brief comment about (2). God's continuing creation is multifaceted and may include not only the universe's "moving towards an extended expansion of organization and consciousness," but also the transition from our earthly, finite forms, "outside" of God to an existence "inside" of God. This would be in line with Marjorie Suchocki's view of subjective immortality and Philip Clayton's panentheistic elaboration of this view.[122]

More generally, the question of the relation between science and religion have led me to the conclusion that the relation is *doctrine-specific*.[123] Cosmology may be (weakly) relevant to the Christian doctrine of creation, but not to the Christian doctrine of salvation and forgiveness. When it comes to soteriology, the model of independence or separation seems more adequate. Moreover, sciences beyond natural science are relevant to other areas of Christian doctrine. Historical science and literary science are relevant to the doctrines of the Bible and biblical revelation, and social science to the doctrine of the church. John Macquarrie argues that philosophical anthropology is largely relevant to Christian hope, and is quite silent on cosmology in his work on human nature. Analogues of doctrine-specificity is found in other religious traditions. In one of his later works, the American ethicist James Gustafson has explored the intricate intersections between ethics, science and theology.[124]

The doctrine-specificity of the relation between science and religion may also shed some light on some of the empirical findings in Chapter 3, where I explored Swedish clergy and their opinions concerning science and religion. One result was puzzling. Many respondents who affirmed one alternative about science and religion being a unity, because they are entirely different perspectives (model of independence) also affirmed a second alternative that science and religion albeit different activities can come into contact with each other (model of contact). One possible explanation could be that the respondents were thinking about certain religious beliefs when choosing the first independence-alternative, and other beliefs when choosing the contact-alternative. In other words, *the question of the relation between science and religion is doctrine-specific*. The first alternative may direct attention to the belief that is true that Jesus is God incarnate has nothing to

---

122. Clayton, "Eschatology as Metaphysics."
123. Suggested in passing by Russell, "Eschatology in Science and Theology," 547.
124. Gustafson, *Intersections*.

do with physics or chemistry. On the other hand, there are other religious beliefs for which it is true that their truth-value is (at least partly) determined by certain scientific results and theories. For example, the belief that God created the world is affected by evolutionary theory, because evolutionary theory explains—at least in part—how God created the world.

One final note: natural and other sciences are relevant to religious beliefs and doctrines in different ways, but there are large religious fields outside the sciences altogether. In the introduction to Part 3, I referred to a minister in the Church of Sweden criticizing the view that equals religion with a number of truth-claims. "Imagination, intuition the feeling for things that cannot be touched, is the religious attitude." The analysis in Part 2 on twentieth century Swedish theology shows that Swedish theologians up until about the 1970s are generally more inclined to a model of independence and separation. The reduction of religion to the atheoretical dimension may be unjustified. At the same time, it is true that a religion which cannot offer "a feeling for things that cannot be touched," cannot live very long.

Appendix 1

Worldview-studies (2006–2017)[1]

	Date	Group	Method	Content
Study 1 (ch. 1)	2006 (May 9–18)	National sample (n=503), 18–74 years of age	Telephone interviews	Biologization and worldviews
Study 2 (ch. 1)	2007 (April–May)	Teachers, students, scholars in five regions (n=36)	Postal survey	Biologization and worldviews
Study 3 (ch. 2)	2009 (May 26–June 4)	National sample (n=502)	Telephone interviews	Biologization, philosophy, literature, politics, and physics
Study 4 (ch. 5)	2009 (May 7–25)	Teachers, students, and researchers in literature, political science and physics in five Swedish regions (n=233)	Web questionnaire	Biologization, literature, politics, and physics

1. *Study 1–5* has been supported by the Swedish Research Council. *Study 6–8* has been supported by the John Templeton Foundation.

	Date	Group	Method	Content
Study 5	2012 (May)	Swedish youth, 18 years old (n=1196)	Classroom questionnaire	Biologization, environmentalism, society, and worldviews
Study 6 (ch. 2, and 3)	2015 (February 26–April 4)	Ministers in the Church of Sweden and pastors in the Uniting Church (n=1492)	Web questionnaire	Biologization, science, and religion
Study 7 (ch. 5)	2015 (May)	Ministers and pastors (n=149)	Web questionnaire	Biologization, science, and religion
Study 8 (ch. 1)	2017 (May 17 – June 7)	National sample (n=503), 18–74 years of age	Telephone interviews	Biologization and worldviews

Appendix 2

Design of telephone-interviews on (1) May 9th, until May 18th, 2006 (Study 1),

(2) May 17th until June 7th, 2007 (Study 2), and (3) May 26th until June 6th 2009 (Study 3, replication of Study 1) on biologism, ecologism and religious belief

Introduction

Hello, my name is John Doe, and I´m calling from Samhällsinformation AB in Stockholm. We are executing a study for Uppsala University and we would like to pose a number of questions to . . . . (see list of sample)

I will now read you a number of statements about worldviews in the general meaning, i.e. how you view environment, biology and technology in different respects. Respond throughout on a 5-point scale, where the figure 5 means that you fully agree with the statement and the figure 1 that you fully disagree.

(Scale 1-5+=cannot answer)

A
(Technology)

1. Technology will be able to solve the problems with our limted resources of nature.
2. Humans would be much better off if they lived a simple life without so much technology.
3. Scientific research creates more problems than solutions.

4. The development of nuclear power is the only alternative to the use of coal and hydropower.

(Biology and ecology)

5. Humanity is approaching a self-inflicted environmental disaster.
6. Human beings have certain unique qualities which no other creatures have.
7. Biology gives us the complete answer to the meaning of life.
8. We must respect the order of nature irrespective of how this affects human welfare.
9. God created human beings at a specific moment.
10. Human beings have evolved through a goaldirected process which is realizing a certain plan.
11. Human beings have emerged through a random biological process.

(Enviroment)

12. The balance in nature is vary fragile and is easily disturbed byu human influence.
13. The earth is like a speceship with limited space and limited resources.
14. Plants and animals exist only for the purpose of human beings.
15. Changes in the environment for human purposes creates serious problems.
16. There are no limits for growth in a country like Sweden.
17. Humanity exists to govern over the rest of nature.
18. I encounter the holy primarily in nature.

(Morality and religion)

19. Christian belief in creation is irreconcialable with the biological doctrine of evolution.
20. Humans consist only of body and matter.
21. Biology teaches that only the strongest survive.
22. Life has no other meaning than to further the human species.
23. Nature is filled by a spiritual force permeating all life.

24. What becomes of a human being is mostly dependent on her biological heritage.
25. There is a god, a supreme power or force.

B

[only 2017—replicating items in a studies made in 1999 and 2000]

I am now going to pose some questions about your view on religious and science. I ask you to respond on a 10-point scale, where the figure 10 means that you fully agree with the statement and the figure 1 that you fully disagree.

1. The more science advances, the more difficult will it be to believe in a God which created the world.

   1 2 3 4 5 6 7 8 9 10 0

2. When religious faith and science contradict each other, religion is always right.

   1 2 3 4 5 6 7 8 9 10 0

3. Which of the following alternatives accords best with how you view science and religion. Science and religion . . .
   a. Have no points of contact at all.
   b. Complement each other, express different aspects of one and the same reality.
   c. Stand in opposition to each other.

4. Which of the following statements suits yourself best?
   a. I believe in a God to which you can have a personal relationship.
   b. I believe in an impersonal higher power, force or energy.
   c. I believe that God is something within each human being and not something beyond our existence.
   d. I do not at all believe in a God, supernatural power or force.
   e. I do not know what to believe (not read).

C

I will now describe some different personality types and ask you to answer how much you think that these different persons are like you. We are now using a 6-point scale, where the figure 6 means that your resemble the person very much and the figure 1 that you do not do it all.

Scale 1-6 plus 0 = do not know.

> V80 To think of new ideas and be creative is important for this person, who wants to do things in his or her own way.
>
> V81 It is important for this person to be rich and have lots of money and expensive things.
>
> V82 It is important for this person to live in a safe environment and to stay away from dangerous things.
>
> V83 To have a good time and enjoy yourself is important to this person who likes to spoil his or her self.
>
> V84 It is important for this person to help human beings in the neighbourhood and see to that the have a good life.
>
> V85 To be very successful is important to this person, who wants people to acknowledge the achievements you have accomplished.
>
> V86 For this person it is important with adeventures and risks ant to live an exciting life.
>
> V87 It is important for this person to always behave correctly and to evade to do things which people would consider wrong.
>
> V88 To really care for nature and to protect the environment is important for this person.
>
> V89 Traditions are important for this person. To follow the customs which you learned from your religion or from your family.

Finally I have some background questions:

1. What is your age?
    a. 18-34 years
    b. 35-54 years
    c. 55-74 years

2. Which is your highest completed education?
   a. a. Primary school or equivalent
   b. b. Gymnasium or equivalent
   c. University
3. What is your main occupation?
   a. Employed (incl. self employed)
   b. Working at home
   c. Student
   d. Ritered
   e. Sick leave
   f. In search of a job
   g. Other
4. Which type of area do you live in?
   a. Big city (Stockholm,Gothenburg, Malmo)
   b. Minor city
   c. Countryside
   d. No answer

Thank you and conclude

Make note of

1. Man
2. Woman

Appendix 3

Web-based Questionnaire on Science and Religion

Open from February 28, until March 15, 2015 (Study 6)

BACKGROUND

As part of a three-year project in collaboration between the Department of Theology at Uppsala University, the Church of Sweden, the Sigtuna Foundation, the study association Ibn Rushd, Uniting Church in Sweden, and Chalmers University, a web-survey consisting of a questionnaire was delivered to minsters in the Church of Sweden and pastors in Uniting Church in Sweden from February 28, until March 15, 2015. (The original idea was that even imams and rabbis would be included in the survey but since the number of email addresses of these were not sufficient to guarantee full anonymity, the answers of these respondents are not reported here.) The purpose of the survey was to enhance knowledge about the perception of the Swedish clergy on issues about science and religion. It was part of the Cusanus-project, financially supported by the John Templeton Foundation in the U.S. and was also directed to the wider community of believers as well as the general public.

The project leader, Carl Reinhold Bråkenhielm, was responsible for the design of the questionnaire, Ingrid Friberg (Samhällsinformation AB) administered the sample and made the basic analysis of the results together with Jan Nylund.

METHOD

The survey consisted of items with standardized response options (mainly 10-point scale), as well as open questions where the respondent had to answer in their own words. E-mails concerning the web-based questionnaire were distributed to more than 3000 active minsters in the Church of Sweden and active pastors in the Uniting Church in Sweden. Some email addresses were incorrect, some were duplicates, and some people were no longer active as a minister and therefore refrained from answering.

The first e-mail was sent on February 26, 2015. The mail had an attachment with detailed information about the project and an internet-link to the webpage with the questionnaire. A first reminder, to those who had not responded to the questionnaire, expired on March 5 and then reminded it twice, March 12 and March 19. The reminders were only available to people who did not answer the questions at the time of reminder mailings.

1492 persons answered all the questions in the questionnaire. This corresponds to a response rate of about 50 percent, which is a good result in comparison to what is normally the case with web-based questionnaires. A special dropout-study was conducted concerning the percentage of (1) men and women, and (2) minsters and pastors. The study revealed no significant differences.

WEB-SURVEY (TRANSLATED FROM SWEDISH)

Thank you for responding to the questions below!

Respond throughout on a 10-point scale, where the figure 10 means that you fully agree with the statement and the figure 1 that you fully disagree.

I. First some questions about how you view biology and environment in various respects.

1. Human beings have certain unique qualities which no other creatures have.

    1 2 3 4 5 6 7 8 9 10 0

2. Humanity is approaching a self-inflicted environmental disaster.

    1 2 3 4 5 6 7 8 9 10 0

3. We must respect the order of nature irrespective of how this affects human welfare.

    1 2 3 4 5 6 7 8 9 10 0

4. Humans would be much better off if they lived a simple life without so much technology.

   1   2   3   4   5   6   7   8   9   10   0

5. Biology gives us the complete answer to the meaning of life.

   1   2   3   4   5   6   7   8   9   10   0

6. Humanity exists to govern the rest of creation.

   1   2   3   4   5   6   7   8   9   10   0

7. The more science advances, the more difficult is it to believe in a God who created the world.

   1   2   3   4   5   6   7   8   9   10   0

8. Humans consist only of body and matter.

   1   2   3   4   5   6   7   8   9   10   0

9. Nature is filled by a spiritual force permeating all life.

   1   2   3   4   5   6   7   8   9   10   0

10. Life has no other meaning than to further the human species.

    1   2   3   4   5   6   7   8   9   10   0

11. What becomes of a human being is mostly dependent on her biological heritage.

    1   2   3   4   5   6   7   8   9   10   0

OPEN FIELD FOR COMMENTARIES: _____

II. Now, we would like to put some questions about your view on the relationship between science and religion.

1. When religious faith and science contradict each other, religion is always right.

   1   2   3   4   5   6   7   8   9   10   0

2. Natural science and religion can be united because they are two totally different perspectives on reality with different questions and purposes.

   1   2   3   4   5   6   7   8   9   10   0

3. Natural science and religion are two different activities, but they can nevertheless come into contact with each other and affect each other's content.

   1   2   3   4   5   6   7   8   9   10   0

4. It is important to reinterpret the Christian/Jewish/Muslim faith so that it at all points is consistent with the scientific worldview.

   1 2 3 4 5 6 7 8 9 10 0

5. Christian/Jewish/Muslim faith rests on a scientific foundation.

   1 2 3 4 5 6 7 8 9 10 0

6. Religion should to a higher degree influence the direction and methods of natural science.

   1 2 3 4 5 6 7 8 9 10 0

OPEN FIELD FOR COMMENTARIES: _____

III. We continue with some questions about what you would like to know about contemporary science

1. I think I know enough about contemporary science and its relationship to religion.

   1 2 3 4 5 6 7 8 9 10 0

2. General knowledge about science is important, but in my role as a religious leader/minster/pastor/imam it is more important for me to go deeper into my own religious tradition.

   1 2 3 4 5 6 7 8 9 10 0

3. Deeper knowledge in science and its relationship to religion should be of great significance in my work as a religious leader/minister/pastor/imam.

   1 2 3 4 5 6 7 8 9 10 0

OPEN FIELD FOR COMMENTARIES: _____

IV. Here are some more open question, which you yourself can give a response to:

1. In your activity as a religious leader/minister/pastor/imam have you encountered especially important questions about science and its relation to religion?

   _____
   _____
   _____

2. If you were offered the opportunity, would you be willing to participate in courses and conferences where scientific researchers present their science and engage in dialogue about science and religion with representatives of different religions?

   1. Yes, willingly
   2. Yes, maybe
   3. Doubtful
   4. No

   Welcome to comment your answer!

   _____

   _____

   _____

3. Have your view on the relationship between science and religion changed during your activity as a religious leader/minister/pastor/imam?

   1. Yes, to a high degree
   2. Yes, to a certain degree
   3. No, hardly
   4. No, not at all

   Welcome to comment your answer!

   _____

   _____

   _____

4. Do you think that your view on science and religion would change if you were given the opportunity to obtain increased knowledge about science and religion?

   1. Yes
   2. No
   3. Doubtful

Welcome to comment your answer!

_____

_____

_____

V. Finally, we have some background questions:
  1. Which year were you born?
  2. To which of the following groups do you belong?
     1. Minister in the Church of Sweden
     2. Pastor in a free denominational church
     3. Imam in a Muslim community
     4. Other: _____
  3. Which type of area do you live in?
     1. Stockholm
     2. Gothenburg
     3. Malmö
     4. Uppsala
     5. Umeå
     6. Other urban area
     7. Sparsely populated area, less than 500 inhabitants
  4. Which is your highest completed education?
     1. High school
     2. Academic exam (not doctoral exam)
     3. Academic exam (doctoral exam)
     4. Other, what? _____
  5. Are you a man or a woman?
     1. Man
     2. Woman
     3. Other

Would you be interested in receiving continued information about the project Science and religion—a platform for dialogue and education in Sweden?

If you are interested, please note the e-mail address below in the field so that we can reach you with a newsletter.

As a supplement to this survey we will conduct some shorter telephone-interviews with a sample of the respondents that have responded to the questionnaire.

If you want to participate, we ask you kindly to register your notes of contact below. As the responses to the survey, all responses in the telephone-interviews will be treated strictly confidential and not be linked to the responses to the questionnaire.

Thank you for taking your time!

Appendix 4

Cluster-analysis of responses to telephone-interviews

A national sample of the Swedish adult population
(Study 1, May 2006, and Study 2, May-June, 2017)

INTRODUCTION

The purpose of cluster analysis was to (1) examine whether there are population segments ("clusters") with substantially different approach to current worldview items and (2) describe how these segments differ in terms of both philosophical and socio-demographic questions. As input in cluster analysis was all worldview items used in web-based questionnaire (see Appendix 2).

The cluster analysis resulted in a total of 73 percent (2006) and 69 percent (2017) were distributed in 6 clusters, each with a special profile beliefs, values, personality-types. The remaining 27 percent (2006) and 31 percent (2017) persons were placed in a residual group ("cluster 0") without any apparent or distinctive profile.

The final cluster solution from 2006 is stable. The probability that a person repeatedly cluster analysis ends up in the same cluster is 80-90 percent (except for cluster 0, which was 65 percent).

The total population is relatively homogeneous in their approach to many of the questions. The differences between the clusters are relatively small.

FACTOR DESCRIPTION (FOR ITEMS, SEE (SEE APPENDIX 2)

*Factor 1—traditional religion—*is constituted by a correlation between responses to five items, namely item 9 about God's creation, 10 about evolution realizing a plan, 11 about the randomness of evolution (negatively), 14 about naure existing for sake of humans (positively), and 17 about human soverignity (positively). The cluster has close affinity to cluster 2 and shows higher than average affirmation than any other cluster of V81 (the importance of wealth!). Sociodemographic profile (2017): higher than average women over men with university education and on sick leave.

*Factor 2—spiritualism—*emerges from a correlation between responses relating to a spiritual presence in nature, namely item 18, 20 (negatively), 23, and 25. The cluster is also higher than average on V80 (underlining creativity and individualism). Sociodemographic profile (2017): lower than average 18-34 years old, and higher than average women.

*Factor 3—biologism—*is derived from a correlation between responses to question about the significance of biology, i.e. item 21 on biology teaching that only the strongest survive, item 22 on reproduction as the meaning of life, and item 24 about the significance of biological inheritance. This cluster also displays higher than average affirmation of item 7 on biology giving the answer to the meaning of life, and V82, V84 and V87-89 about traditional values of correct behavior. On the other hand, the cluster does *not* display higher than average on item 11 about random evolution, yet positive towards item 10 about evolution realizing a certain plan. Furthermore, this cluster is also higher than average on items indicating environmental concern (item 88, 12, 15)as well as showing a certain sympathy for technology and anthropocentrism (item 1, 6 and 17). Sociodemographic profile (2017): higher than average men, higher than average primary school education only and lower than average university education, lower than average from big cities and higher than avarage from small cities.

*Factor 4—anti-environmentalism—*is constituted by a correlation between items on the environment showing low concern (item 5, 12, 13, and 15) and lower than average on V 84 and V87-89 (in stark contrast to cluster 1-3). No significant sociodemographic profile (2017).

*Factor 5—skepticism—*arises from a correlation between responses to item 18 about holiness in nature (negatively), 20 about humans being only matter (positively), 23 about the ensoulment of nature (negatively) and 25 about God (negatively). In contrast to cluster 4, cluster 5 is more environmentally concerned (items 5, 12, 13, and 15), but display a strong affirmation of the irreconciability between science and the Christian doctrine of creation (item 19) and the randomness of the evolutionary process (item 11). Sociodemographic profile: young, well-educated men in big cities.

*Factor 6—environmental pessimism—*shows high environmental concern (higher than average on item 12, 15, and 15), but with stronger than average belief in human uniqueness (item 6) and the randomness of the evolutionary process (item 11). No significant sociodemographic profile (2017).

Appendix 5

Cluster-analysis of web-survey among Swedish clergy

(Study 6, February- March, 2015)

INTRODUCTION

The purpose of cluster analysis was to (1) examine whether there are population segments ("clusters") with substantially different approach to current worldview items and (2) describe how these segments differ in terms of both philosophical and socio-demographic questions. As input in cluster analysis was all worldview items used in web-based questionnaire (see Appendix 1).

The cluster analysis resulted in a total of 1139 (76 percent) persons were distributed in 6 clusters, each with a special personality profile of human values and beliefs. The remaining 353 (24 percent) persons were placed in a residual group ("cluster 0") without any apparent or distinctive profile.

The final cluster solution is stable. The probability that a person repeatedly cluster analysis ends up in the same cluster is 82 percent. In most cases, the stability value was between 70–90 percent.

The total population is relatively homogeneous in their approach to many of the questions. The differences between the clusters are relatively small.

FACTOR DESCRIPTION

*Factor 1—anthropocentrism—*is constituted by a correlation between responses to five different items in the web survey (appendix 1), namely item 6 (in section 1) + item 1 (in section 2) + item 1 (in section 1) + item 2 (in section 3) and item 4 (section 2). It is named "anthropocentrism" after the responses to the question "humanity exists to govern over the rest of creation," i.e. the question that differentiates most clearly between the different clusters.

*Factor 2—anti-science—*is derived from a correlation between responses to question about the significance of science, namely item 10 (in section 1) + item 5 (in section 1) + item 7 (in section 1) + item 8 (in section 1) and item 11 (section 1).

*Factor 3—environmentalism—*emerges from a correlation between responses about environmental concern, namely item 3 (in section 1) + item 2 (in section 1) + item 4 (in section 1) + item 9 (in section 1).

*Factor 4—dialogue (between science and religion)—*is constituted by a correlation between two questions about the relationship between science and religion, namely item 2 (in section 2) + item 3 (in section 2). Item 3 expresses a model of contact, while item 2 expresses a model of independence.

*Factor 5—integration—*arises from a correlation between two responses to two other questions about the relationship between science and religion, namely item 5 (in section 2) and item 6 (in section 2).

*Factor 6—deeper knowledge—*has to do with the perceived need for more knowledge about science and its relationship to religion, namely item 1 (in section 3) and item 3 (in section 3).

Appendix 6

Results of Study 1 & Study 8 (Percent)

A

	2006	2017

1. Technology will be able to solve the problems with our limted resources of nature.

	2006	2017
1–2) Disagree	23	27
1. Disagree completely	8	10
2.	15	16
3.	41	38
4.	25	23
5. Agree completely	10	11
4–5) Agree	35	34
Mean value	3, 2	3, 1
Don´t know	1	

2. Humans would be much better off if they lived a simple life without so much technology.

	2006	2017
1–2) Disagree	33	34
1. Disagree completely	13	13
2.	20	21
3.	31	32
4.	22	21
5. Agree completely	13	12
4–5) Agree	35	34
Mean value	3	3
Don´t know		0

	2006	2017
3. Scientific research creates more problems than solutions.		
1–2) Disagree	60	64
1. Disagree completely	32	42
2.	28	22
3.	24	23
4.	12	8
5. Agree completely	3	4
4–5) Agree	15	12
Mean value	2, 3	2, 1
Don´t know	1	0
4. The development of nuclear power is the only alternative to the use of coal and hydropower.		
1–2) Disagree	52	65
1. Disagree completely	27	40
2.	25	25
3.	19	23
4.	16	8
5. Agree completely	10	3
4–5) Agree	26	11
Mean value	2, 5	2, 1
Don´t know	3	1
5. Humanity is approaching a self-inflicted environmental disaster.		
1–2) Disagree	6	4
1. Disagree completely	11	8
2.	25	21
3.	31	31
4.	27	36
5. Agree completely	58	67
4–5) Agree		
Mean value	3, 6	3, 9
Don´t know	0	0
6. Human beings have certain unique qualities which no other creatures have.		
1–2) Disagree	6	9
1. Disagree completely	2	4
2.	4	6
3.	14	16
4.	24	27
5. Agree completely	56	48
4–5) Agree	80	75
Mean value	4, 3	4, 1
Don´t know	1	

	2006	2017
7. Biology gives us the complete answer to the meaning of life.		
1–2) Disagree	47	45
1. Disagree completely	22	21
2.	25	24
3.	30	36
4.	13	13
5. Agree completely	7	5
4–5) Agree	20	18
Mean value	2, 6	2, 6
Don´t know	3	0
8. We must respect the order of nature irrespective of how this affects human welfare.		
1–2) Disagree	8	10
1. Disagree completely	3	3
2.	5	7
3.	20	27
4.	34	35
5. Agree completely	38	27
4–5) Agree	72	63
Mean value	4	3, 8
Don´t know	1	1
9. God created human beings at a specific moment.		
1–2) Disagree	70	78
1. Disagree completely	56	69
2.	14	9
3.	13	12
4.	5	4
5. Agree completely	8	6
4–5) Agree	13	10
Mean value	1, 9	1, 7
Don´t know	3	1
10. Human beings have evolved through a goaldirected process which is realizing a certain plan.		
1–2) Disagree	49	52
1. Disagree completely	29	33
2.	20	19
3.	27	25
4.	11	12
5. Agree completely	8	8
4–5) Agree	19	20
Mean value	2, 5	2, 4
Don´t know	5	2

	2006	2017

11. Human beings have emerged through a random biological process.

	2006	2017
1–2) Disagree	36	29
1. Disagree completely	20	18
2.	16	11
3.	19	23
4.	23	22
5. Agree completely	18	24
4–5) Agree	41	46
Mean value	3	3, 2
Don't know	4	2

12. The balance in nature is vary fragile and is easily disturbed by human influence.

	2006	2017
1–2) Disagree	4	6
1. Disagree completely	1	1
2.	3	5
3.	10	15
4.	31	33
5. Agree completely	54	47
4–5) Agree	85	79
Mean value	4, 3	4, 2
Don't know	0	0

13. The earth is like a speceship with limited space and limited resources.

	2006	2017
1–2) Disagree	10	8
1. Disagree completely	4	4
2.	6	5
3.	18	15
4.	32	30
5. Agree completely	40	46
4–5) Agree	72	76
Mean value	4	4, 1
Don't know	1	0

14. Plants and animals exist only for the purpose of human beings.

	2006	2017
1–2) Disagree	74	74
1. Disagree completely	51	55
2.	23	19
3.	13	15
4.	8	5
5. Agree completely	5	5
4–5) Agree	13	10
Mean value	1, 9	1, 9
Don't know		

	2006	2017
15. Changes in the environment for human purposes creates serious problems.		
1–2) Disagree	7	8
1. Disagree completely	2	1
2.	5	6
3.	23	27
4.	36	36
5. Agree completely	34	30
4–5) Agree	70	66
Mean value	4	3,9
Don't know	1	
16. There are no limits for growth in a country like Sweden.		
1–2) Disagree	58	51
1. Disagree completely	25	24
2.	33	27
3.	27	32
4.	10	11
5. Agree completely	3	4
4–5) Agree	13	15
Mean value	2,3	2,4
Don't know	2	1
17. Humanity exists to govern over the rest of nature.		
1–2) Disagree	71	70
1. Disagree completely	50	51
2.	21	18
3.	19	18
4.	6	8
5. Agree completely	3	4
4–5) Agree	9	12
Mean value	1,9	2
Don't know		
18. I encounter the holy primarily in nature.		
1–2) Disagree	23	26
1. Disagree completely	15	16
2.	29	29
3.	18	17
4.	10	10
5. Agree completely	28	27
4–5) Agree		
Mean value	2,8	2,7
Don't know	5	1

APPENDIX 6: RESULTS OF STUDY 1 & STUDY 8

	2006	2017

19. Christian belief in creation is irreconcialable with the biological doctrine of evolution.

	2006	2017
1–2) Disagree	39	36
1. Disagree completely	19	18
2.	20	18
3.	28	35
4.	11	11
5. Agree completely	14	16
4–5) Agree	25	27
Mean value	2, 8	2, 9
Don´t know	7	2

20. Humans consist only of body and matter.

	2006	2017
1–2) Disagree	53	49
1. Disagree completely	33	29
2.	20	21
3.	17	24
4.	11	13
5. Agree completely	16	13
4–5) Agree	27	26
Mean value	2, 6	2, 6
Don´t know	2	1

21. Biology teaches that only the strongest survive.

	2006	2017
1–2) Disagree	27	34
1. Disagree completely	14	15
2.	13	18
3.	28	27
4.	28	26
5. Agree completely	17	14
4–5) Agree	45	40
Mean value	3, 2	3
Don´t know		

22. Life has no other meaning than to further the human species.

	2006	2017
1–2) Disagree	49	51
1. Disagree completely	30	28
2.	19	23
3.	22	28
4.	15	13
5. Agree completely	12	7
4–5) Agree	27	20
Mean value	2, 6	2, 5
Don´t know	1	0

	2006	2017

23. Nature is filled by a spiritual force permeating all life.

	2006	2017
1–2) Disagree	25	30
1. Disagree completely	12	14
2.	13	16
3.	27	33
4.	26	20
5. Agree completely	19	16
4–5) Agree	45	36
Mean value	3, 3	3, 1
Don´t know	3	1

24. What becomes of a human being is mostly dependent on her biological heritage.

	2006	2017
1–2) Disagree	38	38
1. Disagree completely	13	13
2.	25	25
3.	35	43
4.	18	14
5. Agree completely	7	4
4–5) Agree	25	18
Mean value	2, 8	2, 7
Don´t know	1	0

25. There is a god, a supreme power or force.

	2006	2017
1–2) Disagree	48	60
1. Disagree completely	33	45
2.	15	15
3.	17	16
4.	16	10
5. Agree completely	16	13
4–5) Agree	32	23
Mean value	2, 7	2, 3
Don´t know	3	1

B

Year	2000	2017

26. The more science advances, the more difficult will it be to believe in a God which created the world.

	2000	2017
1–3) Disagree		29
1. Disagree completely		14
2.		6
3.		9
4.		5
5.		18
6.		5
7.		9
8.		15
9.		3
10. Agree completely		15
8–10) Agree	48	33
Mean value		5, 6
Don´t know		1

27. When religious faith and science contradict each other, religion is always right.

	2000	2017
1–3) Disagree		84
1. Disagree completely		64
2.		14
3.		6
4.		4
5.		8
6.		1
7.		1
8.		1
9.		0
10. Agree completely		1
8–10) Agree		2
Mean value		2
Don´t know		0

28. Which of the following alternatives accords best with how you view science and religion. Science and religion . . .

	1999	2017
a. . . . have no points of contact at all.	24	22
b. . . . complement each other, express different aspects of one and the same reality.	57	56
c. . . . stand in opposition to each other.	19	22

	2000	2017
29. Which of the following statements suits yourself best?		
a. I believe in a God to whom you can have a personal relationship.	19	11
b. I believe in an impersonal higher power, force or energy.	27	24
c. I believe that God is something within each human being and not something beyond our existence.	21	24
d. I do not at all believe in a God. Supernatural power or force.	15	40
e. I don not know what to believe. (Not an explicit alternative in 2017.)	18	1

C

Year	2006	2017
80. Which of the following statements suits yourself best?		
1–2) Does not resemble	5	5
1. Does not at all resemble the person	1	2
2.	4	3
3.	16	14
4.	35	27
5.	29	27
6. Resembles the person very much	16	27
5–6) Resembles	45	54
Mean value	4, 3	4, 6
Don´t know	0	0
81. It is important for this person to be rich and have lots of money and expensive things.		
1–2) Does not resemble	59	66
1. Does not at all resemble the person	23	30
2.	36	36
3.	25	23
4.	12	9
5.	2	1
6. Resembles the person very much	1	0
5–6) Resembles	3	1
Mean value	2, 4	2, 2
Don´t know	1	

	2006	2017

82. It is important for this person to live in a safe environment and to stay away from dangerous things.

	2006	2017
1–2) Does not resemble	8	11
1. Does not at all resemble the person	2	3
2.	6	7
3.	19	22
4.	30	24
5.	30	25
6. Resembles the person very much	12	19
5–6) Resembles	42	44
Mean value	4, 2	4, 2
Don´t know	1	

83. To have a good time and enjoy yourself is important to this person who likes to spoil his or her self.

	2006	2017
1–2) Does not resemble	17	17
1. Does not at all resemble the person	4	5
2.	13	12
3.	26	32
4.	26	26
5.	21	18
6. Resembles the person very much	10	8
5–6) Resembles	31	25
Mean value	3, 8	3, 6
Don´t know	1	

84. It is important for this person to help human beings in the neighbourhood and see to that the have a good life.

	2006	2017
1–2) Does not resemble	1	1
1. Does not at all resemble the person	0	1
2.	1	1
3.	4	6
4.	20	17
5.	37	32
6. Resembles the person very much	37	43
5–6) Resembles	74	75
Mean value	5	5, 1
Don´t know	1	

85. To be very successful is important to this person, who wants people to acknowledge the achievements you have accomplished.

	2006	2017
1–2) Does not resemble	30	36
1. Does not at all resemble the person	9	12
2.	21	24
3.	29	33

	2006	2017
4.	25	21
5.	13	8
6. Resembles the person very much	3	2
5–6) Resembles	16	10
Mean value	3, 2	3
Don´t know	1	

86. For this person it is important with adeventures and risks ant to live an exciting life.

1–2) Does not resemble	38	32
1. Does not at all resemble the person	11	10
2.	27	22
3.	32	28
4.	16	25
5.	10	11
6. Resembles the person very much	3	4
5–6) Resembles	13	15
Mean value	3	3, 2
Don´t know	1	

87. It is important for this person to always behave correctly and to evade to do things which people would consider wrong.

1–2) Does not resemble	26	26
1. Does not at all resemble the person	8	9
2.	18	17
3.	25	31
4.	23	23
5.	19	15
6. Resembles the person very much	6	5
5–6) Resembles	25	20
Mean value	3, 4	3, 3
Don´t know	1	

88. To really care for nature and to protect the environment is important for this person.

1–2) Does not resemble	3	4
1. Does not at all resemble the person	0	1
2.	3	3
3.	12	15
4.	29	21
5.	33	29
6. Resembles the person very much	21	31
5–6) Resembles	54	61
Mean value	4, 6	4, 7
Don´t know	1	

	2006	2017
89. Traditions are important for this person. To follow the customs which you learned from your religion or from your family.		
1–2) Does not resemble	22	32
1. Does not at all resemble the person	6	12
2.	16	20
3.	20	23
4.	24	18
5.	22	19
6. Resembles the person very much	11	8
5–6) Resembles	33	26
Mean value	3,7	3,3
Don't know	1	0
Which is your age?		
18–34 year	30	31
35–54 year	39	37
55–74 year	31	32
Which is your age?		
a. Primary school or equivalent	17	7
b. Gymnasium or equivalent	47	45
c. University	35	48
Which is your main occupation?		
a. Employed (incl. self-employed)	69	68
b. Working in home	0	1
c. Student	8	8
d. Ritered	17	18
e. Sick leave	2	3
f. In search of a job	2	2
g. Other	1	1
Which is your type of residence area?		
a. Big city (Stockholm, Göteborg, Malmö)	25	38
b. Smaller city	58	49
c. Countryside	16	12
d. No answer		
Gender		
Man	50	50
Woman	50	50

Bibliography

Albert, David. Review of *A Universe from Nothing*, by Lawrence Krauss. *The New York Times*, March 23, 2012. http://www.nytimes.com/2012/03/25/books/review/a-universe-from-nothing-by-lawrence-m-krauss.html.

Alston, William. *Divine Nature and Human Language: Essays in Philosophical Theology*. Ithaca: Cornell University Press, 1989.

———. *Perceiving God. The Epistemology of Religious Experience*. Ithaca: Cornell University Press, 1991.

Andrae, Tor. *Nathan Söderblom*. Uppsala: Lindblad, 1931.

Anér, Kerstin. "Teilhard de Chardin, en presentation." In *Teilhard de Chardin: Teologi, filosof, naturvetare*, edited by Jarl Hemberg, 12–29. Stockholm: Verbum, 1973.

———. *Framtidens insida. Om en uthärdlig värld för våra barn*. Stockholm: Proprius, 1978.

Augustine. *Enchiridion on Faith, Hope, and Love*. Translated by Albert C. Outler. Philadelphia: Westminster, 1955. http://www.tertullian.org/fathers/augustine_enchiridion_02_trans.htm.

———. *On the Trinity*. Translated by Arthur West Hadden. Edinburgh: T. & T. Clark, 1872. http://www.logoslibrary.org/augustine/trinity/index.html.

Aulén, Gustaf. *Christus Victor: An Historical Study of the Three Main Types of the Idea of Atonement*. Translated by A. G. Herbert. Eugene, OR: Wipf and Stock, 2003.

Aurelius, Erik. "Olikheter ryms i Nya testamentet." In *Kyrkans tidning*, January 2, 2014.

Barbour, Ian. *Issues in Science and Religion*. London: SCM, 1972.

———. *Religion and Science: Historical and Contemporary Issues*. San Francisco: HarperSanFrancisco, 1997.

———. *Religion in an Age of Science*. San Francisco: HarperSanFrancisco, 1990.

Bazeley, Partricia, and Kristi Jackson. *Qualitative Data Analysis with NVivo*. 2nd ed. Los Angeles: SAGE, 2013.

Bejerholm, Lars. *Harald Eklunds religionsfilosofi: Efterlämnade uppsatser—Bibliografi*. Lund: Gleerup, 1964.

Bergson, L'Evolution creatrice. Paris: Alcan, 1908. https://archive.org/details/levolutioncreatrooberguoft.
A Bishops' Letter about the Climate. Uppsala: Church of Sweden, Bishops' Conference, 2014. https://www.svenskakyrkan.se/omoss/biskoparnas-brev-om-klimatet.
Björkman, Jenny, and Arne Jarrick, eds. Religionen tur och retur. RJ:s årsbok 2017/2018. Göteborg: Makadam, 2017.
Bóchenski, Jósef Maria. The Logic of Religion. New York: New York University Press, 1965.
Bohlin, Torsten. Den ofrånkomlige. Stockholm: Sveriges kristliga studentrörelse, 1932.
———. Tro och uppenbarelse. En studie till teologins kris och "krisens teologi." Stockholm: Diakonistyrelsen, 1926.
Boström, Christopher Jacob. Philosophy of Religion. Translated by Victor E. Beck and Robert N. Beck. New Haven, CT: Yale University Press, 1962.
———. Review of Philosophy of Religion, by Ninian Smart. Philosophical Quarterly 14 (1964) 381.
Braithwaite, Richard Bevan. An Empiricit's View of the Nature of Religious Belief. Cambridge: Cambridge University Press, 1955.
Bring, Ragnar. Gud och människan: Ett alternativ och en kritisk kommentar till det i Uppsala utgivna arbetet "Människan och Gud. En kristen teologi." Borås: Norma, 1987.
———. "Kristen tro och vetenskaplig forskning: Några reflektioner med anledning av den populära debatten om 'tro och vetande.'" Svensk teologisk kvartalsskrift 25 (1949) 201–243.
———. Till frågan om den systematiska teologiens uppgift: med särskild hänsyn till inom svensk teologi föreliggande problemställningar. Acta Universitatis Lundensis. Lund: Ohlsson, 1933.
Broad, Charlie Dunbar. Religion, Philosophy, and Psychical Research. New York: Harcourt, Brace, 1953
Brown, David. "Basil Mitchell 1917–2011." Biographical Memoirs of Fellows of the British Academy 12 (2013) 303–21. http://www.britac.ac.uk/sites/default/files/11%20Mitchell.pdf.
Brown, Warren S. "Conclusion: Reconciling Scientific and Biblical Portraits of Human Nature." In Whatever Happened to the Soul? Scientific and Theological Portraits of Human Nature, edited by Warren S. Brown et al., 213–28. Minneapolis: Fortress, 1998.
Bruce, Steve. Choice and Religion: A Critique of Rational Choice Theory. Oxford: Oxford University Press, 1999.
———. From Cathedrals to Cults: Religion in the Modern World. Oxford. Oxford University Press, 1996.
Bråkenhielm, Carl Reinhold. "Christian Tradition and Contemporary Society." Concilium 256 (1994) 23–34.
———. "Environmentalism as a Civil Religion: the Case of Sweden." In Constellations of Value: European Perspectives on the Intersections of Religion, Politics and Society, edited by Christoph Jedan, 49–66. Münster: LIT, 2012.
———. How Philosophy Shapes Theories of Religion: An Analysis of Contemporary Philosophies of Religion with Special Regard to the Thought of John Wilson, John Hick, and D Z Phillips. Lund: Gleerup, 1975.
———. "Livsåskådning och vetenskap." Religion och livsfrågor 1 (2017) 6–8.

———. "Religion och vetenskap." *Filosofisk tidskrift* 3 (1980) 41–48.
———. *Verklighetsbilder*. Nora: Nya Doxa, 2009.
Bråkenhielm, Carl Reinhold, ed. *Världsbild och mening: En empirisk studie av livsåskådningar i Sverige*. Nora: Nya Doxa, 2000.
Bråkenhielm, Carl Reinhold, and Mats G. Hansson. *Livets grundmönster och mångfald: En bok om genetic, etik och livsåskådning*. Stockholm: Liberutbildning, 1995.
Bråkenhielm, Carl Reinhold, et al. *Att bekänna människans värde: En bok från Bekännelsearbetets arbetsgrupp Människosyn och etik*. Stockholm: Verbum, 1992.
Bultmann, Rudolf. "Zum Problem der Entmythologisierung." In *Kerygma und Mythos*, edited by Hans Werner Bartsch, 2:179–208. Hamburg: Herbert Reich Evangelischer, 1952.
Capps, Walter H. *Religious Studies: The Making of a Discipline*. Minneapolis: Fortress, 1995.
Carroll, John W. "Laws of Nature." *Stanford Encyclopedia of Philosophy* (Fall 2016 ed.), edited by Edward N. Zalta. https://plato.stanford.edu/archives/fall2016/entries/laws-of-nature/.
Carroll, Sean. "Does the Universe Need God?" In *The Blackwell Companion to Science and Christianity*, edited by J. B. Stump, and Alan G. Padgett, 185–97. Oxford: Wiley & Sons, 2012.
Casanova, José. *Public Religion in the Modern Sphere*. Chicago: Chicago University Press, 1994.
Catechism of the Roman Catholic Church. Vatican City: Libreria Editrice Vaticana, 1993. http://www.vatican.va/archive/ENG0015 /_INDEX.HTM.
Chignell, Andrew. "The Ethics of Belief." In *The Stanford Encyclopedia of Philosophy Archive*, edited by Edward N. Zalta. Winter 2016 ed. https://plato.stanford.edu/archives/win2016/entries/ethics-belief/.
Clark, Kelly James. "God, Chance, and Purpose." PPT-presentation. http://slideplayer.com/slide/2905727/.
———. *Religion and the Sciences of Origins: Historical and Contemporary Discussions*. London: Palgrave Macmillan, 2014.
Clayton, Philip. "Eschatology as Metaphysics under the Guise of Hope." In *World without End: Christian Eschatology from a Process Perspective*, edited by Joseph A. Bracken, 128–49. Grand Rapids: Eerdmans, 2005.
———. "God in Process, World in Process: Constructive Theology and the New Engagement with the Sciences." Zygon Center for Religion and Science. November 9, 2015. http://zygoncenter.org/event/advanced-seminar-2015-purpose-process-science-theology-7/.
Clifford, William. *The Ethics of Belief: William K. Clifford, Lectures and Essays*. Edited by Leslie Stephen and Frederick Pollock. London: Macmillan, 1886. http://people.brandeis.edu/~teuber/Clifford_ethics.pdf.
Climate Feedback. "Scientists Explain What New York Magazine Article on 'The Uninhabitable Earth' Gets Wrong." Climate Feedback. https://climatefeedback.org/evaluation/scientists-explain-what-new-york-magazine-article-on-the-uninhabitable-earth-gets-wrong-david-wallace-wells/.
Coyne, Jerry A. *Faith Versus Fact: Why Science and Religion Are Incompatible*. New York: Viking, 2015.
Craig, William Lane. "God and Cosmology: The Existence of God in Light of Contemporary Cosmology." Debate with Sean Carroll. New Orleans

Baptist Theological Seminary. March 2014. http://www.reasonablefaith. org/god-and-cosmology-the-existence-of-god-in-light-of-contemporary-cosmology#ixzz4cKkg8rKf.

———. "What Is the Relation between Science and Religion?" Reasonable Faith. https://www.reasonablefaith.org/writings/popular-writings/science-theology/what-is-the-relation-between-science-and-religion/.

———. "#336 'Honesty, Transparency, Full Disclosure' and the Borde-Guth-Vilenkin Theorem." Reasonable Faith. September 23, 2013. https://www.reasonablefaith.org/writings/question-answer/honesty-transparency-full-disclosure-and-the-borde-guth-vilenkin-theorem#ixzz4hRNJj0rk.

Craig, William Lane, and Quentin Smith. *Theism, Atheism, and Big Bang*. Oxford: Clarendon, 1993.

Cullberg, John. *Das Du und die Wirklichkeit: Zum ontologischen Hintergrund der Geminschaftskategorie*. Uppsala: Uppsala universitets årsskrift, 1933.

———. *Den fördolde Guden: Studier till den bibliska gudsföreställningen*. Stockholm: Verbum, 1968.

———. "Religion och vetenskap än en gång." *Svensk teologisk kvartalsskrift* 4 (1931) 399–405.

———. *Religion och vetenskap: Till frågan om den systematiska teologins vetenskapliga grundläggning*. Stockholm: Svenska Kyrkans Diakonistyrelses, 1930.

———. Review of *Den fördolde Guden*, by Hampus Lyttkens. *Årsbok för kristen humanism* 31 (1969) 127–33.

———. *Tro och verklighet*. Stockholm, 1955.

Danielsson, Ulf. *Mörkret vid tidens ände: En bok om universums mörka sida*. Stockholm: Fri tanke, 2015.

Davie, Grace. *Europe: The Exceptional Case: Parameters of Faith in the Modern World*. London: Darton, Longman and Todd, 2010.

———. *Religion in Modern Europe: A Memory Mutates*. Oxford: Oxford University Press, 2000.

Dawkins, Richard. *The Blind Watchmaker*. London: Penguin, 1991.

———. *The God Delusion*. London: Bantham, 2006.

———. "God vs. Science, Richard Dawkins and Francis Collins interviewed by D. Cray." *Time Magazine*, November 5, 2006. Republished at http://inters.org/Dawkins-Collins-Cray-Science.

de Vries, Paul. "Naturalism in the Natural Sciences." *Christian Scholar's Review* 15 (1986) 388–96.

Dell Inc. *Statistics: Methods and Applications*. Dell, 2013. https://i4iam.files.wordpress.com/2013/09/statistics-methods-and-applications.pdf.

Dennett, Daniel. *Darwin's Dangerous Idea. Evolution and the Meaning of Life*. New York: Simon and Schuster, 1996.

Domning, Daryl P. *Original Selfishness: Original Sin and Evil in the Light of Evolution*. Burlington, VT: Ashgate, 2006.

Draper, Paul "Evolution and the Problem of Evil." In *Philosophy of Religion: An Anthology*, edited by Michael Rea and Louis P. Pojman, 271–82. 7th ed. Stamford, CT: Cenage Learning, 2014.

———. "Natural Selection and the Problem of Evil." *The SecularWeb*. https://infidels.org/library/modern/paul_draper/evil.html.

Drees, Willem B. *Beyond the Big Bang: Quantum Cosmologies and God*. La Salle: Open Court, 1990.

———. "Cosmology as Contact between Science and Theology." *Revista Portuguesa de Filosofia* 63 (2007) 533–53.

Ecklund, Elaine Howard. *Science vs. Religion: What Scientists Really Think*. Oxford: Oxford University Press, 2010.

Ecklund, Elaine Howard, and Christopher Scheitle. "Religious Communities, Science, Scientists, and Perceptions: A Comprehensive Survey." A paper prepared for presentation at the annual meeting of the American Association for the Advancement of Science. February 16, 2014. http://www.aaas.org/sites/default/files/content_files/RU_AAASPresentationNotes_2014_0219%20(1).pdf

Ecklund, Elaine Howard, et al. "Religion among Scientists in International Context: A New Study of Scientists in Eight Regions." *Socius: Sociological Research for a Dynamic World* 2 (2016) 1–9. http://journals.sagepub.com/doi/pdf/10.1177/2378023116664353.

Eddington, Arthur. *New Pathways in Science*. New York: Macmillan, 1935.

Ehnmark, Erland. "Nathan Söderbloms ställning till problemet tro och vetande." *Vår Lösen* 3 (1945) 83–98.

Eklund, Harald. "Kristendom och vetenskap. Några reflexioner." *Ad Lucem* 36, no. 2 (1945) 10.

———. "Nystestamentlig och modern verklighetssyn." *Svensk exegetisk årsbok* 13 (1948) 1–8.

———. "On the Logic of Creeds." *Theoria* 22 (1956) 75–82.

———. Review of *Troslärans perspektiv*, by Axel Gyllenkrok. *Svensk teologisk kvartalsskrift* 4 (1948) 293–94.

———. *Tro, erfarenhet, verklighet*. Stockholm: Natur och Kultur, 1956.

———. *Troslärans perspektiv*. Åbo: Åbo Akademi, 1944.

Emmet, Dorothy Mary. *The Effectiveness of Causes*. London: Palgrave Macmillan, 1984.

"The Falsification Principle." https://revisionworld.com/a2-level-level-revision/religious-studies/philosophy-religion/attributes-god/falsification-principle.

Finnish Society for Scientific Information. *Finnish Science Barometer 2007: A Study of the Finns' Attitudes towards Science and Their Opinions on Scientific and Technological Progress*. Translated by Malcolm Hicks. Helsinki: Yliopistopaino, 2008. http://www.sci.fi/~yhdys/tb3/Finnish%20Science%20Barometer%202007.pdf.

———. *Summary of Finnish Science Barometer 2013: A Study of the Finns' Attitudes towards Science and Their Opinions on Scientific and Technological Progress*. Helsinki: Finnish Society for Scientific Information, 2013. http://www.tieteentiedotus.fi/files/Sciencebarometer_2013_netsummary.pdf.

Fogelklou, Emilia. *Barhuvad*. Stockholm: Bonnier, 1951.

———. *Medan gräset gror: En bok om det växande*. Vol. 2. Stockholm: Bonnier, 1911.

Forell, Urban. "Begreppet 'Guds verksamhet' i ljuset av naturvetenskapliga förklaringstyper." In *Religion, erfarenhet, verifikation*, edited by Urban Forell and Hampus Lyttkens, 33–51. Lund: Studentlitteratur, 1970.

———. "Gud och rummet." *Vår Lösen* 2 (1964) 64–71.

———. *Wunderbegriffe und logische Analyse: logisch-philosophische Analyse von Begriffen und Begriffsbildungen aus der deutschen protestantischen Theologie des 20. Jahrhunderts*. Göttingen: Vandenhoeck & Ruprecht, 1967.

Friedrichsen, Anton. "Epilog." *Svensk exegetisk årsbok* 13 (1948) 116–23.
Fries, Martin. *Metafysiken i modern svensk teologi*. Stockholm: Natur och kultur, 1948.
Fuller, Steve, Mikael Stenmark, and Ulf Zackariasson, eds. *The Customization of Science: The Impact of Religious and Political Worldviews on Contemporary Science*. Basingtoke, UK: Palgrave Macmillan, 2014.
Funk, Cary, and Becka A. Alper. "Religion and Views on Climate and Energy Issues." *Pew Research Center*, 2015. http://www.pewinternet.org/2015/10/22/religion-and-views-on-climate-and-energy-issues/.
Gerholm, Tor Ragnar. *Fysiken och människan: en introduktion till den moderna fysikens världsbild*. Stockholm: Aldus, 1962.
Gilkey, Langdon. *Maker of Heaven and Earth: The Christian Doctrine of Creation in the Light of Modern Knowledge*. Garden City, NY: Doubleday, 1959.
Grant, Edward. *The Foundations of Modern Science in the Middle Ages: Their Religious, Institutional, and Intellectual Contexts*. Cambridge: Cambridge University Press, 1996.
Grantén, Eva-Lotta. *Patterns of Care: Relating Altruism in Sociobiology and the Christian Tradition of Agape*. Lund Studies in Ethics and Theology. Lund: Centre for Theology and Religious Studies, Lund University, 2004.
———. *Utanför paradiset: Arvsyndsläran i nutida luthersk teologi och etik*. Stockholm: Verbum, 2013.
Green, Joel B. "Bodies—That Is, Human Lives: A Re-Examination of Human Nature in the Bible." In *Whatever Happened to the Soul? Scientific and Theological Portraits of Human Nature*, edited by Warren S. Brown et al., 149–74. Minneapolis: Fortress, 1998.
Gregersen, Niels Henrik, and Ulf Görman, eds. *Design and Disorder: Perspectives from Science and Theology*. London: T. & T. Clark, 2002.
Gould, Stephen Jay. "Nonoverlapping Magisteria." *Natural History* 106 (1997) 16–22. http://www.blc.arizona.edu/courses/schaffer/449/Gould%20Nonoverlapping%20Magisteria.htm.
Gustafson, James. *Intersections. Science, Theology, and Ethics*. Cleveland: Pilgrim, 1996.
Gustafsson, Bengt. s.v. "den stora smällen." *Nationalencyclopedin (NE)*. http://www.ne.se/uppslagsverk/encyklopedi/lång/stora-smällen.
———. *Svarta hål*. Stockholm: Fri tanke, 2015.
Gustafsson, Göran. *Tro, samfund och samhälle. Sociologiska perspektiv*. Örebro: Libris, 2000.
Gyllenkrok, Axel. *Systematisk teologi och vetenskaplig med särskild hänsyn till etiken*. With an English Summary. Uppsala: Uppsala Universitets Årsskrift, 1959.
Gyllenkrok, Axel, and Anders Jeffner. "Theological and Ideological Studies." In *Faculty of Theology at Uppsala University*, edited by Helmer Ringgren, 119–29. Stockholm: Almqvist & Wiksell, 1976.
Görman, Ulf. "Svenskars uppfattningar om relationen mellan naturvetenskap och religion." *Svensk teologisk kvartalsskrift* 76 (2000) 34–38.
———. "Vänlig växelverkan mellan naturvetenskap och religion? Diskussion av ett förslag från Viggo Mortensen." *Svensk teologisk kvartalskrift* 66 (1990) 50–59.
Hägerström, Axel. *Filosofi och vetenskap*. Stockholm: Ehlins, 1957.
———. *Die Philosophie der Gegenwart in Selbstdarstellungen*. Vol. 7. Leipzig: Meiner, 1929.

---. *Das Prinzip der Wissenschaft. Eine logisch-erkenntnistheoretische Untersuchung.* Vol. 1, *Die realität*. Uppsala: Skrifter utgivna av Kungliga Humanistiska Vetenskaps-Samfundet i Uppsala, 1908.

---. *Religionsfilosofi*. Stockholm: Natur och kultur, 1964.

Haikola, Lars. "Naturvetenskapen—en ifrågasatt världsbildsskapare." *Vår lösen* 5–6 (1993) 295–302.

---. *Religion as Language-Game: A Critical Study with Special Regard to D.Z. Philips*. Lund: Gleerup, 1977.

---. "Skapelse och försyn i ljuset av modern vetenskap." *Svensk teologisk kvartalsskrift* 67 (1991) 122–29.

Hall, Thor. *Anders Nygren*. Makers of the Modern Theological Mind. Waco, TX: Word, 1978.

---. "The Nygren Corpus: Annotations to the Major Works of Anders Nygren of Lund." *Journal of the American Academy of Religion* 47 (1979) 269–89.

Halldén. *Universum, döden och den logiska analysen*. Stockholm: Almqvist & Wiksell, 1961.

Halvorson, Hans, and Helge Kragh. "Cosmology and Theology." *Stanford Encyclopedia of Philosophy*. Edited by Edward N. Zalta. Summer 2017 ed. https://plato.stanford.edu/archives/sum2017/entries/cosmology-theology.

Hamberg, Eva. "Svenskarnas Gud." In *Gud: Sju Lundateologer föreläser*, edited by Göran Bexell, 23–36. Stockholm: Östlings Symposion, 1997.

Hanes, Jonathan. "The Presupposition of Science-Based Atheism." *SCAST Online Journal* 11 (2015). http://www.iscast.org/journal/Hanes_J_2015-11-17_Presupposition_of_Atheism.pdf.

Haught, John F. *God after Darwin: A Theology of Evolution*. Boulder, CO: Westview, 2000.

Hawking, Stephen. *A Brief History of Time: From Big Bang to Black Holes*. New York: Bantam, 1988.

---. "Quantum Cosmology." In *Three Hundred Years of Gravitation*, edited by Stephen Hawking and Werner Israel, 631–51. Cambridge: Cambridge University Press, 1987.

Hedenius, Ingemar. *Att välja livsåskådning*. Stockholm: Bonniers, 1951.

---. *Tro och vetande*. Stockholm: Bonniers, 1949.

---. "Hedenius, Ingemar." In *Filosofiskt lexikon*, edited by Alf Ahlberg, 63–66. Stockholm: Natur och Kultur, 1963.

Hemberg, Jarl. "Axel Hägerströms skäl att vara ateist." *Svensk teologisk kvartalsskrift* 40 (1964) 172–82.

---. *Religion och metafysik: Axel Hägerströms och Anders Nygrens religionsteorier och dessa inflytande i svensk religionsdebatt*. Acta Univeristatis Upsaliensis. Studia Doctrinae Christianae Upsaliensia 4. Stockholm: Diakonistyrelsen, 1966.

---. "Naturvetenskaplig världsbild och kristen tro II." *Tro och liv* 6 (1980) 3–10.

---. "Världens ursprung och den logiska analysen." *Svensk teologisk kvartalsskrift* 43 (1967) 104–16.

Hemberg, Jarl, ed. *Teilhard de Chardin: Teolog, filosof, naturvetare*. Stockholm: Verbum, 1973.

Hemberg, Jarl, Ragnar Holte, and Anders Jeffner. *Människan och Gud: en kristen teologi*. Lund: Liber, 1982.

Hermansson, Karin. *Vetenskap att tro på?* Stockholm: VA-rapport, 2009.

Hick, John. *An Interpretation of Religion: Human Responses to the Transcendent.* London: Macmillan, 1991.

———. *Philosophy of Religion.* 2nd ed. Englewood Cliffs, NJ: Prentice Hall, 1973.

Hill, Jonathan P. *National Study of Religion & Human Origins.* BioLogos, 2014. https://biologos.org/uploads/projects/nsrho-report.pdf.

Holmstrand, Ingemar. *Karl Heim on Philosophy, Science and the Transcendence of God.* Acta Universitatis Upsaliensis; Studia Doctrinae Christianae Upsaliensia. Uppsala: Almqvist & Wiksell, 1980.

Holmström, Folke. *Uppenbarelsereligion och mystik: En undersökning av Nathan Söderbloms teologi.* Stockholm: Diakonistyrelsen, 1937.

Holte, Ragnar. *Människa, livstolkning, Gudstro: Teorier och metoder inom tros- och livsåskådningsvetenskapen.* Nora: Doxa, 1984.

Hope, Aimie L. B., and Christopher R. Jones. "The Impact of Religious Faith on Attitudes to Environmental Issues and Carbon Capture and Storage (CCS) Technologies: A Mixed Methods Study." *Technology in Society* 38 (2014) 48–59.

Horst, Steven. *Cognitive Pluralism.* Cambridge, MA: MIT Press, 2016.

Hume, David. *Dialogues concerning Natural Religion.* Edited with an Introduction by Henry D. Aiken. New York: Hafner, 1945.

———. *A Treatise on Human Nature.* Edited by L. A. Selby-Bigge. Oxford: Clarendon. http://psycnet.apa.org/index.cfm?fa=browsePB.chapters&pbid=12868.

Höjer, Henrik. "Den svenska upplysningen—fanns den?" *Forskning & Framsteg* 4 (2006) 50–54. http://fof.se/tidning/2006/4/den-svenska-upplysningen-fanns-den.

Images of Sweden Abroad—A Study of the Changes, the Present Situation and Assessment Methods. Stockholm: Regeringskansliet, 2003. http://www.regeringen.se/contentassets/36ac21ea67094813b336115917e1bec5/images-of-sweden-abroad.

Ingemyr, Mikael, ed. *VoF-undersökningen 2015.* Stockholm: Föreningen Vetenskap och folkbildning, 2015.

Jackelén, Antje. "Creativity through Emergence: A Vision of Nature and God." In *Understanding Darwin and Darwinian Understanding*, edited by Anne Runehov and Charles Taliaferro, 69–95. Copenhagen University Discussions in Science and Religion 2. Copenhagen: Copenhagen University Network of Science and Religion, 2013.

———. "Eskatologi—ett naturvetenskapligt ämne?" *Svensk teologisk kvartalsskrift* 76 (2000) 43–47.

———. "Mellan ett reduktionstiskt och teistiskt förflutet och framtidens dynamiska komplexitet." *Svenskt teologisk kvartalsskrift* 85 (2009) 146–52.

———. *Tidsinställningar: Tiden i naturvetenskap och teologi.* Lund: Arcus, 2000.

———. *Time and Eternity: The Question of Time in Church, Science, and Society.* Translated by B. Harshaw. Philadelphia: Templeton Foundation, 2005.

———. "What Theology Can Do for Science." *Theology and Science* 6 (2008) 287–303.

———. *Zeit und Ewigkeit: Die Frage der Zeit in Kirche, Theologie und Naturwissenschaft.* Neukirchen-Vluyn: Neukirchener, 2002.

Jastrow, Robert. *God and the Astronomers.* 2nd ed. New York: Norton, 1992.

Jeffner, Anders. *Biology and Religion as Interpreting Patterns of Human Life.* Oxford: Harris Manchester College, 1999.

———. *Butler and Hume on Religion: A Comparative Analysis.* Stockholm: Diakonistyrelsen, 1966.

———. *Filosofisk religionsdebatt.* Stockholm: Verbum, 1966.

———. *I vetandets gränsmarker: Nio essäer om religion och verklighetstolkning.* Stockholm: Fri tanke, 2017.

———. *Kriterien christlicher Glaubenslehre: Eine prinzipielle Untersuchung heutiger protestantischer Dogmatik im deutschen Sprachbereich.* Göttingen: Vandenhoeck & Ruprecht, 1977.

———. *Livsåskådningsforskning.* Uppsala: Teologiska institutionen, 1976.

———. "A New View of the World Emerging among Ordinary People." In *Christian Faith and Philosophical Theology*, edited by Gijsbert van den Brink, Luco J. Van den Brom, and Marcel Sarot, 137–45. Kampen: Kok Pharos, 1992.

———. "Sjuk av jungfrufödelsen: Teologi och filosofi hos Emilia Fogelklou." In *Emilia Fogelklou läst idag. Nio essäer*, edited by Anders Jeffner, 35–46. Stockholm: Royal Academy of Letters, 2010.

———. *The Study of Religious Language.* London: SCM, 1972.

———. "Teologin inför vetenskapens utmaningar." In *Modern svensk teologi—Strömningar och perspektivskiften under 1900-talet*, 136–86. Stockholm: Verbum, 1999.

———. *Theology and Integration: Four Essays in Philosophical Theology.* Acta Universitatis Uppsaliensis. Uppsala: Uppsala Universitet, 1987.

———. "Torsten Bohlin som teolog." *Kyrkohistorisk årsskrift* (1991) 21–27.

Johansson, Jens. Review of *Har nutida fysik religiös betydelse?* by Åsa Nordén. *Filosofisk tidskrift* 3 (2001) 53–62.

John Paul II. "Message to the Reverend George V. Coyne, S.J., Director of the Vatican Observatory." In *Interdisciplinary Encyclopedia of Religion and Science*, edited by Giuseppe Tanzella-Nitti and A. Strumia. *Interdisciplinary Encyclopedia of Science and Religion.* http://inters.org/John-Paul-II-Coyne-Vatican-Observatory.

Jonson, Jonas. *Nathan Söderblom: Called to Serve.* Translated by Norman E. Hjelm. Grand Rapids, Eerdmans, 2016.

Jonsson, Åke. *Skapelseteologi: En studie av teologiska motiv i Gunnar Edmans texter.* Nora: Nya Doxa, 1999.

Jonsson, Kjell. *Vid vetandets gräns: Om skiljelinjen mellan naturvetenskap och metafysik i svensk kulturdebatt 1870–1920.* Arkiv avhandlingsserie 26. Lund: Arkiv, 1987.

Jonsson, Ulf. *Med tanke på Gud. En introduktion till religionsfilosofin.* Skellefteå: Artos, 2004.

Jordan, Jeffrey. *Pascal's Wager: Pragmatic Arguments and Belief in God.* Oxford: Oxford University Press, 2006.

———. "Pragmatic Arguments and Belief in God." In *Stanford Encyclopedia of Philosophy*, edited by Edward N. Zalta. Winter 2014 ed. https://plato.stanford.edu/entries/pragmatic-belief-god/.

Kegley, Charles W. *The Philosophy and Theology of Anders Nygren.* Carbondale: Southern Illinois University Press, 1970.

Kelly, Thomas. "Evidence." In *Stanford Encyclopedia of Philosophy*, edited by Edward N. Zalta. Winter 2014 ed. https://plato.stanford.edu/entries/evidence/.

Kim, Jaegwon. "The American Origins of Philosophical Naturalism." *Journal of Philosophical Research* 28, Issue Supplement (2003) 83–98. http://fewd.univie.ac.at/fileadmin/user_upload/inst_ethik_wiss_dialog/Kim__J._2003._The_American_origins_of_Phil_Naturalism.pdf.

Kind, Amy. "Qualia." *The Internet Encyclopedia of Philosophy.* http://www.iep.utm.edu/.

Knutsson Bråkenhielm, Lotta. *Religion—evolutionens missfoster eller kärleksbarn? Kognitionsvetenskaplig religionsforskning och dess relevans för religiösa trosföreställningars rationalitet*. Uppsala: Uppsala Universitet, 2016.

Krauss, Lawrence. *A Universe from Nothing: Why There is Something Rather Than Nothing*. New York: Free, 2012.

Kurtén, Tage. "Basic Trust—The Hidden Presence of God." *Studia theologica* 2 (1994) 110–24.

Laeyendecker, Leo. "The Church as a Cognitive Minority." *Concilium* 6 (1971) 71–81.

Lakatos, Imre. *The Methodology of Scientific Research Programmes: Philosophical Papers*. Vol. 1. Edited by John Worrall and Gregory Currie. Cambridge: Cambridge University Press, 1978.

Larson, Edward J., and Larry Witham. "Scientists Are Still Keeping the Faith." *Nature* 386 (1998) 435–36. http://www.nature.com/nature/journal/v386/n6624/pdf/386435a0.pdf.

———. "Leading Scientists Still Reject God." *Nature* 394 (1998) 313. http://www.nature.com/nature/journal/v394/n6691/full/394313a0.html.

Lemberg, Robert. *Jag, du och verkligheten. Religiös kunskap och teologi som vetenskap i John Cullbergs religionsfilosofi*. Åbo: Åbo Akademi, 2000.

Leslie, John, and Robert Lawrence Kuhn, eds. *The Mystery of Existence: Why Is There Anything at All?* Chichester: Wiley-Blackwell, 2013.

Lindbeck, George. *The Nature of Doctrine: Religion and Theology in a Postliberal Age*. Louisville:: Westminster John Knox, 1984.

Linde, A. D. "Initial Conditions for Inflation." *Physics Letters B* 162 (1984) 281–86.

Lindfelt, Mikael. *Teologi och kristen humanism: Ett perspektiv på Torsten Bohlins teologiska tänkande*. Åbo: Åbo Akademi, 1996.

Lindroth, Hjalmar. *Filosofiska och teologiska essayer*. Uppsala: Almqvist & Wiksell/Gebers, 1965.

———. *Kyrklig dogmatik: Den kristna trosåskådningen med särskild hänsyn till det eskatologiska motivet och den frälsningshistoriska grundsynen*. Vols. 1–2. Uppsala: Almqvist & Wiksell International, 1975.

———. "Religion och vetenskap." *Svensk teologisk kvartalsskrift* 3 (1931) 269–85.

———. *Tron och vetenskapens gräns: Kritiska synpunkter på den moderna Upsalafilosofien*. Uppsala: Lindblads, 1933.

———. *Verkligheten och vetenskapen: En inblick i Axel Hägerströms filosofi*. Uppsala: Lindblads, 1929.

Lloyd, Michael. "Are Animals Fallen?" In *Animals on the Agenda: Questions about Animals for Theology and Ethics*, edited by Andrew Linzey and Dorothy Yamamoto, 147–60. London: SCM, 1998.

Lorrimar, Victoria. "Are Scientific Research Programmes Applicable to Theology? On Philip Hefner's Use of Lakatos." *Theology and Science* 15, no. 2 (2017) 188–202.

Löwdin, Per-Olov. "Människan och hennes psyke i den moderna kvantteorins världsbild." *Forskning och Praktik* 6 (1971) 121–25.

Lundborg, Johan. *När ateismen erövrade Sverige: Ingemar Hedenius och debatten kring tro och vetande*. Nora: Nya Doxa, 2002.

Lyttkens, Hampus. *The Analogy between God and the World: An Investigation of Its Background and Interpretation of Its Use by Thomas of Aquino*. Uppsala: Uppsala universitets årsskrift, 1953:5

———. "Kan kristna trosutsagor verifieras genom religiös erfarenhet?" In *Religion, erfarenhet, verifikation,* edited by Urban Forell and Hampus Lyttkens, 52–70. Lund: Studentlitteratur, 1970.

———. *Religiös samhällsanalys: Åtta artiklar om samhällsförändring och kristen tro.* Lund: Religio 19, 1985.

MacIntyre, Alasdair. *Difficulties in Christian Belief.* London: SCM, 1959.

MacKay, Donald. "'Complementarity' in Scientific and Theological Thinking." *Zygon* 9 (1974) 225–44.

Macquarrie, John. *In Search of Humanity: A Theological & Philosophical Approach.* London: SCM, 1984.

Månsson, Bengt-Åke. *Vetenskap, evighet och religion.* [Kristianstad: B.-Å. Månsson], 1996.

Marc-Wogau, Konrad. *Studier i Axel Hägerströms filosofi.* Stockholm: Prisma, 1968.

Martin, Michael. "Justifying Methodological Naturalism (2002)." The Secular Web, https://infidels.org/library/modern/michael_martin/naturalism.html.

Mastin, Luke. "Important Scientists. Lemaître, Georges (1894–1966)." Physics of the Universe. 2009. http://www.physicsoftheuniverse.com/scientists_lemaitre.html.

McInerny, Ralph, and John O'Callaghan. "Saint Thomas Aquinas." In *The Stanford Encyclopedia of Philosophy,* edited by Edward N. Zalta. Spring 2015 ed. http://plato.stanford.edu/archives/spr2015/entries/aquinas/.

McLeod, Hugh. *The Religious Crisis of the 1960s.* Oxford: Oxford University Press, 2007.

McMullin, Ernan. "How Should Cosmology Relate to Theology?" In *The Sciences and Theology in the Twentieth Century,* edited by Arthur Peacocke, 17–57. Notre Dame: University of Notre Dame, 1981.

Meier, John P. *A Marginal Jew: Rethinking the Historical Jesus.* Vol. 2, *Mentor, Message, and Miracles.* New Haven: Yale University Press, 1994.

Mindus, Patricia. *A Real Mind: The Life and Work of Axel Hägerström.* Dordrecht: Springer, 2009.

Mitchell, Basil. *Faith and Criticism.* Oxford: Oxford University Press, 1994.

———. "Theology and Falsification. C." In *New Essays in Philosophical Theology,* edited by Antony Flew and Alasdair MacIntyre, 103–5. London: SCM, 1963.

Moltmann, Jürgen. *God in Creation: An Ecological Doctrine of Creation: The Gifford Lectures 1984–1985.* London: SCM, 1985.

Moritz, Joshua M. *Science and Religion: Beyond Warfare and toward Understanding,* Winona: Anselm Academic, 2016.

Mortensen, Viggo. *Teologi og naturvidenskap: Hinsides restriktion og ekspansion.* København: Munksgaards, 1989.

Murphy, Nancey. "Human Nature: Historical, Scientific, and Religious Issues." In *Whatever Happened to the Soul? Scientific and Theological Portraits of Human Nature,* edited by Warren S. Brown et al., 1–29. Minneapolis: Fortress, 1998.

———. "Nonreductive Physicalism: Philosophical Issues." In *Whatever Happened to the Soul? Scientific and Theological Portraits of Human Nature,* edited by Warren S. Brown et al., 127–48. Minneapolis: Fortress, 1998.

Nagel, Ernest. "Can Logic Be Divorced from Ontology?" *Journal of Philosophy* 26 (1929) 705–12.

———. "Naturalism Reconsidered." *Proceedings and Addresses of the American Philosophical Association* 28 (1954–1955) 5–17.

Nagel, Thomas. "The Core of 'Mind and Cosmos.'" *The New York Times*, August 18, 2013. http://opinionator.blogs.nytimes.com/2013/08/18/the-core-of-mind-and-cosmos/?_r=0#more-148050.

———. *Mind and Cosmos: Why the Materialist Neo-Darwinian Conception of Nature Is Almost Certainly False*. Oxford: Oxford University Press, 2012.

———. *Mortal Questions*. New York: Cambridge University Press, 1979.

———. *Secular Philosophy and the Religious Temperament: Essays 2002–2008*. Oxford: Oxford University Press, 2009.

———. "What Is It Like to Be a Bat?" *Philosophical Review* 83 (1974) 435–50. http://organizations.utep.edu/portals/1475/nagel_bat.pdf.

Newport, Frank. "In U.S., 42% Believe Creationist View of Human Origins." *Gallup*. http://www.gallup.com/poll/170822/believe-creationist-view-human-origins.aspx?g_source=humans+evolved%2c+with+God+guiding+2014&g_medium=search&g_campaign=tiles.

Nordén, Åsa. *Har nutida fysik religiös betydelse?* Stockholm: Podium, 1999.

Nordgren, Anders. *Evolutionary Thinking: An Analysis of Rationality, Morality and Religion from an Evolutionary Perspective*. Stockholm: Almqvist & Wiksell, 1994.

Nordin, Svante. *Den Boströmska skolan och den svenska idealismens fall*. Lund: Doxa, 1981.

Nowotny, Helga, Peter Scott, and Michael Gibbons. *Re-Thinking Science: Knowledge and the Public in an Age of Uncertainty*. Malden, MA: Polity, 2001.

Nygren, Anders. *Agape and Eros*. Part 1. A Study of the Christian Idea of Love and Part 2. The History of the Christian Idea of Love. Translated by Philip S. Watson. London: SPCK, 1953.

———. *Anders Nygren's Religious Apriori*. Edited by Walter H. Capps and Kjell O. Lejon. Introduction by Walter H. Capps. Linköping Studies in Religion and Religious Education 2. Linköping: Linköping University Electronic Press, 2000.

———. *Filosofi och motivforskning*. Stockholm: Diakonistyrelsen, 1940.

———. *Meaning and Method: Prolegomena to a Scientific Philosophy of Religion and a Scientific Theology*. Translated by Philip S. Watson. London: Epworth. 1972.

———. *Mening och metod: Prolegomena till en vetenskaplig religionsfilosofi och en vetenskapslig teologi*. Åbo: Åbo Akademi, 1982.

———. *Det religionsfilosofiska grundproblemet*. Vols. 1–3. Lund: Gleerupska universitetsbokhandeln, 1921.

———. *Religiöst apriori: dess filosofiska förutsättningar och teologiska konsekvenser*. Lund: Gleerupska, 1921.

———. Review of *Meaning and Method*, by Paul R. Clifford. *Religious Studies* 9, no. 4 (1973) 496.

———. "Tro och vetande: Ur en diskussionsinledning." *Vår lösen* 5 (1945) 166–73.

Nürnberger, Klaus. "Eschatology and Entropy: An Alternative to Robert John Russell's Proposal". *Zygon* 46 (2012) 970–96.

O'Halloran, Nathan. "Cosmic Alienation and the Origin of Evil: Rejecting the 'Only Way' Option." *Theology and Science* 13 (2015) 43–63.

Page, Don. "Does God Love the Multiverse?" In *The Blackwell Companion to Science and Christianity*, edited by J. B. Stump, and Alan G. Padgett, 198–215. Oxford: Wiley & Sons, 2012.

Paley, William. *Natural Theology*. 2nd ed. Oxford: Vincent.

Pappas, William Matthew. "Is There a Belief in God and Immortality among Eminent Psychology Scholars?" PhD diss., University of Texas at Austin, 2007.
Pederson, Ann M., and Mary Solberg. "Cyberflesh: An Embodied Feminist Pedagogy for Science and Religion." Paper presented at the American Academy of Religion, Annual Meeting, San Francisco, November 1997.
Peters, Ted. *God as Trinity: Relationality and Temporality in the Divine Life.* Louisville: Westmisnter John Knox, 1993.
Pettersson, Thorleif. "Svensken och religionen." In *Svenskt kynne,* edited by Leif Lewin, 7–23. Uppsala: Universitetsbiblioteket, 2000.
Philipson, Sten M. *Med naturen som referenspunkt: om livsåskådningar i miljörörelsen.* Lund: Doxa, 1984
Pincock, Christopher. "Ernest Nagel's Naturalism: A Microhistory of the American Reception of Logical Empiricism." In *Analytic Philosophy: An Inerpretive History,* edited by Aaron Preston, 160–74. Oxford: Routledge, 2017. http://pincockyilmazer.com/chris/nagelle.pdf.
Pitts, J. Brian. "Why the Big Bang Singularity does not Help the Kalām Cosmological Argument for Theism." *British Journal for the Philosophy of Science* 59 (2008) 675–708.
Plantinga, Alvin. *Knowledge and Christian Belief.* Grand Rapids: Eerdmans, 2015.
———. *Where the Conflict Really Lies. Science, Religion and Naturalism.* Oxford: Oxford University Press, 2011.
Policy för fred och omstallning till en hallbar varld. Equmeniakyrkan, 2013. http://equmeniakyrkan.se/wp-content/uploads/2013/05/Policy-folder-A3-EQ.pdf.
Polkinghorne, John. *The Faith of a Physicist: Reflections of a Bottom-Up Thinker.* Princeton, NJ: Princeton University Press, 1994.
Pollock, John. *Contemporary Theories of Knowledge.* 1st ed. Towota, NJ: Rowman and Littlefield, 1986.
Price, H. H. "Faith and Belief." In *Faith and the Philosophers,* edited by John Hick, 3–25. London: Macmillan, 1964.
Questionnaire Design. U.S. Survey Research. Pew Research Center. http://www.people-press.org/methodology/questionnaire-design/open-and-closed-ended-questions/.
Rad, Gerhard von. *Genesis: A Commentary.* Philadelphia: Westminster, 1972.
Rasmusson, Arne. "A Century of Swedish Theology." *Lutheran Quarterly* 21, no. 2 (2007) 125–62.
Religion and Science: Highly Religious Americans Are Less Likely Than Others to See Conflict between Faith and Science. Pew Research Center Report, 2015. http://www.pewinternet.org/2015/10/22/science-and-religion/.
Repstad, Pål. Review of *Is God Back? Redonsidering the New Visibility of Religion,* edited by Titus Hjelm. *Journal of Contemporary Religion* 31 (2016) 433–35.
Ricoeur, Paul. *Time and Narrative.* Vols. 1–3. Translated by Kathleen McLaughlin and David Pellauer. Chicago: University of Chicago Press, 1984, 1985, and 1988.
Runehov, Anne. *Sacred or Neural? Neuroscientific Explanations of Religious Experience.* Uppsala: Uppsala Universitet, 2004.
———. *Sacred or Neural? The Potential of Neuroscience to Explain Religious Experience.* Göttingen: Vandehoek & Ruprecht, 2007.
Ruse, Michael. "Atheism and Science." In *The Customization of Science: The Impact of Religious and Political Worldviews on Contemporary Science,* edited by Steve

Fuller, Mikael Stenmark, and Ulf Zackariasson, 73–88. Basingtoke, UK: Palgrave Macmillan, 2014.

———. *Can a Darwinian be a Christian? The Relationship between Science and Religion.* Cambridge: Cambridge University Press, 2001.

Russell, Bertrand. *The Analysis of Mind.* London: Allen & Unwin, 1949.

———. *The Scientific Outlook.* London: Allen & Unwin, 1931.

———. *Why I Am Not a Christian and Other Essays on Religion and Related Subjects.* New York: Touchstone, 1957.

Russell, Robert John. *Cosmology: From Alpha to Omega.* Minneapolis: Fortress, 2008.

———. "Dialogue, Science and Theology." In *Interdisciplinary Encyclopedia of Religion and Science*, edited by Giuseppe Tanzella-Nitti and A. Strumia. *Interdisciplinary Encyclopedia of Science and Religion.*. http://inters.org/dialogue-science-theology.

———. "Eschatology and Scientific Cosmology: From Deadlock to Interaction." *Zygon* 46 (2012) 997–1014.

———. "Eschatology in Science and Theology." In *The Blackwell Companion to Science and Christianity*, edited by J. B. Stump, and Alan G. Padgett, 543–45. Oxford: Wiley & Sons, 2012.

———. "The Groaning of Creation. Does God suffer with all Life?" In *The Evolution of Evil*, edited by Gaymont Bennett et al., 120–40. Göttingen: Vandenhoek & Ruprecht, 2008.

———. *Time in Eternity: Pannenberg, Physics and Eschatology in Creative Mutual Interaction.* Notre Dame: University of Notre Dame Press, 2012.

Scarr, Sandra. "Theories for the 1990s; Development and Individual Differences." *Child Development* 63 (1992) 1–19.

"Scientific Method and Religion." Wikipedia. https://en.wikipedia.org/wiki/Scientific_method_and_religion.

Simonsson, Tord. *Face to Face with Darwinism: A Critical Analysis of the Christian Front in Swedish Discussion of the Later Nineteenth Century.* Lund: Gleerup, 1958.

———. "Livsåskådning och vetenskap." In *Att välja ståndpunkt: Orienterande essäer om livsåskådning, religion och etik*, edited by Jarl Hemberg and Anders Jeffner, 64–77. Stockholm: Diakonisstyrelsens, 1965.

Söderblom, Nathan. *Jesu bärgspredikan och vår tid.* Stockholm, 1898.

———. *The Living God: Basal Forms of Personal Religion.* The Gifford Lectures, delivered in the University of Edinburgh in the year 1931. With a biographical introduction by Dr. Yngve Brilioth. London: Oxford University Press, 1933.

———. *The Nature of Revelation.* Translated by Frederic E. Pamp. London: Oxford University Press, 1933.

———. *Religionsproblemet inom katolicism och Protestantism.* Vol. 2. Stockholm: Gebers, 1910.

Southgate, Christopher. *The Groaning of Creation. God, Evolution, and the Problem of Evil.* Louisville: Westminster John Knox. 2008.

Stark, Rodney, and Roger Finke. *Acts of Faith: Explaining the Human Side of Religion.* Berkeley: University of California Press, 2000.

Stenger, Victor J. *God—the Failed Hypothesis: How Science Shows That God Does Not Exist.* Amherst, NY: Prometheus, 2008.

Stenmark, Mikael. "Competing Conceptions of God: The Personal God versus the God beyond Being." *Religious Studies* 51 (2015) 205–20.

———. "The Customization of Science: An Introduction to the Debate." In *The Customization of Science: The Impact of Religious and Political Worldviews* on Contemporary Science, edited by Steve Fuller, Mikael Stenmark, and Ulf Zackariasson, 1-18. Basingtoke: Palgrave Macmillan, 2014.

———. *How to Relate Science and Religion: A Multidimensional Model*. Grand Rapids: Eerdmans, 2004.

———. "Naturvetenskap och religion." In *Religionsdidaktik—mångfald, livsfrågor och etik i skolan*, edited by Malin Löfstedt, 79-95. Lund: Studentlitteratur, 2011.

———. "Rationality and Different Conceptions of Science." In *The Evolution of Rationality: Interdisciplinary Essays in Honor of J. Wentzel van Huyssteen*, edited by F. Leron Shults, 47-72. Grand Rapids: Eerdmans, 2006.

———. *Rationality in Science, Religion, and Everyday Life: A Critical Evaluation of Four Models of Rationality*. Notre Dame: University of Notre Dame Press, 1995.

———. "Relativism—Pervasive Feature of the Contemporary Western World?" *Social Epistemology* 29 (2015) 31-43.

———. "Science and a Personal Conception of God: A Critical Response to Gordon D. Kaufman." In *Journal of American Academy of Religion* 71 (2003) 175-81.

———. *Scientism: Science, Ethics and Religion*. London: Palgrave, 2001.

———. "Ways of Relating Science and Religion." In *The Cambridge Companion to Science and Religion*, edited by Peter Harrison, 278-95. Cambridge: Cambridge University Press, 2010.

Stenmark, Mikael, Ulf Zackariasson, and Steve Fuller, eds. *The Customization of Science: The Impact of Religious and Political Worldviews on Contemporary Science*. London: Palgrave Macmillan, 2014.

Stoeger, William. "Cosmology." In *Interdisciplinary Encyclopedia of Religion and Science*, edited by G. Tanzella-Nitti, and A. Strumia. *Interdisciplinary Encyclopedia of Science and Religion*. http://inters.org/cosmology.

Suchocki, Marjorie. *The End of Evil: Process Eschatology in Historical Context*. Eugene, OR: Wipf and Stock, 1988.

Sundén, Hjalmar. *Religionen och rollerna: Ett psykologiskt stadium av fromheten*. Andra upplagan. Stockholm: Svenska kyrkans diakonistyrelses, 1960.

Sundkler, Bengt. *Nathan Söderblom: His Life and Work*. Lund. Gleerup, 1968

Sunstein, Cass R. *Laws of Fear: Beyond the Precautionary Principle*. Cambridge: Cambridge University Press, 2005.

Swinburne, Richard. "Arguments to God from the Observable Universe." In *The Blackwell Companion to Science and Christianity*, edited by J. B. Stump, and Alan G. Padgett. 117-29. Oxford: Wiley-Blackwell, 2012.

Tanzella-Nitti, Giuseppe. "The Pius XII—Lemaître Affair (1951-1952) on Big Bang and Creation." *Interdisciplinary Encyclopedia of Science and Religion*. http://inters.org/pio-xii-lemaitre.

Taylor, A. E. *Elements of Metaphysics*. London: Methuen, 1961.

Tegmark, Max. *Our Mathematical Universe: My Quest for the Ultimate Nature of Reality*. New York: Knopf, 2014.

Teilhard de Chardin, Pierre. *The Phenomenon of Man*. Translated by Bernard Wall. New York: Harper & Bros., 1959.

Thistelton, Anthony C. *Hermeneutics: An Introduction*. Grand Rapids: Eerdmans, 2009.

Thomas Aquinas. *De Aeternitate Mundi*. In *An Aquinas Reader*, edited by Mary T. Clark. 178-85. London: Houghton ans Stockton, 1972.

Thunberg, Anne-Marie, ed. *Tradition i rörelse: Ett sekels kultursamtal i Vår Lösen*. Stockholm: Carlssons, 2000.

Thurfjell, David. *Det gudlösa folket: De postkristna svenskarna och religionen*. Stockholm: Molin & Sorgenfrei, 2015.
Tipler, Frank. *The Physics of Christianity*. New York: Doubleday, 2007.
Torely, Vincent. "Is Methodological Naturalism a Defining Feature of Science?" Uncommon Descent, Dec. 14, 2015. https://uncommondescent.com/intelligent-design/is-methodological-naturalism-a-defining-feature-of-science-part-one/.
Trillhaas, Wolfgang. *Dogmatik*. Dritte verbesserte Auflage. Berlin: de Gruyter, 1972.
Trow, Martin, et al. *Carnegie Commission National Survey of Higher Education: Faculty Study*. Ann Arbor, MI: Inter-university Consortium for Political and Social Research. 1975. http://doi.org/10.3886/ICPSR07501.v1.
Uddenberg, Nils. *Det stora sammanhanget: moderna svenskars syn på människans plats i naturen*. Nora: Nya Doxa, 1995.
van Inwagen, Peter. *Problem of Evil*. Oxford: Oxford University Press, 2008.
Vilenkin, Alex. Letter to William Lane Craig. 2013. http://www.reasonablefaith.org/honesty-transparency-full-disclosure-and-bgv-theorem#ixzz4hRNJjork.
———. *Many Worlds in One: The Search for Other Universes*. New York: Hill and Wang, 2006.
Wallace-Wells, David. "The Uninhabitable Earth." *The New York Magazine*, July 9, 2017. http://nymag.com/daily/intelligencer/2017/07/climate-change-earth-too-hot-for-humans.html.
Waller, Margit. *Axel Hägerström: Människan som få kände*. Stockholm: Natur & Kultur, 1961.
Ward, Keith. "Cosmology: From Alpha to Omega: Key Themes and Critiques." *Theology and Science* 9 (1999) 261–63.
Wheeler, John Archibald. "Beyond the End of Time." In *Black Holes, Gravitational Waves, and Cosmology: Introduction to Current Research*. New York: Gordon and Breach, 1974.
Wiles, Maurice. *The Remaking of Christian Doctrine*. Hulsean Lecture. London: SCM, 1974.
Wilson, Edward O. *Consilience: The Unity of Knowledge*. New York: Knopf, 1998.
———. *On Human Nature*. Cambridge: MA: Harvard University Press, 1978.
Wilson, John. *Philosophy and Religion: The Logic of Religious Belief*. London: Oxford University Press, 1961.
Wingren, Gustaf. *Creation and Law*. Translated by Ross MacKenzie. Eugene, OR: Wipf and Stock, 2003.
———. *Öppenhet och egenart: Evangeliet i världen*. Lund: LiberLäromedel, 1979.
———. *Tolken som tiger: Vad teologin är och vad den borde vara*. Stockholm: Gummessons, 1981.
Wisdom, John. "Gods." *Proceedings of the Aristotelian Society* 45, no. 1 (1945) 185–206. https://doi.org/10.1093/aristotelian/45.1.185.
Wittgenstein, Ludwig. *Philosophical Investigations*. Translated by G. E. M. Anscombe. Oxford: Blackwell, 1968.
World Values Survey. Wave 7: 2010–2014. http://www.worldvaluessurvey.org/WVSOnline.jsp.
Wrede, Gösta. "A Hjalmar J Lindroth." *Svenskt biografiskt lexikon* 23 (1980–1981) 575. http://sok.riksarkivet.se/sbl/artikel/10651.
Wright, Georg Henrik von. *Explanation and Understanding*. London: Routledge, 1971.
———. *Vetenskapen och förnuftet*. Stockholm: Bonniers, 1986.
Zackariasson, Ulf. "A Problem with Alston's Indirect Analogy-Argument from Religious Experience." *Religious Studies* 42 (2006) 329–41.

Author Index

Albert, David, 239n41, 239n43, 297
Al-Haytham, Ibn, 197
Alper, 16n24, 302
Alston, William, 172, 172n1, 176189,
    189n38, 200, 200n71, 202,
    202n77, 202n78, 202n79,
    203, 206, 206n3, 213n23, 214,
    214n25, 215nn27–29, 216,
    216nn31–32, 217n33, 218n36,
    219–22
Andrae, Tor, 113n3, 297
Anér, Kerstin, 142n32
Aristotle, 143, 198
Armstrong, David, 192, 194
Atkins, Peter, 63
Augustine, St. 178, 235, 236, 236n36,
    237, 251, 251n80, 252, 252n83,
    297
Aulén, Gustaf, 253, 253n84
Aurelius, Erik, 260, 260n111, 262, 297

Bacon, Francis, 197
Bacon, Roger, 197, 198
Barbour, Ian, xi, 78, 78n1, 97, 98n13,
    146, 170, 205n1, 209, 209nn9–
    10, 297
Barth, Karl, 96, 211
Bazeley, Patricia, 297
Bejerholm, Lars, 118n3, 118nn4–5, 297

Bergman, Ingmar, 5
Bergson, Henri, 80, 87, 87n11, 95n4,
    297
Billing, Einar, 83, 220
Björkman, Jenny, 169n1, 298
Bóchenski, Jósef, 124, 124n27, 298
Bohlin, Torsten, 80, 93, 95, 96, 96n6–n7,
    97, 97nn8–9, 97n12, 102, 114–
    16, 172, 203, 206, 210, 211n17,
    212–13, 218, 218nn37–38, 219,
    219nn40–41, 220, 220n44, 221,
    221n23, 222
Borde, Arvin, 232, 233n29, 300
Boström, Christopher Jacob, 79, 87, 89,
    89nn21–22, 90, 90n23, 92, 148,
    298
Braaten, Carl, 254n88
Braithewaite, R.B., 134, 208, 208n6
Bring, Ragnar, 80, 93, 96, 107, 108n5,
    111, 112, 112n20, n22, 113,
    113n28, 114, 116, 117, 298
Broad, C.D., 114, 214n26, 298
Brown, David, 242n51 , 298
Brown, Harold I., 148
Brown, Warren, 298, 302, 307
Bruce, Steve, 24, 24n6, 73, 74n27, 298
Bråkenhielm, Carl Reinhold, 5n3, 9n13,
    n15, 15, 26n13, 32nn29–31,
    78n2, 125, 153n74, 272, 298, 299

Buber, Martin, 80, 99
Bultmann, Rudolf, 101, 223, 224, 224n4, 258, 299
Butler, Joseph, 133, 134, 134n7, 135, 135nn8–11, 136, 136n12, 137n14, 304

Capps, Walter, 136n6, 110, 110n12, 289
Carroll, John W., 257n103
Carroll, Sean, 231, 231n26, 299
Casanova, José, 21, 21n1, 299
Chignell, 201n74, 299
Clark, Kelly James, 152n69, 239, 239n40, n42, 240, 240nn44–45, 241, 242nn48–49
Clayton, Philip, xi, 225, 263, 263n122, 299
Clifford, Paul R., 110n14, 111n18.
Clifford, William, 201, 201n73
Collins, Francis, 181n19, 182n19, 300
Coyne, Jerry, 199, 199n68, 200, 200n70, 201n61, 202, 299
Craig, William Lane, 162 162n34, 197, 197n62, 198, 231, 231n23, nn25–26, 233, 233n29, 234–35, 237–39, 239n38, 240, 299–300
Cullberg, John, 80, 93, 97–98, 98nn14–16, 99, 99nn17–18, 99n20, 100, 100nn21–24, 101, 101nn26–27, 102, 102nn29–30, 102, 103, 104, 110, 114, 115, 116, 211, 300
Cullmann, Oscar, 223–24

d'Aquili, Eugene, 166
Danielsson, Ulf, 232, 232n28, 234, 300
Darwin, Charles, 25, 62, 132, 175, 183, 195, 195nn55–57, 199, 232, 300, 303, 304
Davie, Grace, 24, 73, 300
Dawkins, Richard, 14, 132, 132n3, 145, 181–82, 182n19, 183, 183nn23–24, 184, 184n25, 232, 300
de Vries, 187, 187n33, 188, 300
Dennett, Daniel, 194, 195, 195nn55–57, 196
Domning, Daryl, 179, 179n14, 300
Draper, Paul, 177, 177n6, 178, 178n9, 300
Drees, Willem, 254, 254n88, 255n89, 300

Dunlap, Riley, 9
Dyson, Freeman, 159, 160, 228, 255, 255n89

Ecklund, Elaine, 25n11, 63, 64, 64nn7–9, 65, 65nn11–12, 66, 66nn14–15, 68, 72, 73, 73n24, 75, 75nn30–31, 76, 301
Eddington, Sir Arthur, 261, 262n120, 301, 17
Ehnmark, 88n15, 89n18, 301
Einstein, Albert, 91n27, 160, 230, 237, 257n103
Eklund, Harald, 81, 104, 105, 105n40, 105nn42–44, 106, 116, 118, 118nn1–5, 119, 119nn6–7, 119nn9–10, 122n19, 123, 180–81, 181n17, 297, 301
Eklund. J. A., 83
Elijah, 184–85
Emmet, Dorothy, 127n40, 301

Ferré, Frederick, 8n10
Finke, Roger, 24n6, n8, 26, 310
Flew, Antony, 134, 241, 307
Fogelklou, Emilia, 81, 85, 93, 94, 94n2, 95, 95n4, 103, 301, 306
Forell, Urban, 81, 125, 125n32, n33, 126, 127, 127n38, 128, 128n44, 132n2, 301
Friberg, Ingrid, 272
Fries, Martin, 83, 83n7, 301
Fuller, Steven, 145n43, 142n70, 226n11, 302, 309, 311
Funk, 16n24, 302

Galileo, Galilei, 137, 175, 198, 199
Geijer, Erik-Gustaf, 85
Gerholm, Tor Ragnar, 140, 140n25, 302
Gibbons, Michael, 1, 1n1, 308
Gilkey, Langdon, 18, 302
Gould, Steven Jay, 75n32, 107–8, 145, 172, 205, 206, 206n4, 207, 207n5, 208–9, 209n8, 221, 224n6, 302
Grant Edward, 18n66, 302
Grantén, Eva-Lotta, 82, 163, 163nn37–40, 164, 164nn41–45, 165, 165nn46–47, 302
Green, 28n18, 302

Gregersen, Niels Henrik, 131n53, 302
Gustafson, James, 263, 263n124, 302
Gustafsson, Bengt, 92, 230n20, 243n53, 302
Gustafsson, Göran, 302
Guth, Alan, 232, 233n29, 300
Gyllenkrok, Axel, 122n19, 133n5, 301, 302
Göransson, N. J., 89n16
Görman, Ulf, 40n5, 41, 41n6, 42, 131, 131n53, 132, 132n1, 142, 302

Haikola, Lars, 81, 128, 128n42, n43, 129, 129nn46-48, 130, 130n49, 131, 132, 137, 155, 303
Hall, Thor, 303
Halldén, Sören, 136n13, 140, 140n26, 303
Halvorson, Hans, 255n89, 262n120, 303
Hamberg, Eva, 10n17, 303
Hanes, Jonathan, 199n69, 303
Hansson, Mats G., 153n74, 299
Hare, Richard, 241
Hartle, James, 236
Hartshorne, Charles, 249
Haught, John, 303
Hawking, Stephen, 162, 171, 235, 235n33, n34, 235n33, 236, 236n36, 238, 263n34, 303
Hedenius, Ingemar, 39, 39n2, 40, 51, 80, 96, 97, 97n8, 111, 112, 112n23, 113-14, 114nn30-32, 115, 115nn33-35, 116-17, 187, 187n37, 188, 200, 200n72, 201-3, 203nn81-82, 303
Hefner, Philip, 164, 225, 306
Heidegger, Martin, 138
Hemberg, Jarl, 81, 83, 83n7, 93, 93n39, n40, 98, 99n18, 111, 111n19, 122, 133, 139-40, 140n25, 140n27, 141, 141nn28-30, 142, 142n32, 297, 303, 310
Hermansson, Karin 25, 26, 42, 131, 303
Hick, John, 33, 125n31, 185, 186n30, 198n66, 208nn6-7, 221n52, 298, 303-4
Hilbert, David, 235
Hill, Jonathan P., 75n32, 304
Hjelm, Norman, 305
Hjelm, Titus, 11n19, 21, 25

Holmstrand, Ingmar, 125n33, 126nn35-37, 304
Holmström, Folke, 85n2, 304
Holte, Ragnar, 41n8, 81, 142, 142n33, 143, 143n34, 143nn36-37, 144, 144n39, 303-4
Hook, Sidney, 199
Hope, Aimie L. B., 37n40, 304
Horst, Steven, 7, 7n8, 8, 8n9, 60, 144, 144n38, 170, 170n2, 171, 171nn3-4, 304
Hume, David, 133-34, 134n7, 135, 135nn8-11, 136n12, 136, 137n14, 169, 182, 182n22, 183, 188-89, 194, 195n54, 304
Hägerström, Axel, 79-80, 90, 90n25, 91, 91nn26-27, 91n29, 92, 92nn31-35, 93, 93n39, 94-96, 98-100, 100n22, 102, 103, 111-14, 139, 191n43, 302-3, 306-7, 312
Höjer, Henrik, 304

Ingemyr, Mikael, 5n3, 304

Jackelén, Antje, 78, 82, 156, 157nn9-12, 158, 158nn13-17, 159, 159nn18-25, 160, 160nn26-30, 161, 229, 255, 255n91, n93, 304
Jackson, Kristi, 297
James, William, 97, 203
Jarrick, Arne, 69, 298
Jastrow, Robert, 230, 304
Jeffner, Anders, 2, 6, 6n5, 7nn6-7, 8, 8nn11-12, 9, 14, 14n21, 15, 15n25, 30, 30nn24-25, 31n26, 32n28, 34, 34n35, 35, 35n36, 69, 69n18, 81, 83, 83n9, 88n14, 94nn1-2, 95, 96n6, 97n8, 99, 122, 133, 133nn4-5, 134, 134nn6-7, 135, 135nn8-11, 136, 136nn12-13, 137, 137nn14-15, 138, 139nn18-22, 139, 139nn23-24, 161, 161n31, 167n55, 173, 181n18, 184n26, 185, 186n30, 194, 194n53, 211n16, 213-14, 214n24, 219n41, 221n51, 227-28, 238n37, 242, 242n50, 243n52, 251, 253n84, 302-5, 310

Jesus Christ, 44, 88, 96, 98, 101, 102, 104, 118, 128, 203, 210–12, 215–16, 257–59, 259n110, 260, 260n114, 261, 263, 307
Jones, Christopher R., 37n40
Jonson, Jonas, 85n1, 305
Jonsson, Kjell, 120, 120n13, 121n15, 305
Jonsson, Ulf, 167n56, 305
Jonsson, Åke, 82, 161, 161n32, 305
Jordan, Jeffrey, 97, 97nn10–11, 305

Kant, Immanuel, 90, 109, 142–43, 206, 208
Kaufman, Gordon, 150–51, 311
Kegley, Charles W., 108n4, 305
Kelly, 38, 101, 102, 15, 21
Kelly, Thomas, 176nn2–3, 305
Kepler, 198
Kierkegaard, Søren, 80, 95, 96, 149, 211
Kim, Jaegwon, 190, 191, 191n42, n44, 192, 192nn47–48, 193, 193n51, 199305
Kind, Amy, 248, 248n67, 305
Knutsson Bråkenhielm, Lotta, 167–68, 305
Kragh, Helge, 255, 255n89, 262, 303
Krauss, Lawrence, 232, 232, 232n27, 233, 245, 245n56, 297, 306
Kristiansson, Lars, 143n36, 144
Kuhn, Robert Lawrence, 306
Kuhn, Thomas, 132, 148
Kurtén, Tage, 113, 113n25, 306

Laeyendecker, Leo, 21n2, 306
Lakatos, Imre, 225, 225n9, 226, 306
Larson, Edward J., 62n4, 63, 63n5, 68, 306
Larsson, Stieg, 5
Leibniz, Gottfried, 90, 114
Lemberg, Robert, 99n20, 100n22, 306
Leuba, James, 62, 63, 64
Lindbeck, George, 173n6, 306
Linde, Andrej, 234,, 234n32, 306
Lindfelt, Mikael, 96, 96n5, n7, 97nn8–9, 211, 21nn17–18, 212, 212nn19–22, 213, 220n46, n48, 306
Lindgren, Astrid, 5

Lindroth, Hjalmar, 80, 93, 100, 100nn24–25, 101–2, 102nn32–34, 103, 103n35, n37, 104, 104nn38–39, 105, 105n41, 106, 119, 119n8, 123, 123n24, 210, 210n15, 211, 258, 258n104, 306, 312
Lloyd, Michael, 179, 179n13, 306
Locke, John, 142, 143, 197
Lorrimar, Victoria, 225, 225n10, 306
Luke, St., 118
Lundborg, Johan, 40n4, 97, 97n8, n12, 111, 112, 112nn21–22, 115n34, 306
Lyttkens, Hampus, 81, 99, 123, 123nn24–25, 124, 124n26, 124n28, 124n30, 125, 128, 300, 301, 306
Löwdin, Per Gunnar, 143, 143n35, 306

MacIntyre, Alasdair, 126, 126n37, 189, 189n39, 307
MacKay, Donald, 41n8, 307
Macquarrie, John, 161, 253, 254n86, 258, 258n105, 263, 307
Marc-Wogau, Konrad, 91, 91nn26–28, 92nn35–36, 93n39, 307
Martin, Michael, 187n36, 307
Mastin, Luke, 230n21, 307
McLeod, Hugh, 39, 39n1, 40n3, 307
McMullin, Ernan, 243, 244n54, 307
Meier, John P., 260n114, 307
Mindus, Patricia, 90n25, 91n27, 307
Mitchell, Basil, 241, 241n47, 242, 242n51, 298, 307
Mithen, Steven, 165
Mlodinow, Leonard, 171
Moltmann, Jürgen, 159, 161, 223n1, 307
Moritz, Joshua M., 186n31, 307
Mortensen, Viggo, 41n8, 132n1, 302, 307
Murphy, Nancey, 28, 29, 29nn20–21, 225, 307
Månsson, Bengt-Åke, 108n5, 307

Nagel, Ernest, 92n30, 190, 190n40, 191, 191n43, 192, 193, 193n50, 194, 194n52, 307, 309

Nagel, Thomas, 29n22, 172, 227, 246, 246nn57–60, 247, 247nn61–64, 248, 248nn65–69, 249, 249nn70–72, 250, 250nn74–77, 251, 251n78, 252, 252nn81–82, 253–55, 307, 308, 309
Nasr, Sayyed Hossein, 152
Newberg, Andrew, 166
Newport, Frank, 61, 61n2, 308
Newton, Isaac, 129, 137, 197
Nielsen, Kai, 199n67
Nordén, Åsa, 82, 161, 162, 162n33, 162nn35–36, 305, 308
Nordgren, Anders, 82, 154, 154n1, 155, 155nn2–3, n5, 156, 156nn6–7, 308
Nordin, Svante, 92n33, 308
Norström, Vitalis, 87
Nowotny, Helga, 1, 1n1, 2, 308
Nygren, Anders, 41n1, 75n28, 80, 81, 96, 98, 107, 107n3, 108, 108nn4–5, 109, 109n7, 110, 110nn10–11, 110nn13–14, 111, 111nn15–19, 112, 113, 113n24, 113nn26–28, 114–17, 133, 139, 163, 172, 205–6, 208–9, 209n11, 210, 210nn12–14, 211–12, 218, 222 303, 305, 308
Nürnberger, Klaus, 258–59, 259nn108–9, 308

O'Halloran, Nathan, 178, 178nn1–12, 179, 179n14, 180, 308
Occam, 136, 192, 194, 243, 259

Paley, William, 181, 182, 182nn20–21, 183–85, 308
Pannenberg, Wolfhart, 164, 310
Pappas, William Matthew, 63n6, 308
Pascal, Blaise, 97, 97nn10–11, 305
Paul, St., 180, 184–85, 260, 262
Peacocke, Arthur, 128, 161, 307
Pederson, Ann M., 57, 57n5, 309
Persinger, Michael, 165–66
Peters, Ted, 164–65, 225,261, 261n119, 309
Pettersson, Thorleif, 33, 309

Phalén, Adolf, 91n27, 92n36, 114
Philipson, Sten, 8n13, 309
Pincock, Christopher, 191, 309
Pitts, J. Brian, 231n26, 309
Plantinga, Alvin, 89, 89n17, 102n31, 128, 145n46, 152, 173,177, 177nn7–8, 189, 197–98, 198n63, 309
Polkinghorne, John, 128, 160, 160n26, 161, 255, 255n91, 309
Pollock, Frederick, 299
Pollock, John, 176n2, 309
Price, H. H., 309

Rahner, Karl, 178
Rasmusson, Arne, 83, 83n8, 108n6, 309
Repstad, Pål, 11n19, 309
Ricœur, Paul, 309
Rosenqvist, C. G., 121
Runehov, Anne, 82, 109, 165, 165n49, 166, 166nn50–54, 167, 304, 309
Ruse, Michael, 203, 203n83, 309
Russell, Bertand, 114, 186, 186n32, 198, 198nn64–66, 228, 234, 250, 250n73, 310
Russell, Robert John, 78n1, 114, 165, 172, 178, 178n12, 224, 224nn7–8, 225–28, 228n16, 229, 256, 256nn94–97, 257, 257nn98–102, 258, 258n106, 259, 259n110, 260, 260n115, 261, 261n119, 308, 310
Rydberg, Viktor, 85

Sabatier, Paul, 89
Sagan, Carl, 63
Scarr, Sandra, 26
Scheitle, 65, 66, 17
Scheler, Max, 117
Scott, Peter, 1, 24
Sellars, William, 52
Simonsson, Tord, 78, 81, 119, 120, 121, 122, 123, 26
Solberg, 57, 25
Soskice, Janet Martin, 23
Southgate, Christopher, 165, 165n48, 178, 178n10, 310

Stark, Rodney, 24n6, n8, 310
Stenger, Victor, 239, 239n39, 310
Stenmark, Mikael, 29, 30n23, 43n16, 78, 78nn2–3, 79, 79nn4–5, 82, 115, 115n36, 130n51, 133, 144, 144n41, 145, 145nn42–43, 145nn46–47, 146, 146nn48–49, 147, 147n50–54, 148, 148n55–n58, 149, 149nn60–62, 150, 150nn63–65, 151, 151n66, 152, 152nn70–72, 153, 153n73, n75, 160, 160n29, 167, 167n58, 170, 172–73, 201n76, 226, 226nn11–13, 227nn14–15, 228, 229n19, 302, 309, 310, 311
Stenström, Thure, 40
Stoeger, William, 311
Strawson, Galen, 249
Suchocki, Marjorie, 262, 262n121, 263, 311
Sundén, Hjalmar, 217, 311
Sundkler, Bengt, 85n1, 89n19, 27
Sunstein, Cass R., 34n34, 311
Swinburne, Richard, 127, 127n41, 184, 244, 245n55, 311
Söderblom, Nathan, 74, 77, 78, 79, 80, 85, 85n1, 86, 86n3, n5, 87, 87nn7–10, 87n12, 88, 88nn13–16, 89, 89nn18–20, 90, 90n24, 93–96, 101, 107, 110, 114, 116, 141n31, 212n20, 219, 219n40, 220, 297, 310, 311

Tanzella-Nitti, Giuseppe, 310, 311
Taylor, A.E., 251, 311
Tegmark, Max, 234, 234n31, 311
Teilhard de Chardin, Pierre, 142, 142n32, 179, 179n15, 297, 303, 311
Thistelton, Anthony, 122n18, 311
Thomas Aquinas, 28n19, 123, 196, 196n59, 307, 311
Thunberg, Anne-Marie, 139, 139n24, 311
Thurfjell, David, 10n18, 312

Tibell, Gunnar, 139
Tipler, Frank, 160, 160n92, 117, 312
Torely, Vincent, 187n34
Tracy, David, 209
Trillhaas, Wolfgang, 312

Uddenberg, Nils, 8n13, 36, 36n39, 312

van Huysteen, J. Wentzel, 148n65, 164, 165, 311
van Inwagen, Peter, 185, 185n29, 186n29, 312
van Liere, K. D., 9
Vikner, Pontus, 85
Vilenkin, 94, 95, 16, 28
von Rad, Gerard, 196, 196n60, 309
von Wright, Georg Henrik, 127, 127n39, 132, 132n2, 312

Wallace-Wells, David, 253, 253n85, 254, 254n87, 299, 312
Waller, Margit, 91n26, 312
Ward, Keith, 90, 28
Wheeler, 57, 58, 28
Whitehead, Alfred North, 197, 249
Wiles, Maurice, 81, 128, 220n50, 261, 261nn116–18, 312
Williams, Patricia, 164
Wilson, Edward O., 29, 30, 30n3, 79n4, 173, 312
Wilson, John, 23, 238, 312
Wingren, Gustaf, 81, 105–6, 117, 122n19, 139, 223, 223nn2–3, 224, 224n5, 312
Wisdom, John, 117, 28
Witham, 62, 62n4, 63n5, 306
Wittgenstein, 78, 98, 111, 113, 132, 134, 135, 154, 312
Wrede, Gösta, 102n32, 105n41, 119n9, 312

Zackariasson, Ulf, 178n43, 152n70, 215, 215n28, n30, 217, 217n33, 218, 226n11, 255, 255n28, n30, 302, 309, 311, 312

Subject Index

a priori, 50, 75, 98, 107–10, 141, 208, 210, 222, 308
absolute, 87–88, 90, 100, 166, 190–91
Adam and Eve, 124, 169
aesthetics, 109
agnosticism, 62–63, 184
altruism, 82, 163–64, 302
ambiguity, 102–3, 185
ambiguous object, 194, 242–44, 245, 251, 262
analogy, 182–83, 214, 215n30, 217, 217n33, 219, 242–43, 306, 312
anthropocentrism, 17, 19, 55–56, 59, 224, 280, 282
anti-environmentalism, 12–13, 280
argument from design, 183
argument from the origin of science, 198
atheism, 4, 10, 14–16, 40–42, 51, 64–66, 75n32, 93n39, 94, 132, 162, 162n34, 169, 181, 183, 187, 199n69, 203, 203n83, 238, 246, 300, 309
    neo-atheism, xi, 14
    methodological, 187
    science-based, 199n69, 303
authority, 105
    absolute, 200

Biblical, 122, 200
    in Christian belief, 122
    of religious worldviews, 123
    subjective, 166
axiom of reasonable demand, 148

basic, 32, 126
basic attitude/mood, 2, 6–7, 8, 13, 32, 126, 133, 253, 306
basic value-system, 6
belief-norm, 133–34
beliefs
    apocalyptic, 36
    Christian, 10, 14, 17, 19, 41, 45, 54–55, 68–69, 79–80, 113, 116–17, 121–22, 124–25, 126n37, 128, 140, 189n39, 196–97, 200, 206, 217, 219, 221–23, 268, 289, 307, 309
    as cognitive-proposiational, 173n6
    comprehensive, 6, 7–8, 19, 44, 92, 139, 238
    everyday, 66, 87, 105, 114, 144, 148, 148, 237, 311
    in afterlife, see immortality
    natural, 136
    rational, 148
BGV Theorem, 232–33, 312

SUBJECT INDEX

Big Bang-theory, 47, 82, 156, 162, 162n34, 172, 195-96, 226, 229, 230-31, 231n26, 231-37, 244-45, 254n88, 255n89, 261, 300, 303, 309, 311
Big Crunch, 195, 237
biocentrism, 19
biologism, 2, 4, 8-9, 12-13, 16, 20, 22, 31, 38, 42, 53-54, 56, 59, 71, 136, 138, 161, 229, 267
biologization, 3, 4, 8, 9, 17, 20, 31, 69, 70
biologization, xiii, 3-4, 8-9, 12, 20, 29, 31, 69-71, 265-66
biology, vii, ix, x, chapter 1, 2 and 3, 70, 71, 138, 138n18-22, 139, 143, 150, 161, 161n31, 166, 232, 267, 268, 273-74, 280, 286, 289, 304
  sociobiology, 163, 302
bodily resurrection, see immortality
Buddhism, 110

Cambridge school, 114
causes, 127n40, 301
  natural/immanent, 125-26, 126n37, 167
  Center for Theology and the Natural Sciences (CTNS), 157n9
  chance, 3, 18, 20, 66, 129-30, 141, 149, 152n69, 229, 231, 238-40, 244, 246, 299
chaos theory, see physics
chemistry, 5, 44, 87, 242, 246, 264
Christian belief, see belief
Christianity, 14, 31, 36, 39, 41-42, 68, 81, 96, 105, 110, 112, 114-15, 119-20, 152-53, 203, 206, 210-212, 219, 226, 253, 255, 299 308, 310, 311, 312
Church of Sweden, ii, iv, x, xiii, xiv, 22, 22n5, 23, 28, 33, 35-37, 45-54, 58-59, 79, 80, 85, 121, 157n9, 169, 264, 266, 272-73, 277, 298
climate change, 16n24, 36, 36n38, 37n40, 47, 51, 130, 226, 253, 253n85, 254, 298, 299, 302, 312
Cluster analysis (CA), vii, ix, xvii, 3, 11-12, chapter 4, appendix 4 and 5

cognitive feeling, 83, 93, 123, 133, 134, 138, 27, 30, 35, 59, 80
cognitive pluralism (CP), xiv, xvii, 7, 7n8, 60, 144, 144n38, 170n2, 171nn3-4, 304
cognitive science, xviii, 32, 60, 83, 170
cognitive science of religion (CSR), xvii, 29, 30, 47, 167-68
coherence, 134, 261
  with science, 56, 62, 172, 228, 243
complementarity, 41, 41n8, 51, 81, 95, 164, 307
consciousness, 18, 28, 31, 47, 51, 87-88, 90-91, 100, 109-12, 121, 125, 143, 165, 207-8, 219, 226, 236, 238, 246-47, 249-50, 259, 263
  self-consciousness, 90, 102
contexts of experience, 108, 208
contexts of meaning, 80, 108, 110-12, 111n18, 209
Copernican revolution, 137
cosmology, xi, 47, 51, 150, 162, 195, 224nn7-8, 225, 22n17, 229-30, 231n23, 231nn25-26, 232-33, 235, 235n34, 236n36, 238, 239n38, 244n54, 245, 255n89, 256, 257nn98-101, 258n106, 259n110, 260n115, 261-62, 262n120, 263, 299-301, 303, 307, 310, 311-12
creatio continua, 172-73, 223, 229, 238, 245-46, 251-52, 254-55, 263
creatio ex nihilo/out of nothing, 223, 231-33, 237, 239, 244
creatio nova, 173, 223, 254-57, 259, 261-63
creatio originalis, 172, 223, 229, 238, 244, 263
Creative Mutual Interaction (CMI), xvii, 172, 222-25, 227-28, 310
criterion
  for religious truth, 97
  of correlation, 164
  of relevance, 105, 164
  of revelation, 213
  of science, 137
  of simplicity, 156
  pragmatic, 164

cumulative argument, 222
customization of science, 145n43, 152, 152n70, 226, 226nn11–12, 302, 309, 311

Darwinism, 78n3, 119–20, 119n11, 121, 121nn14–17, 310
  neo-darwinism, 172, 227
Dasein, 138
demythologization, 101, 223–24, 253
determinism, 131, 189–90
disagreement
  in attitude, 135
  of gestalt, 135
disaster
  environmental, 11, 15, 23, 36–37, 130, 135, 253, 268, 273, 285
divine action, see act of God under God
doctrine-specific, 263
dogmatics, 95, 100, 102, 104, 122
domain-specific, 7, 11
Doxastic practice, 202
  Christian Mystical Perceptual (CMP), xvii, 202, 214–18
  Mystical perceptual (MP), xvii, 214–15, 219, 220n44
  Religious (RP), xvii, 202
  Sense Perceptual (SP), xvii, 202, 202n79, 203, 214–17
dualism
  of Bergson, 87
  holistic, 28
  radical, 28, 29

ecclesiology, ii, 105
ecologism, 2, 8n13, 9, 14, 16, 267
ecologization, 8n13
economy
  (monetary), 21
  (parsimony), 261
egoism, 12, 13
emergence of universe/life/consciousness/value, 141, 151, 158, 158nn16–17, 159nn18–25, 160, 160n28, 171, 183, 190, 224, 249, 252, 304
empirical science, 134, 138, 140, 181, 184, 202
empiricism, 114, 142, 309

environmental concern, 12–16, 46, 50, 54–55, 57, 59, 280, 281, 283
environmental disaster, see disaster
environmentalism, 8n14, n15, 9, 13, 32n30, 55–56, 266, 280, 283, 298
Epicurean hypothesis, 194–96
epiphenomenalism, 190
epistemic theory
  reliabilism, 167, 214, 219
  evidentialism, 144, 147, 148, 201, 203
epistemological/epistemic imperialism, 200–201, 201n76, 202–3
epistemology, 142, 146, 150, 225, 297, 311
Erlangen school of theology, 212
eschatology
  religious/Christian, 47, 104, 157–60, 165, 179, 196, 225, 228, 22n16, 229, 246, 254, 255–56, 256nn94–95, 257n98, 258, 259nn108–10, 260nn114–15, 261n119, 263nn122–33, 299, 308, 310, 311
  physical/scientific, 160, 228, 255, 21, 22, 90, 117
eternal life, see immortality
ethical nihilism, 90
ethics, xiii, 6, 81, 109, 117, 129, 139, 140, 142, 145, 147, 150, 160, 163, 201n73, 255, 263, 299, 302, 306, 311
  metaethics, 173
European Society for the Study of Science and Theology (ESSSAT), 131, 157n9
evidence (concept/theory of), 116, 125–27, 134, 137, 148, 172, 176, 176nn2–3, 181–83, 185–86, 192–93, 200–201, 203, 213–14, 219, 231, 241, 251, 305
evil, 47, 64, 149, 155, 165–66, 177n6, 178, 178n9, n11, 179, 179n14, 180, 185n29, 203, 228, 241, 252–53, 255, 258, 262, 262n121, 300, 308, 310, 311, 312

evolution (theory of), xii, 2, 87, 87, 121,
124, 138, 141, 147, 182, 207,
246, 289
  ambiguity of, 172, 238, 242, 251, 262
  attitudes to, 15, 17–19, 25, 27,
    27n17, 47, 51, 61–62, 75n32,
    268
  creative, 87, 87n11
  direction of, 18, 150–51, 280
  evolutionary by-product, 9
  evolutionary Platonism, 246n58,
    247
  evolutionary epic, 29, 79n4
  meaning of life, 318
  naturalistic versions of, 155, 247
  evolutionary optimism, 160, 255
  problem of evil, 165, 177, 177n6,
    178, 178n9, 179–80, 300, 310
  randomness of, 15, 18, 151, 280, 281
  relation to religion, 17, 25, 41, 44,
    82, 120, 121, 124, 141, 147–49,
    154, 154n1, 155nn1–3, 155n5,
    156nn6–8, 163–64, 167n57, 182,
    183, 200–201, 207–8, 226, 264,
    289, 303, 305, 308
  theistic interpretation of, 155–56,
    240, 245
existence, mystery of, 140, 141
existentialism, 80, 97, 98, 132, 246
expansionism, 120, 145, 145n46, 152
  experience
    ethical/moral, 134, 138
    existential, 138, 164
    I- thou, 134
    mystical, 138, 213–15, 214n26
    numinous, 215–17
    gestalt, 135
    of impressedness, 96
    of transcendence, 213, 215–16
    personal experience, 212, 213, 219
    propositional and non-
      propositional, 116
    religious, 93, 96, 123–24, 165, 167,
      172, 213–14, 217, 220, 222
explanation, 127n39, 239, 312
  causal, 129, 141, 168
  Hempelian model of, 126
  methodological, 166
  natural, 30, 94, 95, 96, 126, 168, 189,
    232–34, 236
  non-scientific, 110, 194, 242, 248
  reducing, 166
  personal, 127
  scientific, 126, 136, 141, 147, 167,
    189, 193–94, 231, 242–43, 248,
    309
  supernatural, 193
  teleological, 127, 129, 141, 172–73,
    227, 250, 252
  theological, 212
  ultimate, 188

fact, definition of, 134–35
feminism, 226
fields of experience. Se contexts of
  meaning
fine-tuning, 229, 231, 234, 238- 245
Finnish Science Barometer, 26, 26n14,
  27n17, 301
form of life, 104, 119, 210, 214
  fundamental pattern (FP), xv, 227–
    29, 241–45, 251, 262
  existence-implying (EIFP), xvii, 135
  non-existence implying (NIFP), 135
  non-transcending, 228, 243, 245,
    251, 107, 113
  transcending, 161, 172, 227–28, 238,
    243–45, 251, 262

God
  act of, 37n40, 125–26, 126n37, 127,
    192, 229, 246
  as a designer, 136, 182, 239
  as agape, 103, 110, 110n11, 112,
    163–64
  as creator, 81, 90, 120–21, 150–51,
    150, 188, 192, 201, 237, 239–40,
    243, 245, 251, 256, 262
  as ground of being, 238
  as irrational, 104, 210
  as pesonal, 10, 33, 41–42, 49, 62–63,
    127, 138, 149–50, 150n65, 151,
    211, 215, 220, 269, 292, 310, 311
  as providence, 131, 141, 152, 209,
    254–55

concept of, 32, 63, 64, 159, 202, 237, 240
existence of, 29, 62, 97, 99n18, 100, 181, 183–84, 187–88, 195n56, 201, 233, 261, 299, 300
existential relation to, 101, 110
hidden work of, 98, 101, 113n25, 128, 165, 185, 306
immanence of, 32–33, 126, 126n37, 156.
impersonal, 10, 156, 269, 292
kenosis of, 165
Kingdom of, 105, 119, 254, 260, 260n114
knowledge of, 99, 168, 202
mind-oriented models, 156
of the gaps, 126n37, 245
omnipotence of, 178–79
omnipresence of, 33
predstination, 131
pre-knowledge of, 131
presence's of, 88, 113n25, 185
revelation of, 85–86, 86n5, 88, 88n15, 88n16, 88n20, 90, 99–100, 104, 121, 128, 185, 200, 211–13, 219, 219n40, 220, 220n50, 222, 263, 310
space-oriented models, 156
time-oriented models, 156
transcendence of, 103, 125n33, 126nn35–37, 128, 192, 304
trinitarian concept of, 158–59, 251, 251n80, 261n119, 262, 297, 309

hedonism, 12–13
Heilsgeschichte, 223
hermeneutics, 82, 122n18, 157, 160–61, 311
Hilbert's hotel, 235
historical critical method, 85, 259
history, 28, 34, 81, 256
  Christian view of, 104, 115, 211–13, 220, 253
Homo sapiens, 151–52
hope, 6, 13, 138, 160, 252–55, 257, 263, 297, 299
  total, 254, 258, 260
human soverignty, 13, 59

humanism, 96, 96n7, 97nn8–9, 139, 211nn17–18, 212nn19–22, 220n46, 220n48, 246, 300, 306
hypersensitive agency detection device (HADD), xvii, 168
hypothesis
  ad hoc, 124
  auxiliary, 124, 124n29, 225
  creation hypothesis, 196, 242
  failed hypothesis, 310
  non-scientific, 104, 185, 194, 196, 240, 242–43
  naturalistic, 177
  religious, 81, 123, 124, 195
  scientific, 81, 86, 123, 125, 153, 172, 175–76, 180, 183–84, 185, 192–93, 199, 200, 238–40, 243

idealism (philosophical/ontological), 79, 87, 89, 92n36, 308
  absolute, 89
  Boström´s, 79, 89, 92, 120
ideology, 145, 152, 160, 210226
immortality, 62–63, 63n6, 186, 202, 308
  bodily resurrection, 260–61
  subjective, 262–63
induction, 92, 193
integration, model of, 53–54, 78, 97, 129, 154, 163, 209, 253n84, 305
Intelligent Design (ID), xi, 47, 51, 226, 240
International Society for Science and Religion (ISSR), 157n9
intersubjectivity, 98, 143, 193
Islam, 152, 169–70, 173, 197, 231n24
I-Thou/you-philosophy, 80, 97, 101

Jesus Christ, resurrection of, 98, 101–2, 104, 128, 202–3, 257–59, 259n110, 260–62
Judaism, 110

Kalām cosmological argument, 231, 231nn25–26, 309
knowledge, see epistemology

language game, 78, 111, 128n42, 132, 303

law of non-contradiction, 91, 115
law of self-identity, 91
laws, see natural law
linguistic analysis, 97
literalism (biblical), 78, 79
logic, 91, 94, 95, 103, 104, 105, 114, 115,
    119, 127, 141, 34, 50, 53, 59,
    67, 68
logic, 91–92, 92n30, 94, 104, 105, 115,
    119, 119n6, 124n27, 127, 148,
    155
logical connection argument, 127

manicheanism, 179
marxism, 226
materialism, 189, 265
    eliminative/reductive, 29, 31, 166,
        190
    religious evolutionary, 138
metacognition, 171
metaethics, 173
metamodel, of science and religion, 171
metaphor, 180, 207, 215, 219
    untranslatable/irreducible, 136n12,
        173, 180n16, 219n42, 237
metaphysics, 80, 83, 91–94, 97, 102,
    110n14, 141, 142, 158, 242, 251,
    311
    atheistic, 162
    eschatology as, 263, 299
    naturalistic, 193, 193n51, 199, 234
    religion and, 99, 108, 220–21, 110
    process, 262
    teleology as, 251n79
    theistic, 162
    worldview as, 6, 111n18
    Milder form of Worldview-
        Customized Science (mWCS),
        xvii, 226–27
miracle, 47, 62, 88, 101–2, 123, 125–28,
    141, 188–89, 192, 202, 257, 307
Mode-1/Mode-2 society, 2
model
    model of contact, xiii, 43n15, 44, 51,
        78, chapter 9, 144, 154, 157, 170,
        172, 209, 221, chapter 14

model of incompatibility/conflict,
    xiii, 114, 130, 144, 170, chapter
    12
model of independence, xiii, 43–45,
    50–51, 74–75, 86, 99–100, 102,
    104, chapter 8, 120, 129, 134,
    136, 222–64, 283, chapter 13
model of substitution, 29–30, 79n4,
    145, 173, 201n76
mental, 7, 61, 170
monadology, 90
multiverse, 103n36, 192, 195n57, 226,
    234, 240, 244–45, 245n55, 308
mysticism, 99, 138, 215
mythology, 101, 105, 119, 180, 252

naïve realism, 105
National Academy of Science (NAS),
    63, 129
natural laws, 88–89, 101, 126–27, 143,
    189, 234, 252, 255, 256–57,
    257n103, 299
natural selection, 155, 165, 177n6, 183,
    229, 300
naturalism, 76, 136, 139, 147, 153, 177–
    80, 187n33, 190n40, 191n43,
    192–93, 193n50, 194, 194n52,
    227, 241, 300, 305
    evolutionary, 109
    metaphysical (MetaN)/ontological,
        187, 187n35, 190, 193, 195, 199,
        262, 309
    methodological (MN), 187, 187n34,
        n36, 188, 205, 256n97, 307, 312
neo-naturalism, 249, 251–53
nature
    ensoulement of, 10, 55–57, 75, 281
    human, 8, 24, 27, 29n20, 30n23, 31,
        34, 71, 79n4, 135n11, 142, 145,
        298, 302, 304, 307, 312
neo-darwinism, 172, 227
neo-orthodoxy, 97
neuroscience, 165, 165n49, 166–67, 309
neutral monism, 249–50
New Testament, 101, 104–5, 118, 181,
    203, 212, ,216, 258, 260
no-boundary proposal, 236–37, 263

non-interventionist objetive special divine action (NIODA), 257, 259n110
nonoverlapping magisteria (NOMA), xvii, 75n32, 107–8, 145, 172, 206–9, 221, 224n6

Occam's razor, 136, 192, 194, 243, 259
Old Testament, 28, 184
"only way" argument, 165, 178, 179, 180, 252–53, 308
ontology, 80, 90, 92n30, 103, 191n43, 307
optimism, 2, 6, 15, 32, 36, 129, 160, 255
original sin, 82, 163–64, 300
overrider systems, 213, 216–18

panentheism, 263
pantheism, 33
panpsychism, 249
personality-profile/type, 53, 269
pessimism, 2, 6, 10–16, 36, 281
philosophy of science, xiii, 81, 132, 186, 225, 309
physicalism
  reductive, 29, 247, 249
  nonreductive, 29, 29n21
  physics, 4–5, 44, 69, 70, 87–88, 123, 132, 143, 142, 143, 159, 161–62, 178–79, 211, 228, 231, 234, 237–38, 242, 246–47, 249, 252, 255, 255n92, 257, 264, 265
  chaos theory, 131, 159, 227
  modern, 140, 150
  Newtonian, 7, 122, 129, 142
  quantum fluctuations, 232–34, 239, 244
  quantum mechanics, 201, 227
  quantum physics, 91n27, 126n34, 132, 138, 143–44, 158–59, 171, 231, 236–37
  theory of relativity, 92, 130, 158, 160, 227, 230, 237, 243
  thermodynamics, 90, 93, 123
  vacuum, 232–33, 244
Platonism, 35, 110, 246n58, 247
postmodernism, 14, 20
practice

religious, 11, 76, 148–50, 171–72, 197–203
scientific, 149, 153, 189, 200
doxastic, 202–3, 206, 214–15, 218, 220
soteriological, 202
prayer, 12, 47, 63, 65, 68, 124, 141–42, 215–16
predestination, 131
prediction
  scientific, 188, 225, 240, 256–57, 260
presumptionism, 148
primacy of empirically based knowledge, 138, 228
principle of concern, 149
principle of ignorance, 162
principle of intellectual morality, 114–15, 200
problem of evil, see evil

qualia, 191, 248, 248n67, 305

Rational Choice Theory (RCT), xvii, 24, 73, 296
rationalism, 142
rationality, 24, 74, 82- 83, 97, 108, 144, 144n41, 148, 148n55, 148nn57–58, 149–50, 154, 158, 158n14, 167, 167n57, 185, 201, 226n13, 227, 242
reality
  objective/subjective, 98, 101, 104
  realtive, 87, 89, 191
  ultimate, 87–90, 95, 103, 311
reason/reasoning, 40, 51, 83, 103, 104, 114, 136, 168, 170–71, 176n3, 177, 185, 187–89, 196, 200–201, 210, 214, 230
reconciliation
  cosmic, 246
reductionism
  methodological, 166, 188
  ontological, 87, 142, 166
relativism, 139, 144, 144n41, 147, 311
relativity theory, see physics
relevance
  negative, 141, 167–68
  positive, 141, 167–68

Religion aming scientists in international context (RASIC), xvii, 76
religion
  as a priori, 109
  as atheoretical, chapter 8
  concept of, 69, 100, 102, 169
  as evolutionary by-product, 9
  Jeffner´s theory of (JTR), 134–39
  Hägerström´s theory of, 93
  Nygren´s theory of, 81, 107–111
  traditional, 10, 12, 13, 14, 54, 142
religious beliefs
  rebutting of, 171, 176, 176n3
  undercutting of, 171, 176, 176n3, 188, 198
religious experience, 82, 93, 124, 165, 167, 172, 213–14, 217, 220, 222
research program, xi, 2, 225, 225n10, 227, 228, 257, 306
restrictionism, 120, 145, 145n47
reverence for life, 169

science
  customization of, 145n43, 152, 152n70, 226, 226n11–n12, 302, 309–310, 311
  empirical, 134, 138, 140, 181, 184, 202
  method of, 5, 6, 128, 132, 140, 172, 186, 193, 197n61, 197–203, 217, 225, 257, 310
  results and theories, 44, 82, 88, 134, 223, 228, 264
scientism, 82, 108, 115, 115n36, 144, 144n41, 145n42, 147, 147nn50–54, 151, 186, 311
secularization, xii, xviii, 13–14, 21, 21n1, 24, 73–76, 216
self-organization, 132, 155, 160, 250
sense data-theories, 135, 138, 228
singularity, 156, 162, 231n26, 232, 235, 309
skepticism, 4, 12–16, 19–20, 280
soul, 10, 18, 28, 28n19, 29, 33, 34, 55–57, 89, 89n19, 121, 169, 203, 215, 236, 281, 298, 307
  Catholic doctrine of, 207
  transmigration of, 90

space, 80, 92, 92n37, 129, 156, 231–37, 257
spacetime world, 190–92, 194
spatial- temporal context of experience, 79–80, 92, 102–3, 112
spiritualism, 10, 12–14, 16, 20, 53, 191, 280
spontaneous expressions of life, 105
statement
  direct, 173, 173n6
  indirect, 173
  moral, 207–8
  non-symbolic, 238
  symbolic, 122, 136, 180, 238
  Stenmark's bold statement, 149–50
strength of belief-principle, 149
subjectivism, 79, 80, 90, 98, 100, 102, 138
Swedish Association for Science & the Public, 24, 26, 130
Swedish Humanist Association, 14
systematic theology, 107, 111, 122, 125, 142

Taoism, 110
technology, 5, 5n2, 11, 13, 23, 36–37, 55–56, 70, 130, 143, 267, 274, 280, 284, 304
theism, 112, 125, 136, 147, 151, 155–56, 162, 162n34, 172–73, 177–78, 180, 194, 198–99, 220, 240–41, 243, 245n55, 247, 251–52, 300, 309
  empirical, 19, 181, 183–86, 240
theological and Ideological Studies, 81, 133, 302
theological hermeneutics, 82, 122n18, 157, 160–61, 229, 311
theology
  church-theology, 105, 119
  continental, 85
  environmental, 31
  eschatological, 105
  experiential, 212
  Lundensian, 80, chapter 9
  Lutheran, 26, 79
  natural theology, 182nn20–21, 219–20, 220n44, 222, 308

protestant, 216
trinitarian, 159
of time, 158
of the cross, 165
time
  beyond time, 80, 89, 92
  relational understanding of, 159
transcendence, 103, 128, 138, 185, 192, 254, 304l (see also experiences of transcendence)
transcendental deduction, 100–101, 109
Trinity, 158–59, 251 251n80, 261n119, 262, 297, 309
truth/truth-value/truth-claims, 44, 80, 93, 96, 97, 100, 107–8, 110, 112, 113, 114–18, 134, 136, 139, 169, 171–72, 181, 200, 205–6, 208, 213, 215, 218n37, 221, 243, 264, 304
  correspondence theory of, 96, 211
  eternal (analytic) truths and factural truths, 114
  moral truth, 207, 207n5
  objective and subjective, 96, 206, 208, 211
  pragmatic, 97

Uniting Church in Sweden, xiii, 22n5, 23, 26, 29, 33, 35–37, 45–48, 49–51, 55, 272
universals, 190
universe, 58, 156, 194, 242, 243, 246
  ambiguity of, 228, 243–45, 262
  closed, 102, 262
  origin of, 232
  theory of steady state, 230
Uppsalafilosofin, 114

value, 133, 145, 197, 207, 246, 250, 252, 254, 279, 298, 312
  environmental, 15, 55, 57
  experience of, 94–95, 103, 212, 95
  value system, 6–8, 140
  value-judgements, 8
  value/ideology challenge to science, 145, 152

World Values Survey (WVS), 5, 6n4, 67, 67n17, 312
Worldview, 14, 15, 37, 60, 79, 88, 89, 116, 128, 137, 141, 144, 153, 302, 309, 310, 311
  Aristotelian, 129, 137
  biblical, 121, 259
  Christian, 60, 198
  dimensions of, 146
  experiential, 134
  holistic, 130
  inventory of Worldviews (IWV), xvii, 9, 15n23, 17, 226
  mechanistic, 128–29, 130, 132, 143
  medieval, 137, 175
  mythological, 101
  neo-naturalistic, 108, 109, 227, 246–51
  naturalistic, 76
  religious, 79, 217
  scientific, 79, 79n4, 95, 27, 53, 134, 138
  science and, 122–23, 134
  study of, xi-xiii, 2–4, 6–9, 133, appendices
  theistic, 108
  world-view customized science, 226

www.ingramcontent.com/pod-product-compliance
Lightning Source LLC
Chambersburg PA
CBHW070013010526
44117CB00011B/1543